SECOND EDITION

Bioinf[illegible]
Ba[illegible]

Applications in Biolog[illegible]
and Medicine

SECOND EDITION

Bioinformatics Basics

Applications in Biological Science and Medicine

Edited by
Lukas K. Buehler, Ph.D.
Hooman H. Rashidi, MSc.,M.D.

CRC Press is an imprint of the
Taylor & Francis Group, an **informa** business
A TAYLOR & FRANCIS BOOK

Cover Figure Legend: The cover visualizes the emergence of a protein structure from a DNA sequence. The DNA sequence is that of mouse histone protein H3 (*Mus musculus* H3 histone, family 3A; FASTA format; gen bank accession number BC092043.1). The protein dimer is a ribbon representation of the high resolution structure of a DNA binding protein (*Drosophila* TATA Box binding protein associated factor (TAF); Protein Data Bank accession number 1TAF) as obtained by protein crystallization and x-ray diffraction analysis. TAF functions in transcription as a cofactor in RNA polymerase complexes and is structurally homologous to histone H3-H4 dimers (Xie et al., Structural similarity between TAFs and the heterotetrameric core of the histone octamer, *Nature* 380, 316, 1996). Analyzing the structural similarities such as those found between histone proteins controlling chromosome packing and proteins facilitating gene expression is a central theme of bioinformatics.

CRC Press
Taylor & Francis Group
6000 Broken Sound Parkway NW, Suite 300
Boca Raton, FL 33487-2742

First issued in paperback 2019

ISBN-13: 978-0-8493-1283-0 (hbk)
ISBN-13: 978-0-367-39259-8 (pbk)

Library of Congress Cataloging-in-Publication Data

Catalog record is available from the Library of Congress

Visit the Taylor & Francis Web site at
http://www.taylorandfrancis.com

and the CRC Press Web site at
http://www.crcpress.com

Contents

Preface

After decades of research in structure–function relationships, the past decade proved to be tremendously satisfying in its technical advances on genome sequencing (genomics) and protein identification (proteomics). These advancements now necessitate the need of elaborate databases. The presence of public domain databases with billions of data entries requires robust analytical software tools in cataloging and representing this information, especially with respect to their biological significance. The unifying theme is biological relatedness as described by Darwin's Theory of Evolution and its modern "synthesis" of molecular evolutionary processes. The tool to handle this vast amount of data is bioinformatics, and the exponential increase in both computer processing power and disk storage has been instrumental in advancing the age of genomics, proteomics, and biotechnology.

We are very excited to introduce the second edition of the *Bioinformatics Basics* reference book, and hope that it will aid in the field's understanding (especially for the general scientific community). The community's reception of our first edition has fueled our commitment in introducing this second edition. We have received much constructive input and many suggestions and have tried to incorporate these into this edition. Our book is not intended to train bioinformaticians but rather to help the general scientific community in gaining a better understanding of what bioinformatics tools are available to them and how they could be incorporated into their projects or interests.

Our second edition has a more detailed view of the field while staying focused on our original (global basic concept) approach that has popularized the first edition.

Lukas K. Buehler and Hooman H. Rashidi

The Authors

Hooman H. Rashidi, M.Sc., M.D., is currently training in anatomical pathology and laboratory medicine at the Yale University School of Medicine. He completed his preliminary internal medicine training at Harvard University's Beth Israel Deaconess Medical Center. He earned his M.D. from the University of Texas, San Antonio. Prior to this he served as a biological science instructor, researcher, and a bioinformatics consultant. His graduate training focused on computational modeling and sequence motif analysis of EF-hand calcium-binding proteins at the University of California, San Diego. Dr. Rashidi is a graduate of the University of California at San Diego, where he also received his B.S. in biochemistry and cellular biology with highest honors and departmental distinction. He is also a member of the Phi Beta Kappa National Honor Society. He plans on pursuing a career in academic medicine once he completes his medical residency training at Yale.

Lukas K. Buehler, Ph.D., is a bioinformatics consultant and teaches biology at Southwestern College, Chula Vista, California. He is a visiting lecturer at the University of California at San Diego and Riverside and a member of the editorial advisory board of *Pharmaceutical Discovery.* He has an undergraduate degree in biophysics and a Ph.D. in biochemistry from the University of Basel, Switzerland, based on the functional characterization of bacterial porins, under the guidance of professors Hans-Georg Schindler and Jurg P. Rosenbusch. He worked as protein biochemist at the University of California, San Diego (UCSD) and The Scripps Research Institute (TSRI) in La Jolla on pore-forming entities, ranging from peptide nanotubes to antibiotics to cell–cell communication channels. His academic career is complemented by research into target identification of novel ion channels at Johnson & Johnson. He currently works as a consultant in DNA microarray technology.

Acknowledgments

I am grateful to my bright and beautiful wife Kristen for being so supportive. I am also proud to have such gifted parents (Bob and Mitra) and siblings (Hirbod, Jennifer, Hormoz, and Haleh), and appreciate their continued encouragement.

I would also like to thank my mentor Dr. Douglas Smith, University of California, San Diego for his direction and for being a brilliant teacher.

Hooman H. Rashidi

Borries Demeler contributed "Hydrodynamic Methods" in Section 4.2.

Thorsten Forster and Peter Ghazal are the co-authors of "Microarray and Bioarray Technology" in Section 3.4.

Dedication

This book is dedicated to my niece Rhiannon, whom we all love so very much.

Hooman H. Rashidi

To all the students who make teaching science a joy and a challenge.

Lukas K. Buehler

Contributors

Borries Demeler Department of Biochemistry, University of Texas Health Science Center at San Antonio, San Antonio, Texas (Section 4.2, Hydrodynamic Methods)

Thorsten Forster Scottish Centre for Genomic Technology and Informatics, University of Edinburgh Medical School, Edinburgh, Scotland (Section 3.4, Microarray and Bioarray Technology)

Peter Ghazal Scottish Centre for Genomic Technology and Informatics, University of Edinburgh Medical School, Edinburgh, Scotland (Section 3.4, Microarray and Bioarray Technology)

1

Biology and Information

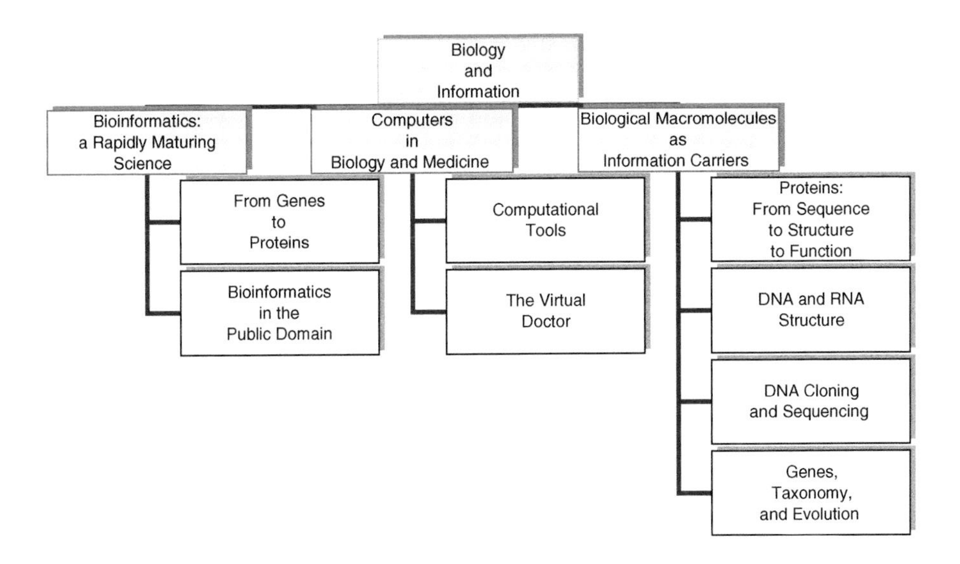

FIGURE 1.1
Chapter overview.

1.1 Bioinformatics—A Rapidly Maturing Science

Bioinformatics is a rapidly growing field within the biological sciences. It dates back to the 1960s following the discovery of the DNA double helix, when cracking the genetic code allowed for the ability to treat genes as strings of information that guide the building of cellular components, the faithful reproduction of an organism's form and function, and its ability to evolve. Today, bioinformatics is driven by the challenge of integrating the large amount of genetic and structural data emanating from biomedical research. Using computational power bioinformaticians catalog and compare genetic and structural information with biochemical, physiological, and medical data furthering our understanding of the cellular organization of life, its diversity, and the fact that all modern organisms are the children of a common ancestral cell.

1.1.1 From Genes to Proteins

Reflecting upon the complexity of living organisms and the many different ways man studies life, the term "bioinformatics" refers to the task of *organizing*, *analyzing*, and *predicting* increasingly complex data arising from modern *molecular* and *biochemical* techniques. For some the meaning extends to the concept of *information flow* within biological systems alluding to the transmission of genetically encoded information transmitted from genes to proteins, from the *blueprint* to the *machinery* of life. Bioinformaticians attempt to understand what it means to say that genes code for physiological traits such as enzyme activity, curly hair, or the susceptibility for disease. Through the creation, annotation, and mining of biological databases they help elucidate the mechanism of genetic complexity and evolutionary relationships among organisms. Bioinformatics depends on the assumption that quantifiable relationships exist not only between the sequence of genes and the structure and function of proteins, but also between the activity of genes and their placement within the genome. Commencing from a single gene, single protein approach to a *global view* of gene activity and protein networks, bioinformaticians make use of database mining techniques to study protein complexes, metabolic pathways, and gene networks.

The bread-and-butter of bioinformatics are *gene sequences* and *protein structures*. So, how can bioinformatics contribute to answering questions in biology and medicine? Genes are the hereditary units of life, and genomes of contemporary organisms are the only molecular source of information of its history (with exceptions of well-preserved paleontological samples of soft tissue). While the accurate copying of genetic information from generation to generation is crucial for the viability of organisms and thus the continuation of life, randomly occurring mutations, i.e., the alterations of gene sequences, are equally central for biological evolution, e.g., the ability of organisms to adapt to environmental changes. Mutations are the agents of change, mostly deleterious, causing disease or outright death of the carrier, but in a few cases also laying the foundation for novel traits that give an organism a reproductive advantage. Many mutations are eliminated, some are retained through natural selection, and as a result, our genomes are full of individual differences (called genetic polymorphism) that constitute the gene pool of a species. Over evolutionary time, the cumulative effect of mutations and natural selection are evident in today's biodiversity and genomes of modern organisms. Bioinformatics characterizes and quantifies this diversity through direct comparison (alignment) of genetic sequences and protein structures.

Evolution can rarely be studied in real time, for evolutionary changes take many generations to take hold. Thus bioinformatics has greatly advanced evolutionary biology through phylogenetic analysis of the molecular tree of life, complementing traditional systematics such as the Linnaean *taxonomic system* for naming, ranking, and classifying organisms and *paleontological* studies of fossil records. How we group organisms to correctly reflect evolutionary history is a work in progress. For instance, in the late 1970s, using bioinformatics tools, Carl Woese and coworkers analyzed 16S ribosomal RNA (rRNA) sequences and proposed a three-domain classification, splitting the kingdom Monera into *archaea* and *bacteria* as two independent domains (*urkingdom*) of the tree of life next to the domain *eukarya* that includes the four kingdoms Protista, Fungi, Plantae, and Animalia. In effect, Woese et al. suggested that the two prokaryotic life forms are more different from each other than humans are from amoebae.

But what about studying evolutionary mechanisms through hypothesis immunodeficiency testing? While macroevolution (speciation) leading to the tree of life can only be recreated from contemporary sequences, some types of microevolution can be tracked in real time. Examples are the rapid mutation rates of the *human immunodeficiency virus* (HIV) and *influenza* (flu) strains that help these viruses evade immune detection, or the recurrence of *antibiotic resistance* of pathogenic bacteria. Under strong selective pressure in hospitals and animal husbandry, these examples are rather impressive proof of principles of natural selection, courtesy of Mother Nature.

Studying evolution is of course not the only, and certainly not the most important, use of bioinformatics. The study of mutations in humans, higher animals, and plants by means of bioinformatics advances our understanding of the pathogenesis of *diseases* and emerging *infections*, contributing to a more and more complete picture of human genetics. Mutations serve as genetic markers allowing molecular biologists to track and locate traits, map genomes, or provide forensic evidence in crime scene investigations. In medicine, geneticists not only work with human subjects, but strongly rely on *animal models* to understand and treat human diseases. This trend has become more, not less, important over the last century. Organisms with short generation times that are measured in hours or days instead of years makes the manipulation of genes in the laboratory practicable. *Model organisms* used by biologists include mice (*Mus musculus*), fruit flies (*Drosophila melanogaster*), yeast (*Saccharomyces cerevisiae*), and bacteria (*Escherichia coli*). The development of novel drug compounds is increasingly dependent on animal models of closely related species. Using animals as models of human diseases, scientists test toxicity and efficacy of drug compounds sometimes by mimicking the *genetics of a human disease* through the introduction of genes (transgenic animals), causing mutations through chemical or radiological means, or by inactivating a gene outright (knock-out models). Plants, too, are important targets of genetic manipulation as agricultural crops in the form of genetically modified organisms (GMOs) or providers of myriads of potent drugs to treat diseases.

Although experimentation on humans is limited to treating diseases for ethical reasons, the use of model organisms cannot entirely replace human subjects when studying mental diseases or human behavior. Such studies include linking genes with neurological disorders like dementias, schizophrenia, or depression. With regard to human behavior and consciousness rather than clinical conditions, research becomes controversial when touching on societal norms. Our answers become an issue of biological determinism that forces us to evaluate the essence of being human.

To appreciate the value and limitation of bioinformatics in biology and medicine, one has to understand the source, quality, and biological significance of the data it depends on; the sequences and structures of genes and proteins. Nucleic acid sequences and protein structures are the *raw data* that bioinformaticians use to find relations among genes, genomes, and proteins. The quality and accuracy of the data is crucial in order to correctly interpret their biological significance. Here's the scenario. When scientists publish a report about mutations in the genes (BRCA-1 and -2) involved in the development of breast or ovarian cancers, the *cloning* of these genes is a central first step in obtaining the sequence information of its protein coding and regulatory components. In this process, a chromosomal DNA fragment (for BRCA-1 and -2 on the long arm fragment of human chromosomes 13 and 17) is identified in affected individuals through restriction fragment length polymorphisms (RFLPs). When a mutation—whether single nucleotide polymorphisms (SNPs; pronounced *snips*), insertions, or deletions—within a restriction enzyme-cutting site is established in a population (e.g., a genetic marker with a frequency of more than 1%), cutting chromosomal DNA using restriction enzymes will result in different DNA fragment sizes in individuals with the marker compared to those without the marker. Any gene located near a marker is potentially involved in the disease. To clone a gene, an isolated fragment from the genome is inserted into a small, functionally customized *vector DNA* resulting in *recombinant* DNA to obtain its nucleotide sequence. These cloning vectors are designed to *amplify* a gene by increasing its copy number in a mammalian or bacterial cell system, making the human gene amenable to sequencing and genetic manipulation. Using an alternate method, amplification of DNA can be achieved by the *polymerase chain reaction* (PCR). Both amplification techniques, bacterial cloning and PCR, result in large amounts of identical copies, or clones, of the desired (human) gene. In the search for genes of hereditary diseases, one not only needs samples from individuals carrying the mutated genes causing a disease, but also samples from healthy people to pinpoint the location of the gene in the genome (gene locus; see DNA cloning and sequencing, Section 1.3). Indeed, a large sample of both afflicted and healthy (control) individuals is necessary to establish the suspected correlation of a genetic marker with a disease.

It is worthwhile at this point to pause and ask why recombinant DNA technology—the mixing and matching of animal, plant, bacterial, and viral DNA—is possible in the first place. The answer is simple. The manipulation and the artificial exchange of genes across species boundaries in the laboratory are possible because all living organisms operate with the same genetic code, molecular material, and mechanisms of replication. This astonishing likeness of genetic mechanisms across all forms of life is the single most important evidence of biological evolution as outlined by Charles Darwin's Theory of Evolution. It also forms the basis of the success of bioinformatics in shaping modern biomedical research.

Cloning human DNA by PCR or in a bacterial cell-culture is not only simple and efficient, but also avoids the difficulty of obtaining large quantities of human tissue for the purpose of DNA sequencing. Sequencing determines the linear arrangement of the four bases adenine (A), guanine (G), cytosine (C), and thymine (T) found in the nucleotide building blocks of DNA. Any sequence of dozens to millions of building blocks encodes genetic information. The same recombinant DNA can also be used for gene expression studies and to synthesize large amounts of protein for biochemical and functional analysis. Sequencing and functional studies are of course related. The sequence of a stretch of DNA reveals a great deal about its properties, be it a gene, a regulatory element, or a noncoding element. For proper genes, the sequence reveals information about the structure and function of the protein or RNA that it is coding for. Thus obtaining the sequence of a gene by way of cloning DNA from any organism produces a great wealth of information about its potential biological significance. Unraveling the relationship between a gene's DNA sequence, its chromosomal localization, and its cellular expression is already under way, but remains the biggest challenge for today's life sciences.

Bioinformatics has broader implications, still, when used in combination with traditional scientific disciplines, widening their scope and methodology. Pharmacogenomics is one such novel discipline combining bioinformatics and pharmacology (see Section 5.1.3). The basic idea behind pharmacogenomics is the observation that individuals respond to drugs in different ways (or to food, dust, or microbial pathogens, for that matter). Pharmacogenomics probes the link between the genetic inheritance or make-up of an individual and his or her susceptibility to drugs, their efficacy and toxicity. Understanding and predicting these individual differences using genomic fingerprints could lead to novel drug regimens (personalized medicine) and allow for an individually optimized administration of drugs, increasing effectiveness while minimizing side effects. Bioinformatics is also used to understand the relationship between genes and diseases, such as potassium channel mutations and heartbeat regulation (e.g., Long QT-syndrome) or chloride channel mutations and malabsorption of salts in kidney and intestines or clearance of fluids in the alveolar capillaries of the lung (e.g., cystic fibrosis). Most genetic links to diseases, however, remain unsolved because multiple genes are determining a phenotype obscuring the contribution of individual genes. The BRCA-1 and -2 genes discussed above, for instance, are responsible for only about 10% of all breast cancers.

Yet the biggest advance in bioinformatics comes from analyzing whole genome sequences available in public databases. Having the global view on gene sequences, their structure and distribution on chromosomes, as well as their cell-type specific expression (transcriptome) will bring a more complete and novel understanding of the relationship between the genome and the proteome. A central role in this analysis will be played by gene and protein *networks* and metabolic *pathway reconstructions*. Genome projects are teaching yet another lesson; that higher organisms contain massive amounts of noncoding DNA, the function and purpose of which has been questioned until very recently. Facing a staggering number of sequence information for noncoding genetic material, theoretical biologists will have to forge new venues that will lead to novel experimental approaches. Experimental biologists are just starting to ask the right questions about novel methodology needed to study this portion of the genome of a species. Purely theoretical models stemming from *in silico* analyses of whole genomes will guide experimentalists to probe the genetic material that eludes current evolutionary reasoning. The challenge to biologists is clear. In the classic words of evolutionary biologist Theodosius Dobzhansky (1900 to 1975) "nothing in biology makes sense except in the light of evolution." The promise of bioinformatics then is to understand life itself. Eureka!

But how central is DNA, the workhorse of bioinformatics, really, to understanding life? Is DNA the only form of information inherited? It turns out that it is not possible to assign inherited information to DNA sequences alone, for DNA can neither replicate nor activate (transcribe) itself outside a cellular environment. DNA serves as an information storage device and is a blueprint that has to be "read" and implemented somehow. In cells this is done by proteins, the structures of which are encoded within this very DNA they access. A third type of macromolecule closely related to DNA is ribonucleic acid, or RNA. Its function is to direct and control the biosynthesis of proteins from DNA. None of these three important cellular components can act on its own (some exceptions are known for RNA molecules that act as information storage and catalytic device) as exemplified by viruses. They depend on host cells with a cytoplasmic metabolic machinery.

Many aspects of this dependency of gene expression on cellular organization are not known. For instance, to understand how much protein a gene is able to make compared to its sister gene (*allele*) or in different cell types like liver cells, neurons, or muscle cells, one has to study spatial and temporal control of gene expression. The challenges and possibilities are tremendous. As more and more genome projects are being completed, the wealth of information they generate is transforming the molecular biology laboratory into computer labs where genetic information is freely accessible from public databases, opening up new ways to elucidate the role of genetic networks in complex biological processes like embryonic development, aging, the structure and function of proteins, memory, and the like. From simple phylogenetic analyses to simulating complex biological systems, bioinformatics is driving biology from a purely experimental science to one where theoretical work will play an ever more important role.

Reliance on computer analyses has also become a dependency of sorts. For instance, a majority of biological annotation of gene sequences found in databases comes from automated protocols. These protocols base their predictions on similarity scores from comparing novel sequences with sequences of genes with known structure and/or function. Thus, a large number of automatically generated annotations await manual curation to verify the correctness of the inferred function. Nonetheless, for an estimated 30% of novel genes identified in whole genomes, no similarity to any known gene is found and thus any biological function associated to them would be purely speculative. In other words, for about one third of all putative genes found in whole genomes, we neither know what protein (or structural RNA) they may code for, nor what the function of such a protein might be. This finding has been consistent for any organism sequenced to date, from bacteria to plant to animal, including *Homo sapiens* as well as the best-studied model organisms, e.g., *E. coli* (bacteria), *D. melanogaster* (fruit fly), and *Rattus norvegicus* (rodent).

The need to establish the function for such a large number of novel genes is driving the development of new technology and bioinformatics application. To rapidly produce biological information associated to this large number of orphan gene sequences, molecular biologists use new tools like DNA and protein microarrays to study global gene expression and protein levels in cells and tissues. The systematic detection and annotation of protein levels analyzed by 2-D gel electrophoresis or protein interactions by the yeast two-hybrid system (proteomics) add another layer of physiologically relevant information in the detection of novel and important proteins associated with development, aging, and disease. This information at the level of proteins is fundamentally important in understanding how the genetic blueprint is read and implemented for the development and functioning of a viable organism.

Novel nanotechnologies that strive to miniaturize detection of novel genes are drastically increasing the output of raw data submitted to databases and amenable for bioinformatics analyses. For instance, DNA and protein microarray technologies (see Section 3.4) are leading the way to quickly complement and use the information from genome projects to study activity patterns of genes by measuring the messenger RNA levels on bioarrays containing tens of thousands of probes on an area of less than a square inch. Technological development is labor- and cost-intensive and thus it cannot come as a surprise that commercial technology is an important component of bioinformatics platforms, as are patents on biological molecules, processes, and whole organisms.

1.1.2 Bioinformatics in the Public Domain

During the 20th century, biologists have used chemistry, physics, and genetics to unravel metabolic pathways and to study the structure and function of enzymes and the relationship of these pathways for hereditary diseases.

Not only have techniques changed and influenced the productivity and goals of biochemistry, but today the need for collective approaches to solve the truly overwhelming work ahead changes the way biologists interact and work in their labs.

Traditionally, understanding metabolic pathways, characterizing enzymes, or sequencing genes is pursued by the individual interest of scientists studying one gene or one protein over many years of intense laboratory research. It is often a personal interest in diseases like cancer and heart disease, or the workings of memory that spurs research done by individuals seeking funding by government agencies such as the National Institute of Health (NIH; www.nih.gov), the National Science Foundation (NSF; www.nsf.gov), and hundreds of state agencies, private corporations, and nonprofit organizations and foundations with special interests in individual diseases.

Academic research is powered by individuals who are knowledge driven and need acknowledgment in the form of public records to gain access to grant money. The close relationship between the scientist and his or her accomplishments is fundamental to scientific progress and closely tied to academic freedom, which involves free access to information while avoiding undue economic pressure. For both of these aspects, the Internet has done wonders and stimulates fast and free flow of information. It is this combination that has allowed the Internet to become a vital tool in proliferating and sharing of information collected in centralized public databases.

Databases are rapidly expanding catalogues of biological information. Some are updated daily to accommodate newly submitted data, and to release curated information and make it accessible to the scientific community. In 1998 the genome projects covered 83 species, with 21 projects completed (predominantly microorganisms), and 63 more in progress, by the end of 2003 the numbers of completed projects included 128 (140) bacterial genomes, 16 (17) archaea genomes, 19 (20) eukaryotic genomes, and over 1000 viral, organellar, plasmid, and phage genomes, with 778 alone accounting for viral genomes (see genomes page at EMBL www.ebi.ac.uk/genomes/index.html and Entrez Genome and NCBI www.ncbi.nlm.nih.gov/entrez/query.fcgi?db=Genome). Dozens of additional genome projects are in the works.

These genome projects are powered by automated cloning and polymerase chain reaction (PCR) for DNA amplification, high-throughput sequencing technology, and bioinformatics to reconstruct gap-free, contiguous sequences (contigs) of randomly generated chromosomal fragments (shotgun approach), which eventually leads to the base-by-base sequence of entire genomes. The fragments to be sequenced are no longer chosen based on their known biological significance. The rationale of the genome projects is a discovery approach; to "shoot first" and ask questions later—questions relevant to generate insight into the still larger unknown portion of life's blueprint. For the human genome project the finished "working draft" was announced on June 16, 2000. Since then, many gaps between existing strings of sequences in this draft have been filled, and with close to three billion base pairs the sequencing effort is near completion. More than 95% of their bases of the

TABLE 1.1

Human Genome Reference Sequence and Chromosome Properties

Chrom. Number	Reference Accession	Sequence Length	Determined Bases*	Coverage (%)
1	NC_000001.4	245,203,898	218,712,898	89.2
2	NC_000002.5	243,315,028	237,043,673	97.4
3	NC_000003.5	199,411,731	193,607,218	97.1
4	NC_000004.5	191,610,523	186,580,523	97.4
5	NC_000005.4	180,967,295	177,524,972	98.1
6	NC_000006.5	170,740,541	166,880,540	97.7
7	NC_000007.7	158,431,299	154,546,299	97.5
8	NC_000008.5	145,908,738	141,694,337	97.1
9	NC_000009.5	134,505,819	115,187,714	85.6
10	NC_000010.4	135,480,874	130,710,865	96.5
11	NC_000011.4	134,978,784	130,709,420	96.8
12	NC_000012.5	133,464,434	129,328,332	96.9
13	NC_000013.5	114,151,656	95,511,656	83.7
14	NC_000014.4	105,311,216	87,191,216	82.8
15	NC_000015.4	100,114,055	81,117,055	81.0
16	NC_000016.4	89,995,999	79,890,791	88.8
17	NC_000017.5	81,691,216	77,480,855	94.8
18	NC_000018.4	77,753,510	74,534,531	95.9
19	NC_000019.5	63,790,860	55,780,860	87.4
20	NC_000020.5	63,644,868	59,424,990	93.4
21	NC_000021.3	46,976,537	33,924,742	72.2
22	NC_000022.4	49,476,972	34,352,051	69.4
X	NC_000023.4	152,634,166	147,686,664	96.8
Y	NC_000024.3	50,961,097	22,761,097	44.7
Unplaced	Various	25,263,157	25,062,835	99.2

HGP goals called for determination of only the euchromatin portion of the genome. Telomeres, centromeres, and other heterochromatic regions have been left undetermined, as have a small number of unclonable gaps.

Source: From NCBI; www.ncbi.nlm.nih.gov/genome/seq. With permission.

human genome are sequenced and their location on the chromosomes determined. The official completion of the human DNA reference sequence was published on April 15, 2003 as summarized in Table 1.1. Due to the enormous size of the genome, searching and navigating the human genome is done one chromosome at a time. Each chromosome has an accession number (reference accession) and the estimated sequence length, number of actual base pairs sequenced, and percent coverage are shown.

Even while incomplete, there is a tremendous amount of information available from the daily progress of these ongoing genome projects. Outcomes of this massive flow of information are threefold. First, everyone with an Internet browser has access to this information. Second, data submitted and stored in centralized databases show a varying degree of redundancy. The 12-megabase genome of the common baker's yeast *S. cerevisiae*, completed in 1997, has been oversubmitted by a factor of 2.5. Much of this sequence information is the result of using yeast as a eukaryotic model system in molecular biology.

Although this may sound wasteful in the eyes of economists, for most biologists, redundancy as a byproduct of multiple submissions from different authors provides a *quality control* for both sequence information and biological annotation. Third, why bother with yeast, when the goal is to cure human diseases? Again, because of the close relationship of all living organisms based on the process of evolution, many genes and biochemical pathways studied in yeast help understand and find homologous genes and pathways in humans. New biological disciplines like *comparative* and *functional* genomics are now being used to address medical problems by analyzing gene activity patterns in cells, tissues, and organs. The technical difficulties and demands, however, make this kind of genomic analysis of diseases extremely expensive, at least at the beginning. Because of the potential findings for new drugs, commercial interests are strongly tied into these genome projects, and fights over patent rights for as yet unknown DNA sequences have rightfully been compared to the gold rush times of the 19th century.

Industry, realizing the potential of information handling and selling, played an important role from the beginning of biotechnology in the early 1980s, pushing technological development and playing an increasingly dominant role. The teaming up of the former NIH scientist Craig Venter with the private sector forced the academic community into a competitive race for being first in sequencing the human genome and thus avoiding the risk of losing free access to genome maps owned by private sector companies. The race has accelerated the pace of sequencing, producing a draft some 10 years before the scheduled completion date. In 2000 the race was called a tie for political reasons, thus both saving the public effort face and Venter's initiative, with parallel publications in *Science* and *Nature*.

This race has highlighted the strength of both public and private initiatives and the role public funding of research plays in fostering private enterprise, particularly in the U.S. Industry uses public databases to complement its own research. Borders between academia and industry become more fluid, and people move from one side to the other. The Institute for Genomic Research (TIGR) founded in 1992 by Craig Venter, is a premier nonprofit research institute that has grown with and made use of the Internet from the very beginning. For many scientists TIGR symbolizes the understanding of genome projects, and has given new meaning to the phrase high-throughput sequencing. TIGR has been instrumental in developing automated sequence procedures and data analysis. The public service of institutes like TIGR cannot be overstated. The accessibility of the information produced at TIGR and other organizations over the Internet is truly astounding.

THE INSTITUTE FOR GENOMIC RESEARCH (TIGR)

> TIGR is a not-for-profit research institute with interests in structural, functional, and comparative analysis of genomes and gene products in viruses, eubacteria, pathogenic bacteria, archaea, and eukaryotes (both

plant and animal), including humans. The Institute is located in Rockville, MD, in the greater Washington, D.C. metropolitan area, and is close to the National Institutes of Health, Johns Hopkins University, The University of Maryland, and other research institutes and biotechnology companies. Institute facilities consist of over 50,000 sq. ft. of laboratory and office space on a 12-acre campus. The Institute has a large DNA sequencing laboratory and has modern facilities for bioinformatics, biochemistry, and molecular biology. (From: www.tigr.org/about/)

TIGR has pioneered the development and distribution of techniques necessary for the mass cloning of DNA fragments and sequencing of so-called expression sequence tags (ESTs) (Figure 1.2). The many hundreds of thousands of electronic sequences deposited in public databases are real goldmines for other scientists interested in biologically relevant research.

TIGR was a dry run for Venter's human genome project that he moved entirely onto the private sector side of research founding Celera Genomics Group in partnership with the bioscience instrument manufacturer Perkin Elmer. Many other privately held EST libraries (Incyte, Millennium Pharmaceuticals) have been established to sell genetic information through proprietary databases to the pharmaceutical industry. Their proprietary

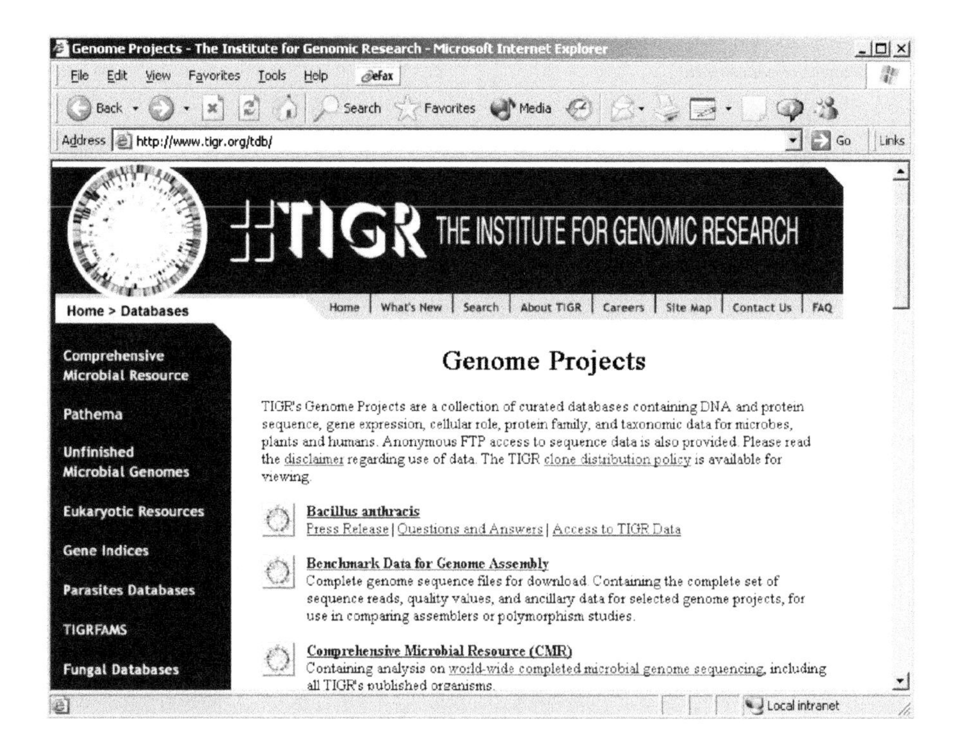

FIGURE 1.2
TIGR genome projects.

databases are built upon public database information by adding their own sequence information. The most valuable asset of these private sector databases, however, is the inclusion of medical data (annotation) suitable for drug discovery, including information on tissue distribution of gene expression or whether the sequences have been obtained from diseased or healthy tissue. Thus functional and therapeutically interesting novel genes can be identified and their use protected by appropriate patents. Incidentally, public databases with an ever-increasing number of freely accessible data have forced many of these information companies out of business or into the business of drug development proper, including Incyte Corp., Celera Genomics Group, Myriad Genetics, Inc., and Millennium Pharmaceuticals.

Another factor facilitating the demise of proprietary databases came at the expense of the rapid growth and utility of public databases integrating raw data with functional annotation and database tools. The problem of freely accessible data is compounded by the meager results of reliably identifying novel drug targets from sequence information alone. The deluge of drugs to cure diseases has not materialized from Genomics and Proteomics without substantial investment into experimental assay development. Predictive biology, it turns out, is still in its infancy.

1.2 Computers in Biology and Medicine

1.2.1 Computational Tools

Experimental science in biology, physics, and chemistry conjures up images of test tubes, colored liquids, microscopes, and men in lab coats with wild hair. Yet daily life in the laboratory can be as banal to a scientist as it is mysterious to the average person. We know scientists experiment, but what does that mean? Experimenting is the testing of an idea by observing and testing a set of specimens over a given period of time. Experimenting obviously requires planning and organization.

Today, computers play a central role in the design, execution, and analysis of experiments. They have not changed scientific thinking or methods, formulation of hypotheses, and falsification of theories and models. But computers, and particularly the personal computer (PC), do affect the dynamics of the modern laboratory in almost every respect. PCs are multifaceted laboratory workhorses and are used for word processing, spreadsheet analysis, presentation, Internet access, and specialized software to control lab equipment. Versatility, speed, increased computational power of personal computers, and local networks (Pentium processors, NT workstations) allow scientists to do their work independently of super computers, such as molecular modeling and multiple-sequence alignment calculations (molecular evolution). Often lab equipment comes with optional computer interfaces that allow the researcher to customize his or her application for specific experimental needs.

Computers count the number of cells in a petri dish, measure the size of cell nuclei in the microscopic cross section of a kidney tumor, record the electrical activity of an isolated neuronal cell that is suspected in chronic pain sensation, and read the sequence information from an electrophoresis gel after having exposed the radioactive-labeled DNA fragments overnight on x-ray-sensitive film. Other indispensable lab support comes from electronic cell-sorting machines, video cameras with online control from computers, product purification by reverse-phase chromatography, digital oscilloscopes, and real-time peak current analysis. Sophisticated computers assist in experimental systems where the human hand and eye would be too slow and inaccurate to guide and record important data. The computer's functions are for control, performance, data manipulation, and storage. Storage capacity and real-time processing of large amounts of data have increased dramatically.

But most computers do not come in the form of PCs. An estimated 1% of all microprocessors in the world are contained in desktop PCs. The other 99% are embedded in airplanes, heating systems, lab instruments, security systems, appliances, wireless telephones, ATM machines, etc. These processors are often referred to as "firmware." These are chips that have their own special functionality prebuilt into them—no programming required.

Modern science, too, requires many machines with embedded processors, such as gas chromatographs, computerized scales, or spectrophotometers. The latter are used to measure light absorption and emission at different wavelengths to determine concentrations, size, or chemical consistency of compounds. Computer control allows real-time measurements to conduct experiments such as monitoring changes in the chemical composition of a test solution. Automated control fractionation of liquid chromatography separating molecular mixtures into individual components according to size or solubility. The microprocessor is controlled through a small window displaying one or more lines of code or command text that can be typed in or selected from a short menu. Essentially they function like an ATM machine where a small display and key pad can be used to access your checking account; you are interacting with a computer that allows you to transfer money, but not to type and edit a letter. However, it is connected to a telephone network and is a terminal of a specialized Internet. During the last 20 years these microprocessors have evolved from simple control circuits to "real" computers allowing storage of large data files and graphics and slowly replacing the need for analog recordings on paper and films.

Precision and high-quality measurements in science were achieved long before the advent of the computer. Today, the precision of an experiment is not determined by computers alone, but by the quality of the instrument used. High-quality alloys must be engineered for the precision cutting of frozen cell samples in electron microscopy, the pulling of a micro-glass pipette for the transfer of a cell nucleus into stem cells to produce knockout mice, or measuring the electrical activity of a single neuron in brain tissue.

In addition, the quality of recordings, whether chemical, electrical, or optical, still depends on the quality of biological specimen preparation. Ramon y Cajal (1852 to 1934), a Spanish neuroanatomist and the 1906 Nobel Laureate in Medicine, exemplifies the scientific excellence of the 19th century. He studied the nervous system and recorded his anatomical studies of the brain in meticulous, accurate drawings. Cajal was the leader in brain research because of his state-of-the art drawings of neurons and groups of neurons based on new staining techniques (the use of dye to color individual cells or groups of cells to show the interconnectedness of neurons) developed by Camillo Golgi, the co-winner of the 1906 Nobel Prize in Medicine. Cajal showed that the vertebrate brain was made up of billions of individual cells and neurons and was not a continuous network of fine arteries. Similarly, new staining techniques combined with photography and computer-aided spectroscopy are used today to construct functional maps of the human brain (see Color Figure 1.5). Staining and sample preparation that make use of a variety of marker molecules are still necessary for single-cell visualization.

The development of fluorescence markers for biological sample preparations in conjunction with high resolution light microscopes, photon detectors, and computer power is revolutionizing cell biology, advancing functional genomics (see Chapter 3), and is giving rise to non- or less-invasive optical brain monitoring techniques, thus replacing the need to implant electrodes to record activity of multiple neurons simultaneously. The interdisciplinary approach—combining chemistry and biology—has energized both scientific disciplines. Material sciences are combining chemistry and biology in ways different from those in classical biochemistry. Nanotechnology, the manipulation of matter at the single molecule level, is a new development and has become a buzzword, indicating that technology is improving the miniaturization of mechanical and electrical devices, with the goal of creating single-molecule machines. The nanometer, the physical dimension of this new technology, is the dimension of the molecular machinery of life itself. Hence, inspiration and creative solutions among biologists, chemists, and engineers are now going hand in hand.

The introduction of the Internet, however, is probably the single most important technical advance that changed how scientists work, publish, and communicate, and has been extremely rewarding to the life sciences. It has dramatically enhanced communicative efforts between the researchers and has stimulated collaborative work in various fields. The presence of data management systems such as NCBI (National Center for Biotechnology Information) and EBI (European Bioinformatics Institute) has enhanced the efficiency of many research efforts around the world, while uniting life scientists from different disciplines. The exponential expansion of biological data necessitates an organized specification of the data by specialized and specific management systems. For example, the biological data pertaining to proteins should be separate from those of polynucleotides (DNA and RNA). The PDB (Protein Data Bank) library[1] is an example of such a data bank where protein data is stored. PDB specifically deals with protein structures. Like most other biological servers, the PDB server will also provide trends and relationships between the stored molecules. In PDB, this information could be retrieved from SCOP (structural classification of proteins).[2] Databases such as SCOP are useful tools in the characterization of macromolecules with respect to each other and their biological systems. The separation of the molecules into specific categories forms only the basis, however important, for the real deal. The data management system should also be able to display *relational information* with regard to the molecule of interest. The information in a particular file should have links to related data on other relevant sites. For example, horse myoglobin's PDB file summary has multiple link options to related information (e.g., related abstracts, etc.) for the myoglobin molecule. The relational information for a given data entry makes it valuable to other entries by displaying potential relationships to other molecules and systems in other servers.

One of the great strengths of the Internet is its redundancy and interactive mode. Most interactive tasks available throughout the Internet run through remote computers and are replacing the need to download appropriate software for local analysis, although this is often desirable and applicable with the manufacturing of more powerful PC systems. PCs are now used to do almost any computing task. Their low prices, large capacity, and ability to interconnect in local and nonlocal networks allow them to form clusters of parallel computing processors. This form of distributed computing has become a low-cost alternative to supercomputing centers. The Internet is a network of workstations connected by switches, routers, and fiberoptic cables. It resembles supercomputers more and more as it provides means of massive parallel computing. In fact, supercomputers are themselves built from parallel integrated PC components. The distinction becomes one of access and control.

Bioinformatics requires complex simulations and 3-D renderings and animations. These are computationally intensive tasks and thus are limited by available hardware that determines processing power, memory, network implementation, and storage capacity. The top listed supercomputer in 2003 was Japan's Earth Simulator (ES), with 35 Tflops used for climate modeling (supercomputers measure processing power in teraflops or 1000 Gigaflops; one trillion floating point operations per second). The ES is based on 5120 (640 8-way nodes) 500 MHz NEC CPUs with 8 Flops per CPU (41 Tflops total; Source: Top500 Supercomputer Sites at www.top500.org). Except for the ES, all top 10 supercomputers are located in the U.S. including the Los Alamos National Laboratory (LANL; 14 Tflops) and the National Center for Supercomputing Applications (NCSA; www.ncsa.uiuc.edu;10 Tflops) at the University of Illinois at Urbana-Champaign.

The NCSA has many partners in the U.S. including the San Diego Super Computer Center (SDSC; Figure 1.3a), a publicly operated service provider

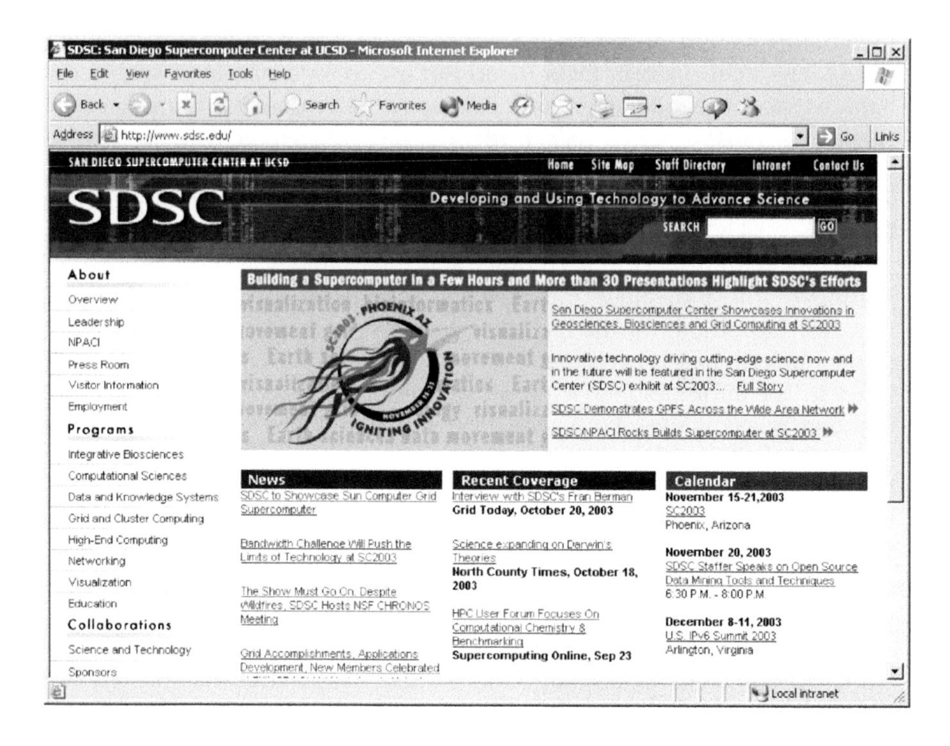

FIGURE 1.3A
San Diego Supercomputer Center at UCSD.

that hosts the PDB Repository. The SDSC provides and supports a wide range of computing resources. Currently, SDSC's production systems are provided by the National Partnership for Advanced Computational Infrastructure's (NPACI) Blue Horizon (Figure 1.3b), a 1152-processor IBM SP capable of 1.7 trillion calculations per second. A 7.9 Tflops system is operational as of April 2004. This computing power is available for use through a peer-review system to academic researchers and students in the U.S. They are also available under special cost-sharing arrangements to commercial and government researchers in the U.S. and abroad. Currently, more than 5100 researchers at more than 240 institutions are using these platforms for their research (source: www.sdsc.edu).

Supercomputing does not just rely on centralized stations, but importantly relies on networking. The TeraGrid Project sponsored by the NSF made 4.5 Tflops of distributed computing power available at the beginning of 2004. The TeraGrid is a multi-year effort to build and deploy the world's largest, most comprehensive distributed infrastructure for open scientific research. The completed *TeraGrid* is scheduled to provide over 20 Tfolps of capability. This newest development has come a long way from older network capability described in this 1997 press release from the Pittsburgh Supercomputing Center:

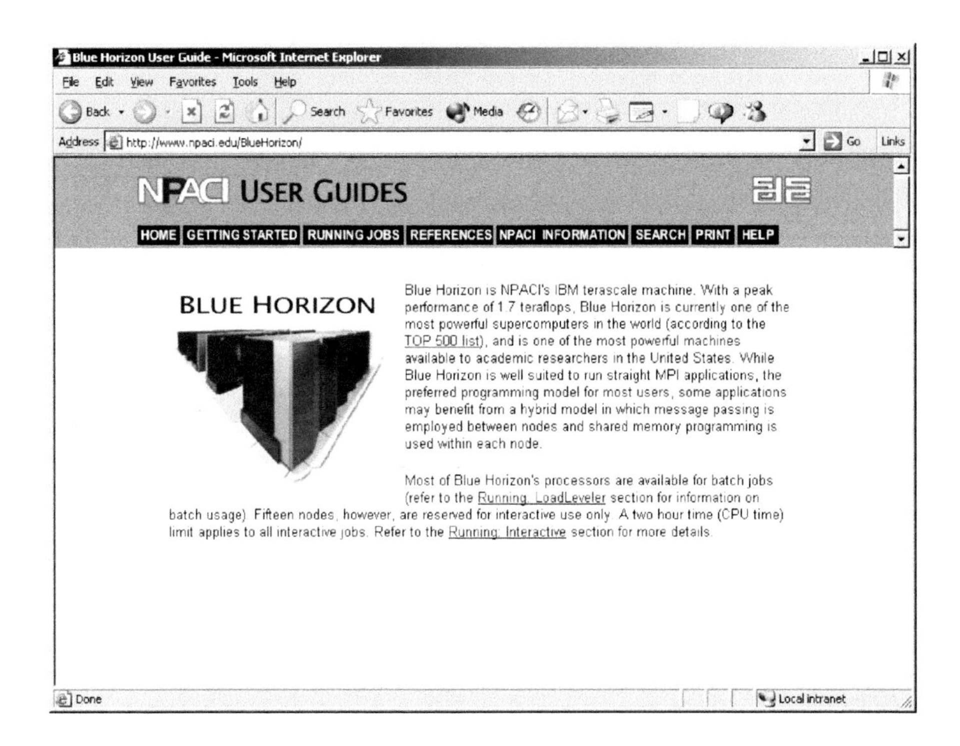

FIGURE 1.3B
Blue Horizon Supercomputer at SDSC.

HIGH-SPEED CONNECTIONS AND PARALLEL COMPUTING

On June 20, 1997 researchers at the Pittsburgh Supercomputing Center and the University of Stuttgart, Germany, linked supercomputers on both sides of the Atlantic via high-speed research networks. This is the first time that high-speed telecommunications networks, such as the very high speed Backbone Network Service (vBNS), have been used for transatlantic metacomputing. Intended as a prototype for international high-performance networking, the project couples Pittsburgh's 512-processor CRAY T3E with another 512-processor T3E at the High Performance Computing Center in Stuttgart (RUS). Linking two or more supercomputers at different locations in this manner, for work on the same computing task, is known as "metacomputing." The Pittsburgh–Stuttgart link creates a virtual system of 1024 processors with a theoretical peak performance of 675 billion calculations a second. The project relies on a series of research networks to create a high-speed transatlantic link between the two centers. Such networks, established during the past few years, allow information to move up to 100 times faster than on the Internet. The vBNS, for example, which connects U.S. supercomputing centers, currently transmits at speeds of up to 622 million bits per second, fast enough to transfer the complete *Encyclopedia Britannica* in less than 10 seconds (from PCS News; Michael Schneider; Pittsburgh Supercomputing Center; www.psc.edu).

The contribution of supercomputers to modern bioinformatics may simply be the access to computational power. Modern PCs are the building blocks of supercomputers, and the latter are no longer single, large, uniquely designed machines. In fact, anyone can potentially put together a supercomputer by integrating a large number of off-the-shelve PCs. Biologists, however, are mostly computer users, not computer scientists or network specialists. Supercomputer centers thus play an important role and fulfill three basic requirements that cannot be achieved using single, though powerful, personal computers. These tasks are:

Storage capacity

Simulation of complex systems

Machine learning

In fact, public database systems are located on supercomputer clusters and provide all the infrastructure and software necessary to store, retrieve, search, and analyze biological information on genes, genomes, proteins, and metabolic networks.

1.2.2 Limitations of Computational Tools

The tools used in bioinformatics are applied mathematics and computer science. Information storage and retrieval, statistical analysis, data fitting, and computer simulation are central tasks, and today's molecular biology would be impossible without them. Computers are essential in processing large amounts of data in a time-efficient manner that is otherwise inefficient through manual processing. Computers, however, need to come with instructions, and the analytical process that goes into the system is the work of the human operator and needs to be included in the overall time it takes to solve a problem with a numerical processor. Human intervention takes time and is error-prone. Thus, many modern processes guided by computers are automated with the future goal of teaching the machine to make decisions (a simple context recognition problem) with the human relegated to supervisory roles. The use of machines to perform human tasks is solved through expert systems. Expert systems are computer-based systems (e.g., robots) that perform narrowly defined tasks requiring enormous computational power. Real-life situations are never totally reproducible, and many decisions that are currently based on human intervention are thought to be solved by neural networks (NNs) that have the designed ability to learn. NNs, "although promising, are difficult to apply successfully to problems that concern manipulation of symbols and memory [and] there are no methods for training NNs that can magically create information that is not contained in the training data" (Warren S. Sarle, Cary, NC; from Neural network FAQ; ftp.sas.com/pub/neural/FAQ.html). Once an algorithm has been set up and successfully implemented, the computer goes through a repetitive task where input and output data may constantly be changed and adjusted to preset values through feedback and feed-forward loops. The goal is to let the machine make the adjustments, a process that today still depends largely on human input.

The real promise of computers, however, was *artificial intelligence* (AI). Regrettably, there is a real lack of progress in this area. Spell checkers are a good example of the pitfalls of simple algorithms that lack analytical "understanding" of language. The computer cannot find a typo that creates a correctly spelled word with a misplaced meaning. Spell checking is in its infancy when it comes to proofreading a text within its *context*. The analytical input of the human mind in recognizing correct spelling, style, and grammar is not satisfactorily solved using computers. And much like proofreading, analysis of scientific data and its subsequent interpretation are aided by computers, but useful only under the strict control of the human mind. Computers are excellent tools for numerical solutions (analysis, simulations), controlling and guiding machines, editing information/text, string searches, finding relationship in data, and managing databases. These are all crucial for bioinformatics.

The idea of biology as becoming a context-driven, that is to say predictable, science is an important aspect of bioinformatics. Predictability has been notoriously difficult in biology, and the role of theory in biology is very different from that of theoretical physics, which usually takes a leading role in research. Much of the difficulties of theoretical biology are rooted in the complexity of biological systems. Many cellular components and mechanisms remain to be discovered. One of the next big discoveries awaiting biology will be real-time interactions between components in biological networks and cellular structures. Genomics and proteomics provide good examples of the difficulties in predicting the behavior of complex system. The reasons are not conceptual, but lie in the nature of incomplete data set and incomplete or wrong data annotation.

Another big issue is the process of selecting the data that can be analyzed and used for interpretation. Data selection is a process that critically depends on the experience of the experimenter. Automated data selection requires filtering based on preset thresholds and statistical tests for sensitivity and specificity of signal detection. With the ever-increasing number of biological data, computer support of this process is critical, yet computer programs are only as good as their proper applications. The intuition and bias of the scientist is the most important factor in making the right decisions, automated processes are also safeguards against bias in data selection. Computer-aided data analysis can take out a case-by-case bias of the experimenter and read data in a consistent manner, and not in a manner that reflects the experimenter's biased vision.

While automation of data selection and analysis will eventually be done only by computers, interpretation of the results constitutes the most difficult challenge for computer science. Interpretation of biological data is largely a matter of identifying relationships within data. Relationships among components in biological systems can often be reduced to identifying patterns within the data. *Pattern recognition* is one of the most exciting and promising applications in bioinformatics. Yet it is important to reflect on the meaning of patterns and our cognitive abilities to interpret them correctly. Take the way we use dictionaries and other reference books. Quickly looking up a word is based on our physiological ability to recognize patterns of a set of x numbers of letters in alphabetical order. Of course, there are hundreds of alphabets that serve this purpose, as there are hundreds of languages. However, a language that creates words from a limited set of symbols is an optimal system for learning pattern recognition and may explain why books are such a useful, widespread, and stable form of communication and information storage. This pattern recognition may simply be referred to as analog pattern recognition as compared to digital, computer-based string searches. Unlike humans, a computer search does not depend on knowing an alphabet, because search strings check an entire document. By contrast, we are slow readers but can see information and relation in a visual representation. We create efficiency by creating order (patterns), not through speed of task. A randomized dictionary would be completely useless to humans, but not for a computer. That's why we invented the convention of saying that the letter H always comes before the letter J. With increasing amounts of data being stored, the inefficiency and errors (missed hits during search) increase in the absence of structured databases. This is why written words are relatively short, and why we see words and not letters when reading sentences.

How would we teach a computer to look up a telephone number exactly the way we do, based on the alphabetical listing of last names first, first names last? We would need to teach it the rules of the alphabet as commands that the machine could interpret as a string of hierarchies and priorities when searching and categorizing information (A before B, B before C, etc.). However, we need to demand not just simple search string patterns, but a task that is much more difficult for a computer to perform: that is, for it not to be precise. If the query string is misspelled, the computer will not find it. If you are looking up a name, however, you may be able to find it even if you are unsure about the correct spelling. In addition, you might find other information that you consider meaningful, even though you did not expect to find it. Computers and minds work very differently. The constant struggle against email spammers is a vivid example of the limited ability to use automated systems to detect unwanted mail (e.g., masking the word "advertisement" as "a%dverti$ement").

Computers can be programmed to find information in scientific projects much the same way as they can be used to find telephone numbers. Science is a human activity where finding relationships between objects is central. This means there is a need for quantification of relationships and numerical solutions for numerical relationships. Strings of information are searched for and compared, and it is the scientist who must define the quality of the string-matching process. Understanding how computers work and how scientists need the information laid out in order to extract meaning determines how successfully computers can be used in science.

→ input →

<u>Interpretation</u>/Man Machine/<u>Computation</u>

← output ←

We prefer a figure over a table of numbers anytime. We create 3-D illustrations in order to understand the meaning of numerical relationships derived from scientific inquiries; for example, we use color to represent almost any physical parameter, e.g., temperature, charge density, height, rigidity, etc.

In this age of computer networks, words like "virtual cell" have new meaning. With the help of computer networks within given hierarchies of interaction, physical–chemical knowledge of cellular dynamics can be modeled mathematically. This is the world needing supercomputers for complex graphical visualization of 3-D objects (molecule structures) and their dynamics (molecular dynamics). In a digitized cell, however, the number of atoms is literally astronomical. In computer simulations for structural movements of proteins, we not only have to include structural information such as size and relative orientation, but also the energy information of the strength of interaction between the atoms in the molecules. These are chemical bonds, dipole–dipole interactions, hydrogen bonds, and ion pairs. To exemplify this complexity and the lack of computational power—as well as the lack of mathematical algorithms—we have to look at the quantum mechanical description of molecules. These systems become so complex that only the simplest molecule—molecular hydrogen—has ever been modeled in quantum mechanical terms alone. All other molecular structures deal with shortcuts that include mixtures of classical mechanics and thermodynamics.

The smallest amino acid, glycine, contains ten atoms. Small proteins contain about 100 amino acids, big ones several thousand. This estimate equals several thousand atoms for a small protein, with several thousand proteins per cell. For the sake of simplicity, assume 10^4 atoms per protein and 10^4 proteins per cell. This corresponds to 100 million (10^8) atoms for all proteins of a cell. The smallest bacterial genome is about 15 million base pairs long. With an average of 70 atoms per base pair, this small genome has about one billion (10^9) atoms. Now we include all metabolites and water molecules and estimate a total of three billion atoms per single bacterial cell. If the position of every atom is defined by an average of five physical parameters, we need a spreadsheet containing 15 billion cells to store this information. We now calculate the dynamics of this system and would like to know where all the atoms are located one trillionth of a second later (a pico second; 10^{-12} s). In order to perform this task, we need 15 billion calculations. Using a 100-MHz computer it would take about 150 seconds to complete the calculation. Calculating

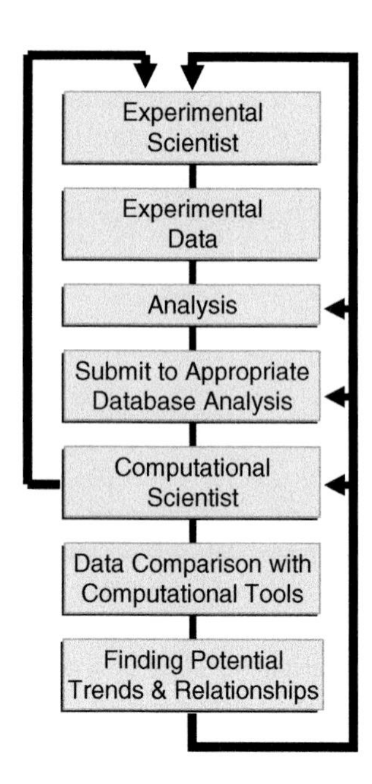

FIGURE 1.4
The marriage between experimental and computational science.

the changes for one second, one picosecond at a time, would take 1.5 million days—or more than 4000 years.

The reason for stressing this point and understanding that the limit of computer generated data depends on the correct input and handling of data by human operators, lies in the acknowledged difficulty of database annotations. It is no secret that the DNA and protein sequence databases contain errors when it comes to the background information on where and how these sequences have been obtained. The success of bioinformatics is directly linked to the complete and reliable annotation of database sequence information with accurate biological data. This process of checking annotation for accuracy is not automated and will not be for a while to come. Instead, it needs many specialists in the relevant fields to go over the information, word by word, to make databases a reliable source of information (Figure 1.4).

Comparing sequences of genes using programs like BLAST[3] is fairly easy. However, to truly understand what the result of such a comparison signifies can be extremely difficult. It depends on the type of sequences compared and the origin of the sequences. Both allude to the fact that besides the mere sequences one needs to know their functional structure and the means and cellular origins from which they have been obtained. In other words, one needs to understand the biology behind the sequence in order to understand the result of a sequence comparison with related or novel sequences. The genome projects produce, and this is of course their goal, mostly sequences for which no biological function is associated. The hope is to extract predictive information so as to design experiments to quickly and reliably corroborate the biology behind the sequence.

References

1. Sussman, J.L. et al., Protein Data Bank (PDB): database of three-dimensional structural information of biological macromolecules. *Acta Crystallogr. D. Biol. Crystallogr.*, 1998, 54(1 [Pt 6]), 1078–1084.
2. Barton, G.J., SCOP: structural classification of proteins. *Trends Biochem. Sci.*, 1994, 19(12), 554–555.
3. Altschul, S.F. et al., Basic local alignment search tool. *J. Mol. Biol.*, 1990, 215(3), 403–410.

1.3 The Virtual Doctor

Computer technology has become an important tool in medicine. But the incredible advancements in biology and medicine over the past half-century are driven by fast-paced technological progress not just by computers, but also optical technologies. Both technologies have been mutually beneficial to each other, evolving into scanning technology employing fast and precise lasers used to read and write molecular information. The same period has, not surprisingly, seen an exponential growth in biological data. The need for efficient and powerful data management systems has become quite obvious. New disciplines in functional magnetic resonance imaging (fMRI), robotic surgery, combinatorial chemistry, high-throughput drug screening assays, and computer assisted drug design (CADD) are gaining rapid influence in medical research. Computer-based technologies promote automation of manufacturing, screening, and analyses and help to prolong and enhance life. Computers are now an intricate part of biology and medicine, and the partnership between these silicon-based computer chips and carbon-based organisms has reached the state of merging man and machine.

Computers and robots are becoming integral partners in advanced medical procedures. The time efficiency and increased accuracy of these procedures are driving their advancement in today's as well as tomorrow's medicine. Emulating software is now used in surgery rehearsals and training at university hospitals and is a training tool for many surgical residents in various specialties and subspecialties. Virtual surgeries could improve the success of certain procedures, but are limited by their ability to anticipate problems that could arise during the actual surgery. This, in turn, will better prepare the surgical staff and will improve the surgery's outcome. Therefore, it is crucial for the emulating software to be as realistic as possible. To ensure realistic volume visualization of the regions displayed, the specification and representations of the anatomical regions in 3-D space must be very accurate. During virtual surgery simulation, the program must be able to detect and quantify changes at the operating level in order to manipulate the data output that enables the operator to visualize the anatomical region and the volume of interest. Noninvasive techniques have greatly facilitated these developments, eliminating the need for surgical procedures or the use of chemicals and x-rays. Two successful noninvasive techniques are fMRI and ultrasound, which make use of the physical properties of specific atoms or molecules shared by all matter. In addition, the development and deployment of expert systems will increase the ease and success of performing routine medical treatments.

Visualization of the human anatomy has been a great challenge for many centuries. Leonardo da Vinci's detailed anatomical drawings (dating from 1489) had important educational, diagnostic, and therapeutic impact during his lifetime. What Leonardo's paintings lacked was *in vivo* representation, which was not made possible until Wilhelm Conrad Roentgen discovered x-rays in 1895. X-rays were used to create the first *in vivo* medical imaging technique. This method was limited to 2-D representation of the cross-sections analyzed, and the information it gathered yielded little information about the organs themselves. Nevertheless, x-rays are very powerful medical tools that are still used today. The development of computerized tomography (CT) in the 1970s and magnetic resonance imaging (MRI) in the 1980s has enhanced cross-sectional imaging of anatomical features. The 2-D images obtained from these techniques have been instrumental in diagnosis and have also helped to improve anatomical teaching tools.

The drawbacks of 2-D images are their lack of volume information and the relationship of the features displayed in 3-D space. 3-D images would add a great deal of realism and relativistic information (tissue–tissue or organ–organ interactions). In a 2-D image, the information obtained is based on a two-coordinate system (x,y). In order to generate 3-D images from these 2-D pictures, information from a third dimension (z) is necessary. Therefore, multiple 2-D images with fixed x,y coordinates and a variable z coordinate are used to generate these 3-D images. Several 3-D reconstruction techniques have been employed to improve the quality and realism of the images portrayed and to allow the observer to get further involved in the displayed image. The "gray level gradient shading,"[1] the "generalized voxel model,"[2] and the reconstruction of the "visible human" model are a few of the most successful projects of the last decade. With the gray-level gradient technique, the computer uses the original tomographic data to compute a smooth surface normal with a high dynamic range. The generalized voxel model allows further exploration of the imaged data displayed. An example of its application is showing different organs in combination with others and allowing selective cutting of the 3-D images displayed. The most recent efforts have been concentrated on constructing the virtual "visible human."[3] In this project, the use of colored, cross-sectional images has enhanced the 3-D quality of the displayed object and has added a great deal of realism to the features portrayed. Computers have greatly enhanced medicine and they are here to stay.

1.3.1 Mapping the Human Brain

Bioinformatics transforms accuracy and speed of genetic analysis of diseases (see Section 5.1). Yet, almost every other aspect in medicine is entering the information age, too. Neuroscience in particular is a discipline with a rapidly increasing need in information processing. *Neuroinformatics* is an emerging medical discipline studying the link between cellular and anatomical structures to disease integrating information from anatomy, neurology, cognitive science, and psychology. The brain is a tremendously complex organ and is central to our understanding of what it means to be human. The early 1990s have seen the beginning of the mapping of the human brain from its architecture (anatomy) to functional networks and integrated atlases combining anatomy with gene expression patterns. The brain mapping initiative has become a large-scale project very much like the human genome project. And the brain, or rather its components—the neurons with their input/output functions—are used as a paradigm of understanding complex systems.

Modeling the way neurons are linked together in networks, the use of *neural networks* has become a form of computational approach to solve nonlinear equations and machine-learning algorithms. The cellular architecture of the brain has become the foundation of the architecture of mathematical modeling of complex systems composed of elements with a high degree of interconnection, simple scalar messages, and adaptive interaction between elements. Neural networks join a growing number of other biologically inspired mathematical models such as genetic algorithms and evolutionary computation.[4] They transcend the usual meaning of bioinformatics assuming a literal meaning, focusing on the flow of information *within* biological systems. Understanding how biological systems create and compute information will ultimately facilitate the merging of biological molecules with electronic circuits.

The brain receives attention from scientists for obvious reasons: our immediate experience with human behavior, consciousness, memory, sleep disorders, sensation, and pain, including that in phantom limbs. Recent computer-aided brain scanning techniques and the success of the Human Genome Project have revitalized attempts to map the human brain both anatomically and functionally. With these maps, it should one day be feasible to colocalize neuronal activity at the level of single sensory neurons as a function of motoneural tasks, learning words, or reading symbols. The newly coined term neuroinformatics promises for neuroscience what is currently happening for the human genome, namely, the integration of physical maps with functional data. Such integration promises mechanistic explanations of genetics from genome analysis and of human behavior from functional brain analysis. The availability of information through computer networks such as the Internet is a vital component of this undertaking. On April 2, 1993, the Human Brain Project was officially announced by the NIH as part of the

1990s *Decade of the Brain*, and is sponsored by the National Institute of Mental Health (NIMH; www.nimh.nih.gov/neuroinformatics). The Human Brain Project[5] is designed as a broad-based, long-term initiative to support research and development of advanced technologies and to open information super-highways (e.g., Internet) to neuroscientists and behavioral scientists.

The Human Brain Project consists of three subprojects: the Multi-Modal Imaging and Analysis of Neuronal Connectivity, or Connectivity Project; the *In Vivo* Atlases of Brain Development, or Atlas Project; and the Goal-Based Algorithms for 3-D Analysis and Visualization, or Algorithms Project. A number of distinct technologies need to be integrated into these projects, such as chemistry, animal models, computation, Web design, networking, and fMRI. These technologies include the design, synthesis, purification, and characterization of novel contrast agents (chemistry core); animal acquisition, care, breeding, surgery, and disposal (animal core); design and maintenance of a stable and effective computational environment; imple-mentation of algorithms associated with goal-directed imaging experiments; dissemination of those results to provide routine hardware and software support for computers, networks, Web page maintenance, and video environments; and finally, the development and maintenance of an imaging facility that uses nuclear magnetic resonance (NMR). To better understand the complexity of the human brain, the ultimate goal of the Brain Project is the "weaving of the informatics and neuroscience components" of brain research into one single unit. According to the California Institute of Technology:

> ...past progress in this Human Brain Project has been fueled by ongoing and dynamic interactions across the various individual Projects and Cores. This is aided by the fact that many of the personnel have expertise and interests that cross the artificial Project/Core boundaries. Work in the informatics components provides the neuroscience components with novel and more efficient ways of collecting, analyzing, and looking at information. This necessitates the computational components learning from the neuroscience components about how the data is collected, as well as what it is about the data that is interesting and important. Thus, we have a win–win situation where each component aids and is in return enriched by the other component. (Source: www.gg.caltech.edu/hbp)

Brain research is divided among myriad specialists, each requiring unique approaches, techniques, and methodologies attracting a wide range of specialists, from molecular biologists and electrophysiologists to cognitive scientists and philosophers. Together they pursue epistemological questions related to consciousness, believing that insights at the molecular level will ultimately reveal the "mystery" of the mind–body problem. While bridging the gap between molecules and consciousness seems difficult enough for methodological reasons, the real challenge in solving the mind–body problem is the lack of a common language among the many specialists, although many scientists believe that such a common language—a unified field—will one day exist.

In addition, there is a conceptual barrier that has yet to be overcome. The complexity of the brain and the effort to understand it are enormous. Feelings and analytical thinking, two central aspects of human consciousness, are elusive properties from a biological point of view. It is true that certain chemicals influence our consciousness, showing that it is firmly rooted in our brain (the biochemical correlate). However, very little is known about potential mechanisms that reveal how consciousness emerges from brain chemistry, including its spatial and temporal patterns of activity. Neurobiology suffers from large gaps in information that still exist among some of the life science disciplines. Saying that the whole is more than the sum of its parts is true, yet no model exists that could explain exactly how such a transcendence of the highest functional level might occur.

The framework of current research into consciousness is that the complex structure and function of the brain has an emergent property, the mind. While simulating the behavior of a complex system with defined components, the simple calculation of input/output functions still misses an essential feature of the mind—experience. In their need to understand functional, biological, and chemical structures, scientists use computers to generate functional "landscapes" in which the desired parameters (behavior) can be visualized. The electrostatic forces between atoms and molecules or action potentials in neurons are good examples. While we are able to visualize their activity and force fields, they remain "things" that are utterly foreign to our own experience and thinking. We have no conscious knowledge of repulsive forces between charged entities. We have limited firsthand understanding of attraction between two bodies based on experience with staying on ground, falling down, or riding a roller coaster. We have experience of hard objects and treat molecules in a similar way, modeling them as pool balls. The difference between a scientific measurement and actual experience can be best described when we define our own senses. Science is capable of analyzing how the eyes function based on electromagnetic radiation converted into action potentials and chemical neurotransmission inside the retina and visual cortex. But our own experience has nothing to do with wavelength, electrical charges moving around, and neurotransmitters binding to their receptors. Yet, this mechanistic description is all we currently have when studying the behavior of an insect, mouse, or bacterium. What we don't have and don't know how to simulate is the experience of the organism during these activities, as exemplified by the "radar" capability of bats in a famous essay written by the philosopher Thomas Nagel.[6]

To overcome our inability to "see" the atomic world and force fields outside of man's experience, we must create images of our own—translations, so to speak—to *see it* with our most powerful sense and to recreate what we think happens. False color imaging is one such powerful tool. The units of a physical parameter are transformed into a color code from red to blue. Because we do not have to read and compare numbers, our minds instantly "see" differences and identities distributed in space. This way, we can "see" temperature gradients (visualized infrared imaging) or molecular oxygen consumption in the brain to localize spots of high metabolic activity. False color imaging is a fascinating and aesthetically satisfying technique because it allows us to instantly transfer abstract mathematical formalism describing numerical relations—and science is all about numerical values—between objects into direct sensory inputs for our visual receptors (eyes). Here again, computational power is instrumental in the implementation of these graphical tools.

An integrated map of the human brain superimposes an anatomical and functional map in both space and time. The types of questions that can be asked depend on the level of resolution. For improved visualization of the spatial arrangements of functional brain units, topological maps, like cartography of the landscape, aim to produce rows of cross sections layered on top of each other. Functional brain mapping makes use of noninvasive techniques such as SPECT/PET (single photon emission computed tomography/positron emission tomography), fMRI, EEG (electroencephalography), MEG (magnetoencephalography), optical imaging, and neuroanatomical tools. These tools are then used to produce maps composed of cross-sectional images of the brain. Cross sections are put back together into a virtual 3-D image reconstruction of the brain.

Spatial resolution in the millimeter range (fMRI) and temporal resolution in the millisecond range (EEG, MEG) need to be correlated by computational means and result in movies of brain activity.[7] These brain maps have a resolution at the neuroanatomical level. The Whole Brain Atlas at Harvard Medical School (www.med.harvard.edu/AANLIB/home.html) is an example of a publicly available real-time map that compares normal brains with those affected by cerebrovascular diseases (strokes), neoplastic diseases (tumors), degenerative diseases (Alzheimer's, Huntington's), and inflammatory or infectious diseases (multiple sclerosis, AIDS-related dementia, Creutzfeld-Jakobs, herpes). These maps reflect the high spatial resolution obtained from hemodynamic or metabolic measurements, such as glucose levels and oxygen consumption with the high temporal resolution of their electromagnetic signals. This correlation is not trivial because the brain is a complex organ with anatomically distinct processing centers that communicate across large distances and several time scales shown in Color Figure 1.5.

However, spatial resolution in millimeters is by no means good enough for molecular studies of brain functions. Although neurons outgrow axons measuring several millimeters to meters (peripheral nervous system), the thickness of their cell bodies must be measured in micrometers or less when studying the functional morphology of their chemical synapses. Thus, neuroanatomical maps like The Whole Brain Atlas can be complemented with molecular details stemming from biochemical, physiological, pharmacological, and molecular biological studies, such as ion channel and receptor distribution (proteome), mRNA-level distribution (transcriptome), synaptic connectivity and plasticity responsible for the enormous complexity of neuronal networks. While the current information content and resolution level of The Whole Brain Atlas can pinpoint anatomical centers of activity for

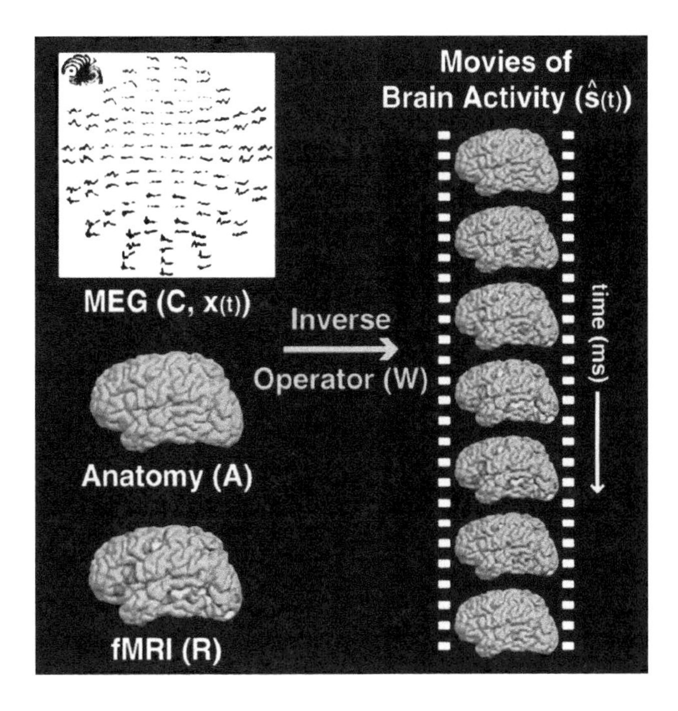

FIGURE 1.5 (See color insert following page 44.)
Spatiotemporal imaging of human brain activity using fMRI. (From Liu, A.K. et al., 1998, *PNAS* 95(15), 8945–8950. With permission.)

motoneuronal tasks, for example, they are not yet able to provide information about electrical activity patterns (firing patterns) and neurotransmitter selectivity of individual neurons or groups of neurons. In other words, the structural details are not yet linked to functional states, the temporal activity of the brain.

Computer-generated virtual images allow the view of objects from a perspective that we are otherwise unable to have. We can now think of ourselves as agents small enough to roam the inside of a body (movies are being made attempting just that, promoting an artistic view of the inside of man). In order to illustrate the relationship between structure and function, imagine a new house you are planning to build with an architect. You can see yourself inside using the space and going from room to room. It is not necessary to create a cross-section map of the house for visualization purposes because we know how its structure and function are related. We are familiar with the sensation of walking into the kitchen and having the smell of dinner awaken our senses, knowing that we are about to eat our favorite foods. We often like a particular food, not only for its taste, but because of the person who cooked it.

We do not have this insight for objects in biology. Understanding a fundamental experience like enjoying eating differs dramatically from scientists trying to understand a virus. As much as we can understand the joy of eating experienced by others, we obviously cannot use an analogy to understand a viral infection from the point of view of the virus. We are left with the mere mathematical description of a process of molecules interacting, virus particles "docking" to cell surfaces. And as much as we are aware of a food when eating it, we are totally oblivious to the process of digestion and nutrient absorption (unless there are problems). Our awareness of metabolic activity is as nonexistent as our ability to feel like a virus. Most of our bodily functions are beyond awareness, and it is a good guess that microbial life happens at a similar unconscious level. We do not feel or think with our liver, but with our brain. If only we could understand the difference.

References

1. Cao, Q.L. et al., Enhanced comprehension of dynamic cardiovascular anatomy by three-dimensional echocardiography with the use of mixed shading techniques. *Echocardiography,* 1994, 11(6), 627–633.
2. Schmidt, R. et al., Visualization of 3-D treatment plans with fast neutrons. *Strahlenther Onkol.*, 1992, 168(12), 698–702.
3. Slavin, K.V., The visible human project. *Surg. Neurol.*, 1997, 48(6), 638–639.
4. The Genetic Algorithms Archive; an archive maintained by Alan C. Schultz at The Navy Center for Applied Research in Artificial Intelligence; http://www.aic.nrl.navy.mil/galist/.
5. Shepherd, G.M., et al., The Human Brain Project: neuroinformatics tools for integrating, searching and modeling multidisciplinary neuroscience data. *Trends Neurosci.*, 1998, 21(11), 460–468.
6. Nagel, T., What is it like to be a bat?, *Philosophical Rev.*, 83, no. 4 (October 1974), 435–450.
7. Liu, A.K., et al., Spatiotemporal imaging of human brain activity using functional MRI constrained magnetoencephalography data: Monte Carlo simulations. *PNAS* 95(15), 1998, 8945–8950.

1.4 Biological Macromolecules as Information Carriers

The biological macromolecules central to life can be divided into four chemical groups: nucleic acids, proteins, carbohydrates, and lipids. Almost all of today's bioinformatics deals only with the first two classes. Thus, the chemical composition and function of proteins and nucleic acids in cells as well as the handling of nucleic acids will be discussed in this chapter. Nucleic acids store the hereditary information in the chromosomes of an organism. Together, chromosomes constitute the genome of an organism.

Genomes are composed of DNA or deoxyribonucleic acid while ribonucleic acid (RNA) serves as intermediary in protein synthesis and regulation. Proteins are amino acid polymers and their amino acid sequence correlates with the sequence of nucleotides found in DNA and messenger RNA (mRNA). Bioinformatics largely deals with the description and comparison of DNA and amino acid sequences. However, the crucial aspect of these sequences is their ability to code for protein structures, which in turn determines protein function. Protein and RNA enzymes are the "tools" used by the cell to read and translate its genomic information into other proteins for performing and controlling cellular processes—metabolism, physiological signaling, energy storage and conversion, and the formation of cellular structures. Thus, the structure–function relationship and the related phenomenon of the protein folding problem are central to information processes in biology.

The identity of proteins is inherited by means of DNA replication from generation to generation and translated via mRNA. This is akin to an information transfer from DNA sequence to amino acid sequence. The information is encoded in the genetic code. This code consists of 64 nucleotide triplets (codons) that define 1 out of 20 amino acids used in protein biosynthesis. The genetic code is the key to properly translating a sequence using four different nucleotides into a sequence using twenty different amino acids. The position of a nucleotide triplet in a gene defines the position of an amino acid in the protein. A unique feature of this process was observed in all bacteria, plants, and animals; translation is only possible from nucleic acid to protein and not vice versa. Nucleic acids serve as templates for protein synthesis, but genes cannot be synthesized from proteins. This is the central dogma of molecular biology and reflects the way all living, cellular organisms reproduce and grow.

Certain groups of viruses contain RNA molecules instead of DNA in their genome. One would expect, according to the central dogma, that these viruses use their RNA to directly guide protein synthesis in their host cells. While some do, it has been found that once inside the host cell, certain viruses use a protein that is able to *reverse*-transcribe the viral RNA genome into a DNA copy (so-called cDNA or complementary DNA). Once inside the host cell, the viral protein *integrase* catalyzes the stable insertion of this cDNA into the host genome. The protein responsible for the synthesis of DNA from RNA templates is called *reverse transcriptase* (RT). Together with a heat-stable DNA polymerase, this enzyme has revolutionized molecular biology in a laboratory process called PCR.

The genetic code is universal and redundant. Knowing this is important in understanding the relationship between the genomic organization of an organism and the structure and compositions of its proteins, cellular structures, and the *diversity of life*. The genetic code is redundant because from a total of 64 possible triple-base combinations (codons; Figure 1.6) available to select among a set of 20 amino acids, 3 codons are reserved for stop or termination signals, while only methionine, the start codon, and tryptophan

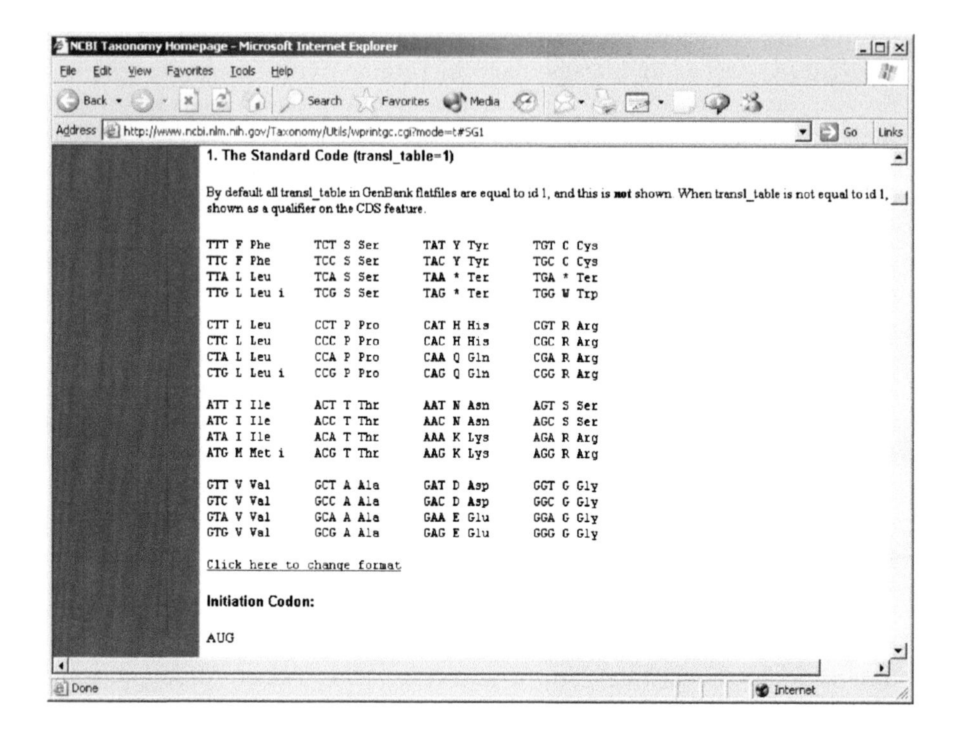

NCBI Taxonomy Homepage - Microsoft Internet Explorer

Address http://www.ncbi.nlm.nih.gov/Taxonomy/Utils/wprintgc.cgi?mode=t#SG1

1. The Standard Code (transl_table=1)

By default all transl_table in GenBank flatfiles are equal to id 1, and this is **not** shown. When transl_table is not equal to id 1, shown as a qualifier on the CDS feature.

TTT F Phe	TCT S Ser	TAT Y Tyr	TGT C Cys
TTC F Phe	TCC S Ser	TAC Y Tyr	TGC C Cys
TTA L Leu	TCA S Ser	TAA * Ter	TGA * Ter
TTG L Leu i	TCG S Ser	TAG * Ter	TGG W Trp
CTT L Leu	CCT P Pro	CAT H His	CGT R Arg
CTC L Leu	CCC P Pro	CAC H His	CGC R Arg
CTA L Leu	CCA P Pro	CAA Q Gln	CGA R Arg
CTG L Leu i	CCG P Pro	CAG Q Gln	CGG R Arg
ATT I Ile	ACT T Thr	AAT N Asn	AGT S Ser
ATC I Ile	ACC T Thr	AAC N Asn	AGC S Ser
ATA I Ile	ACA T Thr	AAA K Lys	AGA R Arg
ATG M Met i	ACG T Thr	AAG K Lys	AGG R Arg
GTT V Val	GCT A Ala	GAT D Asp	GGT G Gly
GTC V Val	GCC A Ala	GAC D Asp	GGC G Gly
GTA V Val	GCA A Ala	GAA E Glu	GGA G Gly
GTG V Val	GCG A Ala	GAG E Glu	GGG G Gly

Click here to change format

Initiation Codon:

AUG

Done Internet

FIGURE 1.6
The genetic code—the standard code. Translation table site for the translation table 1 (includes human) at NCBI's Taxonomy site. A total of 11 codes have been described in nature, including in eukaryotic organelles and the alternative yeast nuclear code. The bacterial code (table 11) is identical to the standard code, except that codons (i) for alanine, leucine, and isoleucine, including GUG and UUG, are documented as alternate start codons in Archaea and Bacteria. Standard start codon is AUG.

are coded for by just one single codon. All other amino acids are coded for by 2 or more codons. This means that the amino acid sequence is more conserved than the DNA sequence. This has implications for the mechanism of evolution because some single-base exchanges do not result in changes in amino acid sequence. These single-point mutations are also called silent mutations because they do not affect the phenotype and therefore are not subject to selective pressure. Such mutations are fairly common particularly in noncoding regions of genomes, and their distribution within a population of individuals is known as single nucleotide polymorphism (SNP), affecting about 1 in every 300 to 500 bases.

Redundancy, however, is often limited because organisms do not make use of the full set of codons, but instead prefer certain codons for a specific amino acid, an organism-specific preference known as codon usage or bias. Codon bias is widespread among microorganisms. For instance, of the six codons for the amino acid arginine, all six are used in human genes varying from 9 to 22%, while in *E. coli* two codons are preferred, with CGC used in 39% and CGU in 49% of all arginines found in proteins.

Codon bias is important for recombinant DNA technology experiments where genes of one organism are cloned and recombined into another organism's genome for manipulation and expression purposes. In some cases, codon bias can result in nonfunctional proteins or affect the level of protein synthesis. Codon bias can also serve as protection against foreign DNA from pathogenic organisms. Unused codons have the effect of stop codons in foreign DNA, effectively inhibiting synthesis of functional proteins necessary for the reproduction of pathogenic organisms.

Codon bias, however, is the result of an otherwise universal use of the genetic code. Except for organellar and some ciliate nuclear DNA, all organisms, including viruses, assign the same codons for the 20 amino acids used for protein synthesis (Figure 1.6). Alternate uses of codons often affect start and termination signals. This means that genes can be transferred between organisms, which are the basis for today's biotechnology industry. It also means that bioinformatics does not have to distinguish sequence information based on its cellular origin. This universality increases the sample base for the statistical analysis of DNA sequences stored in databases, and allows well-studied genes from model organisms such as rat, fruit fly, or *E. coli*, to be easily compared with their human counterparts. DNA sequence similarity allows for the rapid identification, cloning, and sequencing of genes in related organisms. In addition, missing biological information in one organism can be inferred, although with caution, from other species. What "works" in a fruit fly or a nematode may well be compared to human metabolism.

The analysis of lipid and polysaccharide diversity—unlike that of DNA and amino acid sequences—demands a more complex level of analysis, as it cannot be directly deduced from a genetic sequence template. Rather, the lipid molecular diversity and carbohydrate structures found on glycoproteins and glycolipids is the result of spatial and temporal metabolic activity of membrane-bound enzymes in various organelles. Functional genomics and proteomics analyses of gene and protein interaction *networks* will likely point the way in how to deal with the 3-D and temporal expression patterns that lead to the accurate and reproducible formation of carbohydrate (the glycome[1]) and lipid moieties (the lipidome[2]) that serve as receptor and ligands in cellular signaling processes.

The challenges of deciphering the genetics and evolution of carbohydrate and lipid structures are similar to those encountered with epigenetic studies of chromosomal control of gene expression and replication in cells. Here, post-translational modification mechanisms are highly conserved mechanisms of sequence modifications affecting packing parameters of chromatin, an essential part of cell differentiation and growth control.[3]

References

1. Khersonsky, S.M. et al., 2003, Recent advances in glycomics and glycogenetics. *Curr. Top. Med. Chem.*, 3(6), 617–643.
2. Forrester, J.S. et al., 2004, Computational lipidomics: a multiplexed analysis of dynamic changes in membrane lipid composition during signal transduction. *Mol. Pharmacol.*, 65(4), 813–821.
3. Jablonka, E. and Lamb, M.J., 2002. The changing concept of epigenetics. *Ann. NY Acad. Sci.*, 981(1), 82–96.

1.5 Proteins: From Sequence to Structure to Function

Amino acids are the building blocks of all proteins. Understanding the fundamental features of proteins requires a profound grasp of its amino acid substituents.

Amino acids have both an amino and a carboxylic acid group (Figure 1.7). The amino group adds to their basic nature while the carboxylic acid terminus contributes to their acidity. At physiological pH both termini are charged. This means that the amino terminus stays protonated while the carboxyl terminus sustains its deprotonated form. The distinguishing feature of an alpha amino acid is its residue or side chain. The residues are generally known as the R groups, and the characteristic R groups are the distinguishing features of each of the twenty alpha amino acids. Some of these residues are acidic, some are basic, and the remaining residues are relatively neutral.

Acidic residues: glutamate (E) and aspartate (D). These are the conjugate forms of the glutamic acid and aspartic acid residues. At physiological pH they are deprotonated and hold a negative charge.

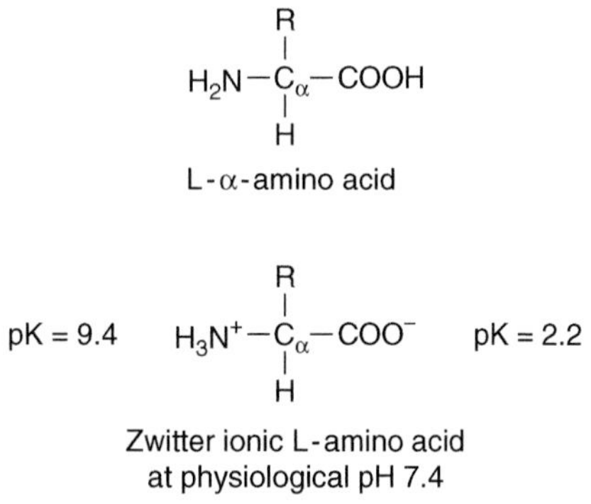

FIGURE 1.7
Chemical structure of an L-α amino acid and its zwitterionic property. An L-α amino acid contains a central alpha-carbon and four chemical substituents. R represents the side chain with chemical and physical properties outlined in the text. NH_2 and COOH represent the basic amino and acid carboxylic groups which are always charged at physiological conditions.

R
$H_2N-C_\alpha-COOH$
H
L-α-amino acid

R
pK = 9.4 $H_3N^+-C_\alpha-COO^-$ pK = 2.2
H
Zwitter ionic L-amino acid
at physiological pH 7.4

Basic residues: lysine (K) and arginine (R). Lysine and arginine are relatively basic at physiological pH. Their basic character promotes their residues to be protonated by their environment and to maintain a positive charge.

Amino acids are also characterized by their relative hydrophobicity. Some are relatively hydrophobic while others tend to prefer polar environments. The relative hydrophobicity of the residue enables us to predict its position within the protein structure. Hydrophobic residues are generally found within the protein core, while the polar residues are predominantly found on the surface of the protein structure, interacting with the aqueous environment. The chemistry concept of "like dissolves like" is also applicable to biological systems. Hence, hydrophobic–hydrophobic interactions in most biological systems are preferred to competing hydrophobic–hydrophilic interactions.

Glutamate, aspartate, lysine, and arginine are charged at physiological pH and are predominantly found on the protein surface, interacting with the polar environment. In general, charged molecules prefer polar environments. This is predominantly due to the polar environment's charge-stabilizing feature (e.g., hydrogen bonding, electrostatic interactions).

Alanine, valine, leucine, isoleucine, phenylalanine, methionine, glycine, cysteine, and tryptophan are widely accepted as relatively hydrophobic and predominantly found in the protein core and other hydrophobic environments. The presence of hydrocarbon chains in these residues adds to their hydrophobic nature.

Asparagine, glutamine, proline, serine, and threonine residues are the uncharged polar moieties that have a tendency of being solvent exposed.

Tyrosine's hydroxyl group adds to its hydrophilic nature, while its aromatic side chain contributes to its hydrophobic characteristic. The dual nature of this residue makes it suitable for either environment.

Histidine is relatively polar. Conformational changes of its ringed side chain contribute to its wide pKa range and add to its amphoteric nature. Histidine can be either protonated or deprotonated, depending on its environment. This feature makes it a suitable Schiff base in the active site of many enzymes (Table 1.2 and Table 1.3).

1.5.1 Molecular Interactions in Protein Structures

1.5.1.1 The Peptide Bond

Amino acids are linked together via peptide bonds. This is basically a condensation reaction that results in the loss of a water molecule (Figure 1.8a). In order to better understand the protein's backbone conformation, one needs to comprehend the nature of the peptide bond and its relationship to the polypeptide's backbone structure. The peptide bond is a special bond that restricts certain angular conformations within the protein (Figure 1.8b).

TABLE 1.2

General Characteristics of Alpha Amino Acids

Residue	1 Letter Code	Hydrophobic	Aromatic	Aliphatic	Small	Polar	Charged
Alanine	A	X	–	–	X	–	–
Arginine	R	–	–	–	–	X	X
Asparagine	N	–	–	–	X	X	–
Aspartate	D	–	–	–	X	X	X
Cysteine	C	–	–	–	X	X	X
Glutamate	E	–	–	–	–	X	X
Glutamine	Q	–	–	–	–	X	–
Glycine	G	X	–	–	X	X	–
Histidine	H	X	X	–	–	X	X
Isoleucine	I	X	–	X	–	–	–
Leucine	L	X	–	X	–	–	–
Lysine	K	X	–	–	–	X	X
Methionine	M	X	–	–	–	–	–
Phenylalanine	F	X	X	–	–	–	–
Proline	P	–	–	–	X	–	–
Serine	S	–	–	–	X	X	–
Threonine	T	–	–	–	X	X	–
Tryptophan	W	X	X	–	–	X	–
Tyrosine	Y	X	X	–	–	X	–
Valine	V	X	–	X	–	–	–

TABLE 1.3

Commonly Used Hydrophobicity Scales

Residue	1 Letter Code	Kyte/Doolittle (1)	Edelman (2)	Eisenberg (3)	von Heijne (12)
Alanine	A	1.8	0.4397	0.26	0.267
Arginine	R	–4.5	–0.7010	–1.80	–2.749
Asparagine	N	–3.5	–1.414	–0.64	–1.988
Aspartate	D	–3.5	–2.588	–0.72	–2.303
Cysteine	C	2.5	1.150	0.04	1.806
Glutamate	E	–3.5	–1.270	–0.62	–2.442
Glutamine	Q	–3.5	–1.656	–0.69	–1.814
Glycine	G	–0.4	–0.8634	0.16	0.160
Histidine	H	–3.2	0.0268	0.40	–2.189
Isoleucine	I	4.5	1.546	0.73	0.071
Leucine	L	3.8	1.517	0.53	0.623
Lysine	K	–3.9	–1.502	–1.10	–2.996
Methionine	M	1.9	1.740	0.26	0.136
Phenylalanine	F	2.8	0.4345	0.61	0.427
Proline	P	–1.6	–1.721	–0.07	–0.451
Serine	S	–0.8	–0.3841	–0.26	–0.119
Threonine	T	–0.7	–0.0078	–0.18	–0.083
Tryptophan	W	–0.9	–0.0638	0.37	–0.875
Tyrosine	Y	–1.3	–0.4585	0.02	–0.386
Valine	V	4.2	0.5056	0.54	0.721

Note: Some are more popular than others, but the important values are those that are consistent in different scales within the table. The positive values represent the hydrophobic residues.

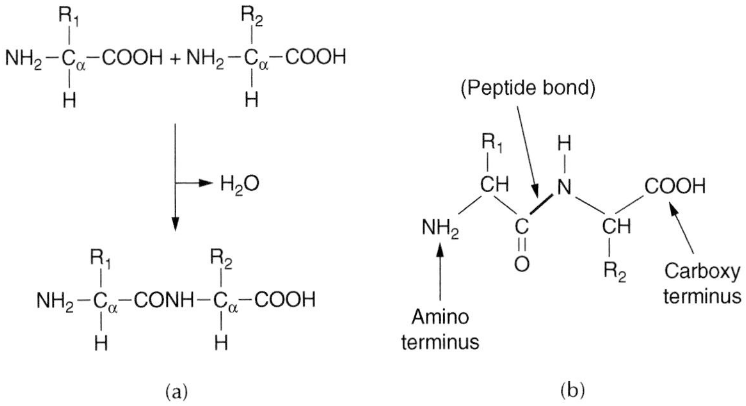

FIGURE 1.8
Formation of the peptide bond. (a) Two amino acids covalently link through their amino and carboxyl group in a condensation reaction releasing one molecule of water and forming a dipeptide (R1 and R2 linked in the same molecular structure). (b) The peptide bond forms a planar structure called the amide plane. The covalent bond between the N–H and C=O group is not flexible unlike the C–C and C–N bonds which allow proteins to fold into complex 3-D structures.

This restriction further limits the total number of possible 3-D conformations for the polypeptide's backbone structure.

1.5.1.2 Characteristics of the Peptide Bond

The peptide bond has a double-bond characteristic that is due mainly to the resonance system present between its substituents (Figure 1.9). The resonance-stabilized nature or double-bond character of the peptide bond contributes to its rigidity and limits its angular rotation. The angular rotation about the peptide bond is also known as omega and is relatively restricted in the polypeptide chain. The double-bond nature of the peptide bond restricts its substituents to a plane, and this planarity restricts most of the

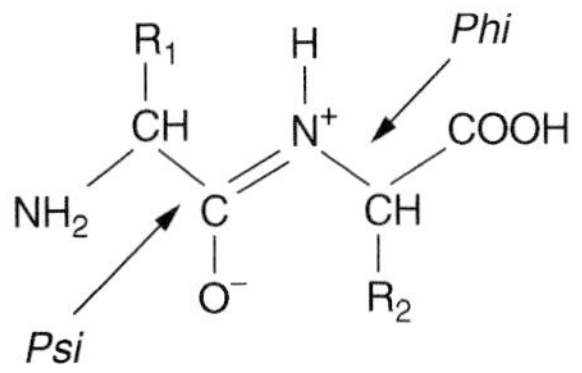

FIGURE 1.9
Electronic resonance character of peptide bond. Evidence for this structure comes from x-ray crystallographic studies of simple peptides showing that the N–C_α bond length is 1.46 Å as expected for a single bond. The C–N peptide bond is 1.33 Å long, only a little longer than the value of 1.27 Å for the average C=N bond length in model compounds. Similar x-ray studies show the six atoms $C_\alpha NHCOC_\alpha$ very close to being coplanar.

omega angles to 180°. This is in part due to the *cis* nature of most dipeptides. In those rare cases where a transconformation is present, the omega angle conforms to a 0° conformation. The *trans*-conformation is a characteristic of the proline residue. Figure 1.9 displays the partial double character of the peptide bond in one of its resonance forms. The phi and psi angles are also marked.

Phi and psi are the two main angular conformations associated with the polypeptide's backbone structure and are used in a 2-D plot to represent a 3-D conformation of the protein backbone. This plot is known as Ramachandran plot and is shown in Color Figure 1.10. Phi is the angular freedom between the alpha carbon and the neighboring N–H group, while psi is the angle between the C-alpha and the attached carbonyl substituent.

The protein's linearly extended conformation is also known as its primary structure. Understanding the primary structure of the protein could yield an insight into the molecule's 3-D structure.

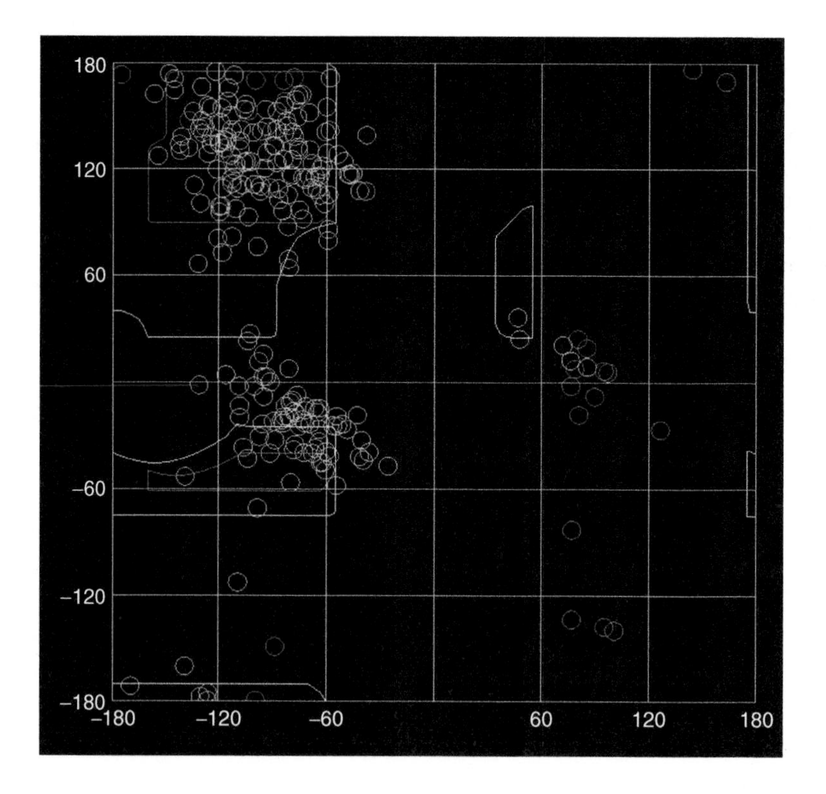

FIGURE 1.10 (See color insert)
Ramachandran plot. Ramachandran Map of 1est.pdb (accession file of Protein Data Bank). Tosylelastase (E.C.3.4.21.11) is a serine protease from porcine pancreas. This enzyme hydrolyzes (cleaves) peptide bonds. The circles in the upper left corner indicate extensive beta strand formation and the group of circles middle-left the presence of two short alpha helical secondary structures. The structure was solved by L. Sawyer et al. in 1976.

The following types of interactions drive protein folding:

- Hydrophobic interactions or hydrophobic effect
- Electrostatic interactions
- Hydrogen bonding
- Conformational entropy
- Van der Waals interactions (packing)
- Covalent bonds (e.g., disulfide bridge)

1.5.1.3 The Hydrophobic Effect and How It Contributes to Protein Folding

The hydrophobic effect is the preference of hydrophilic or polar molecules to stay together and away from hydrophobic molecules. This is readily observed when you mix water with oil. The water molecules are hydrophilic and thus have a tendency to interact with other water molecules through preferred hydrogen bonds and to reduce their association with the nonpolar oil molecules. At room temperature, this effect is predominantly entropically driven.

The hydrophobic effect is widely accepted as the principal driving force (energy source) in the protein-folding pathway (Figure 1.11). The environment around most proteins is an aqueous one. Since water molecules are constantly forming and breaking hydrogen bonds with other water molecules, the presence of nonpolar side chains in proteins would hinder such a

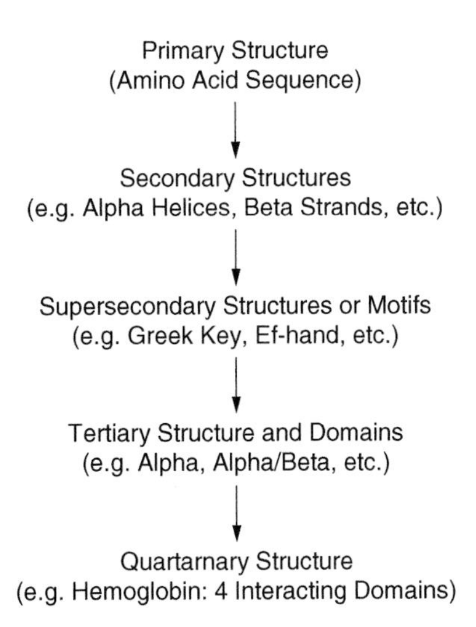

FIGURE 1.11
Protein-folding pathway.

conformational sampling scheme that is entropically favored in the aqueous environment. Hence, the burial of these nonpolar side chains is entropically favorable for an energetically sound coexistence between hydrophobic regions of the protein molecule and its aqueous environment.

How Do We Measure the Relative Hydrophobicityof a Side Chain?

This is usually done by using partition experiments. Early experiments by Fauchere and Pliska obtained a hydrophobicity value for each of the side chains by measuring their concentrations through a modeled compound in a medium that represented the protein core and its aqueous environment.[4] The medium they used was octanol. Octanol's long aliphatic chain and terminal hydroxyl group made it a suitable choice for interacting with the nonpolar and polar side chains. Some of the hydrophobicity scales and values are presented in Table 1.3.

What is the Accessible Surface Area (ASA) and Its Relationship to Hydrophobicity?

The accessible surface area, or ASA, of the solute is defined as the locus of the center of the water probe as it rolls about the surface of the solute (Figure 1.12).[5]

What is the Relationship between the Hydrophobicity of the Solute and Its ASA?

The ASA of nonpolar atoms (Table 1.4) of the extended side chains are to an approximation linearly related to their hydrophobicity.[6] This is not an exact science and the relationship found requires further analysis.

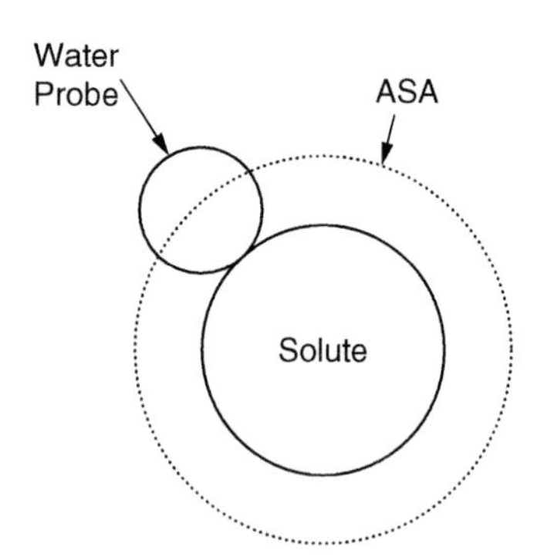

FIGURE 1.12
Solvent-accessible surface area: a solute in water has a specific surface known as the Van der Waals surface. Water molecules will interact with the solute surface and occupy some space, which essentially increases the volume of the solubilized molecule. The apparent surface determined by the center of a spherical water molecule rolled over the Van der Waals surface of the solute determines the water-accessible surface area and represents a useful volume of the solute in modeling drug-receptor interactions.

TABLE 1.4

Nonpolar ASA and Hydrophobicity

Residue	1 Letter Code	Hydrophobicity (8)	Nonpolar ASA (Å²)
Glycine	G	–0.06	33
Alanine	A	–0.20	71
Cystein	C	–0.67	98
Valine	V	–0.61	116
Proline	P	–0.44	120
Isoleucine	I	–0.74	140
Leucine	L	–0.65	143
Methionine	M	–0.71	159
Phenylalanine	F	–0.67	164
Tryptophan	W	–0.45	186
Tyrosine	Y	0.22	135
Threonine	T	0.26	76
Histidine	H	–0.04	96
Serine	S	0.34	48
Glutamine	Q	0.74	53
Asparagine	N	0.69	45
Glutamate	E	1.09	61
Aspartate	D	0.72	50
Lysine	K	2.00	118
Arginine	R	1.34	80

Source: From Miller, S., et al., *Nature*, 1987, 328(6133), 834–836. With permission.

1.5.1.4 Electrostatic Interactions

This generally refers to the electrostatic interactions present between the ion pairs within the protein molecule. Electrostatic interactions are believed to yield specificity to the protein structure. The effects associated with these interactions are, for the most part, governed by Coulomb's Law. The charge of the ion pairs, the distance between the charged groups, and the relative dielectric constants are the key governing features of Coulomb's Law and necessitate maximum interaction between the ion pairs in the protein and its aqueous environment. This is predominantly to maximize a thermodynamically favorable setting for all charged ion pairs. Isolated charges are typically present on the aqueous surface and fully solvated by the water molecules. Disrupting the solvation shells around these isolated charged groups is thermodynamically unfavorable. Therefore, transferring these isolated charged groups to the protein core is less likely to occur. Studies have shown that solvent-exposed charged pairs and isolated groups contribute to the overall stability of the protein molecule. One salt bridge is believed to contribute 0.65 (±0.35) kcal/mol of energy.[8]

1.5.1.5 Hydrogen Bonding

It is not quite clear how much hydrogen bonds contribute to the overall stability of the protein structures, but their periodic arrangement within alpha helices and between beta strands provides a plausible arrangement for the formation of secondary structures. Hydrogen bonds are also involved in side chain–side chain and side chain–main chain interactions.[9] The main reason for a doubt in the contribution of buried hydrogen bonds towards protein stability could be attributed to the competition between protein–solvent and protein–protein hydrogen bonds.

1.5.1.6 Conformational Entropy

The 3-D fold of the protein structure limits the sampling scheme of the backbone and side chain (angular) conformations, which is entropically not favored. The Boltzmann equation could be used to estimate the relative loss of conformational entropy when a side chain is restricted to a single rotamer. The restriction of these rotameric conformations is predominantly due to the burial of the side chain within the protein structure.

1.5.1.7 Van der Waals Interactions (Packing)

The folded structure of most proteins favors close packing of its atoms within the molecule. The core atoms are generally more ordered than those that face the surface of the protein molecule. Upon folding, the core of the protein structure is more solid than its surface or unfolded state. This liquid-to-solid transition is an enthalpy-driven event and believed to contribute around 0.6 kcal/mol per CH_2 substituent.[10]

1.5.1.8 Covalent Bonds (e.g., Disulfide Bridge)

The most significant covalent bond within a protein structure is the disulfide bridge. The disulfide bridge is believed to stabilize the folded structure by restricting certain degrees of freedom of the unfolded chain compared to the same chain without the covalent link. This is entropically driven. Generally, increasing the length of the covalent link corresponds to an increase in stability.[11] This rule is only applicable to a single disulfide bond. The stabilizing effect of the disulfide bond is more profound in smaller proteins.

1.5.2 Protein Functions

Proteins are important biological macromolecules involved in a variety of functions.

1.5.2.1 Enzymes

These are the biological catalysts. Most of the known enzymes are proteins, without which, life as we know it would cease to exist. Their presence is essential to speeding up biological reactions that are otherwise too slow to sustain life. Enzymes are generally substrate specific and their efficiency is typically dependent on the concentration of the substrate in the cell. This dependency, among others, prevents the enzyme from overproducing products that could overstimulate a response and ultimately cause a catastrophic cellular event. The functional specificity of the enzyme is tightly linked to its structure in 3-D space. The 3-D conformation of its active site creates its specificity and differentiates one enzyme from another. Therefore, a better understanding of its structural characteristics enables us to have a better grasp of its functional roles. Understanding structural characteristics of proteins with known structures is essential to finding structure–function relationships for sequences whose structures are yet to be determined. A structure–function understanding of these vital proteins could serve as a powerful tool in gaining control over their activities in the event of a malfunction. A malfunction in the activity of these enzymes is believed to be responsible for a variety of pathogenic events.

1.5.2.2 Regulatory Proteins

These proteins are mainly involved in regulating the activity of other macromolecules within the cell. The concentration of these proteins is responsible for this regulation process. Many of these proteins are involved in a negative feedback regulatory mechanism. In most negative feedback loops, an increase in the concentration of a downstream product hinders the formation of the upstream product. Most feedback regulations are either at the level of transcription (DNA) or at the translation (RNA) level.

1.5.2.3 Storage

Certain ions, metabolites, or small molecules could be complexed with proteins for the sake of storage. For instance, ferritin stores iron in the liver by complexing the iron ion through its heme group.

1.5.2.4 Transportation

Certain proteins act as biological transporters. Transferrin and hemoglobin are transporter proteins that carry iron and oxygen, respectively, throughout the body. A majority of transporters are found in cell membranes facilitating diffusion and active transport of metabolites in and out of cells and organellar compartments.

1.5.2.5 Signaling

Some proteins are specifically involved in the transmission of biological and cellular signals. Many of these proteins are cellular receptors for small molecules and hormones. Binding of the small molecule or hormone to its respective receptor could cause a signal that is ultimately translated into a cellular response.

1.5.2.6 Immunity

Most of the macromolecules involved in our immune system are proteins and polypeptides. Immunoglobulins are great examples of a large family of proteins involved in a variety of immunocellular responses.

1.5.2.7 Structural

A great portion of our proteins have a structural role. These proteins are mainly for mechanical support. Collagen is probably one of the most abundant structural proteins found in all multicellular organisms. It occurs in almost every tissue and is the basis of many cells.

References

1. Kyte, J. and Doolittle, R.F., A simple method for displaying the hydropathic character of a protein. *J. Mol. Biol.*, 1982. 157(1), 105–132.
2. Edelman, J., Quadratic minimization of predictors for protein secondary structure. Application to transmembrane alpha-helices. *J. Mol. Biol.*, 1993. 232(1), 165–191.
3. Eisenberg, D., et al., Analysis of membrane and surface protein sequences with the hydrophobic moment plot. *J. Mol. Biol.*, 1984, 179(1), 125–142.
4. Fauchere, J.L., et al., Amino acid side chain parameters for correlation studies in biology and pharmacology. *Int. J. Pept. Protein Res.*, 1988, 32(4), 269–278.
5. Richards, F.M. and Richmond, T., Solvents, interfaces and protein structure. *Ciba Found. Symp.*, 1977, 60, 23–45.
6. Chothia, C., Hydrophobic bonding and accessible surface area in proteins. *Nature*, 1974, 248(446), 338–339.
7. Miller, S., et al., The accessible surface area and stability of oligomeric proteins. *Nature*, 1987, 328(6133), 834–836.
8. Serrano, L., et al., The folding of an enzyme. II. Substructure of barnase and the contribution of different interactions to protein stability. *J. Mol. Biol.*, 1992, 224(3), 783–804.
9. Baker, E.N. and Hubbard, R.E., Hydrogen bonding in globular proteins. *Prog. Biophys. Mol. Biol.*, 1984, 44(2), 97–179.
10. Nicholls, A., Sharp, K.A., and Honig, B., Protein folding and association: insights from the interfacial and thermodynamic properties of hydrocarbons. *Proteins*, 1991, 11(4), 281–296.
11. Harrison, P.M. and Sternberg, M.J., Analysis and classification of disulphide connectivity in proteins. The entropic effect of cross-linkage. *J. Mol. Biol.*, 1994, 244(4), 448–463.
12. von Heijne, G. *Sequence Analysis in Molecular Biology: Treasure Trove or Trivial Pursuit*? 1987, 98–103, Academic Press, San Diego, CA.

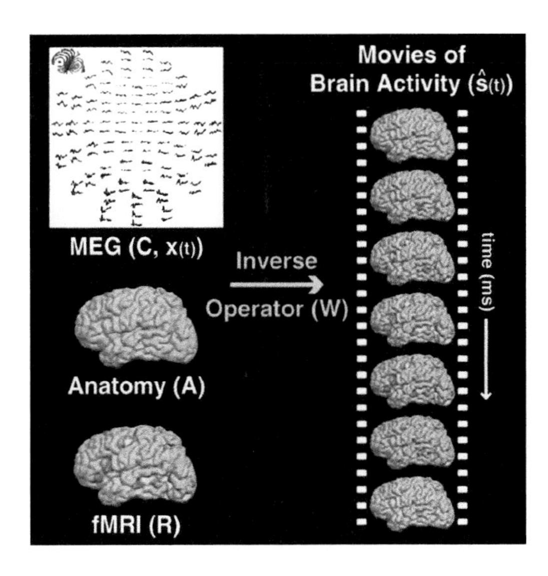

COLOR FIGURE 1.5
Spatiotemporal imaging of human brain activity using functional MRI. (From Liu, A.K. et al., 1998, *PNAS* 95(15), 8945–8950. With permission.)

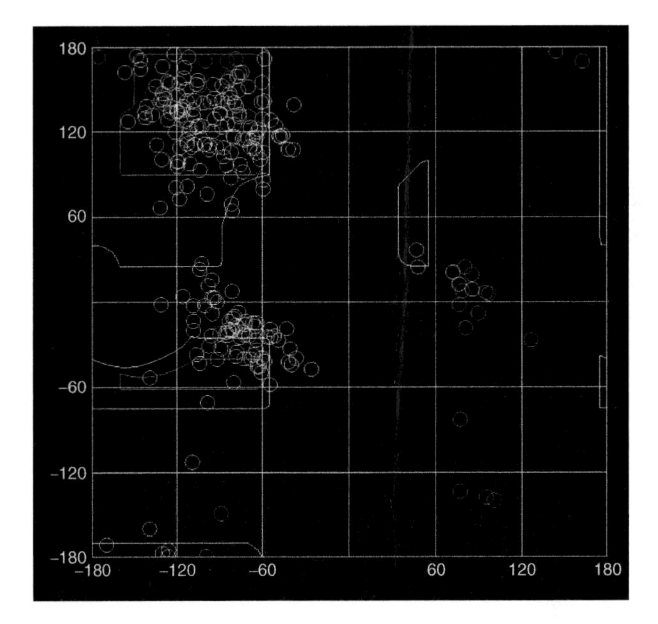

COLOR FIGURE 1.10
Ramachandran plot. Ramachandran Map of 1est.pdb (accession file of Protein Data Bank). Tosyl-elastase (E.C.3.4.21.11) is a serine protease from porcine pancreas. This enzyme hydrolyzes (cleaves) peptide bonds. The circles in the upper left corner indicate extensive beta strand formation and the group of circles middle-left the presence of two short alpha helical secondary structures. The structure has been solved by L. Sawyer et al. in 1976.

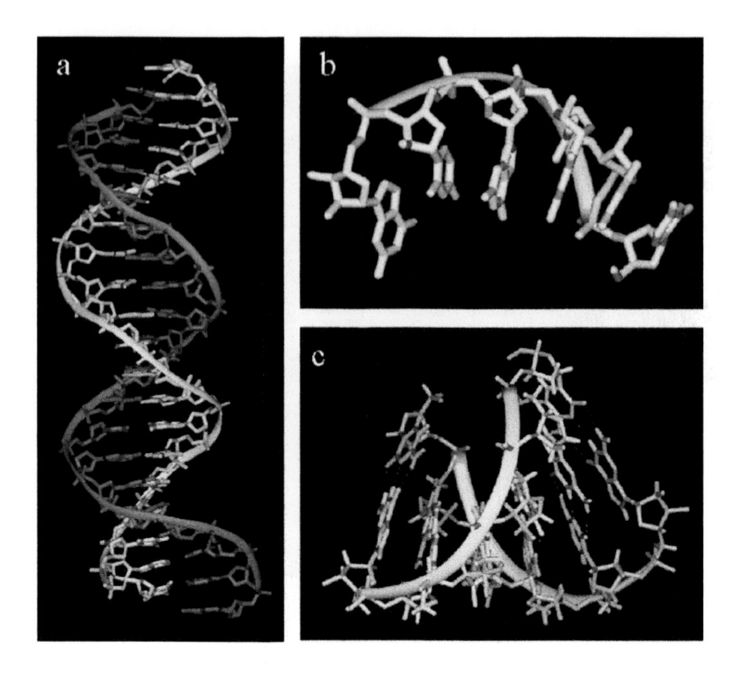

COLOR FIGURE 1.13
DNA and RNA structures. (a) DNA double helix; each single strand is colored differently and the backbone indicated with an idealized solid line respectively. (b) RNA single strand; free base structures pointing downward. (c) RNA double helix with base pairs stabilized through hydrogen (dotted lines). RNA does not form extensive double helical structures, but the more typical stem and loop motifs, where a single RNA strand folds back on itself with the stem forming a short double helix.

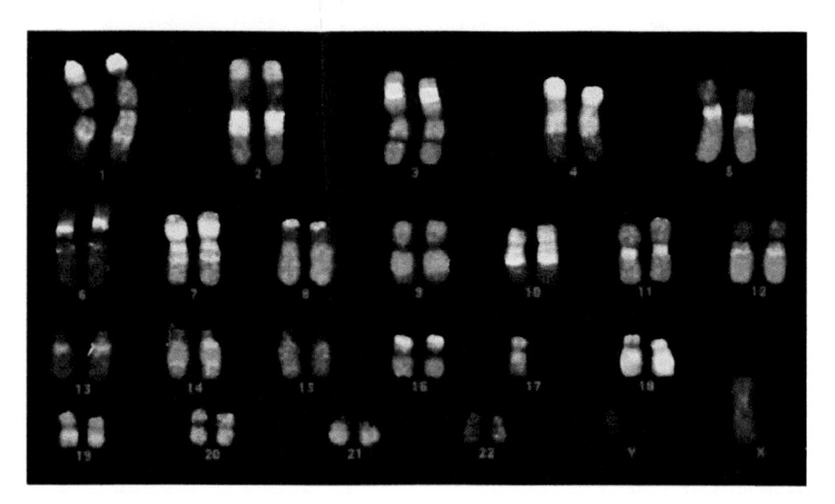

COLOR FIGURE 3.5
Fluorescent *in situ* hybridization (FISH) identification of human chromosomes — chromosome painting. (From Image © Applied Imaging, Hylton Park. With permission.)

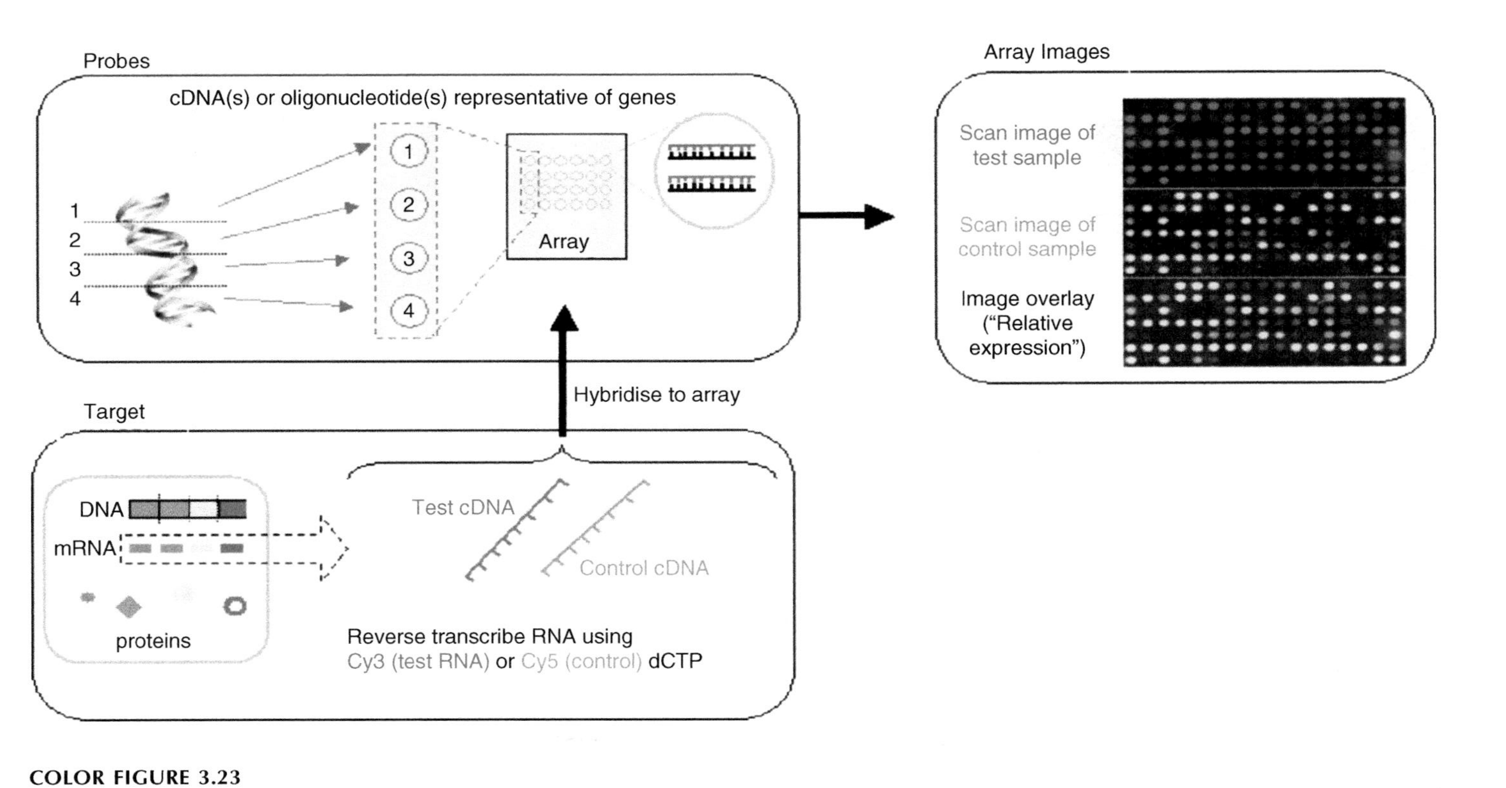

COLOR FIGURE 3.23
Overview of microarray technology as used for DNA readout. The process is very similar for protein arrays, with differences in deposited material and labeling protocols.

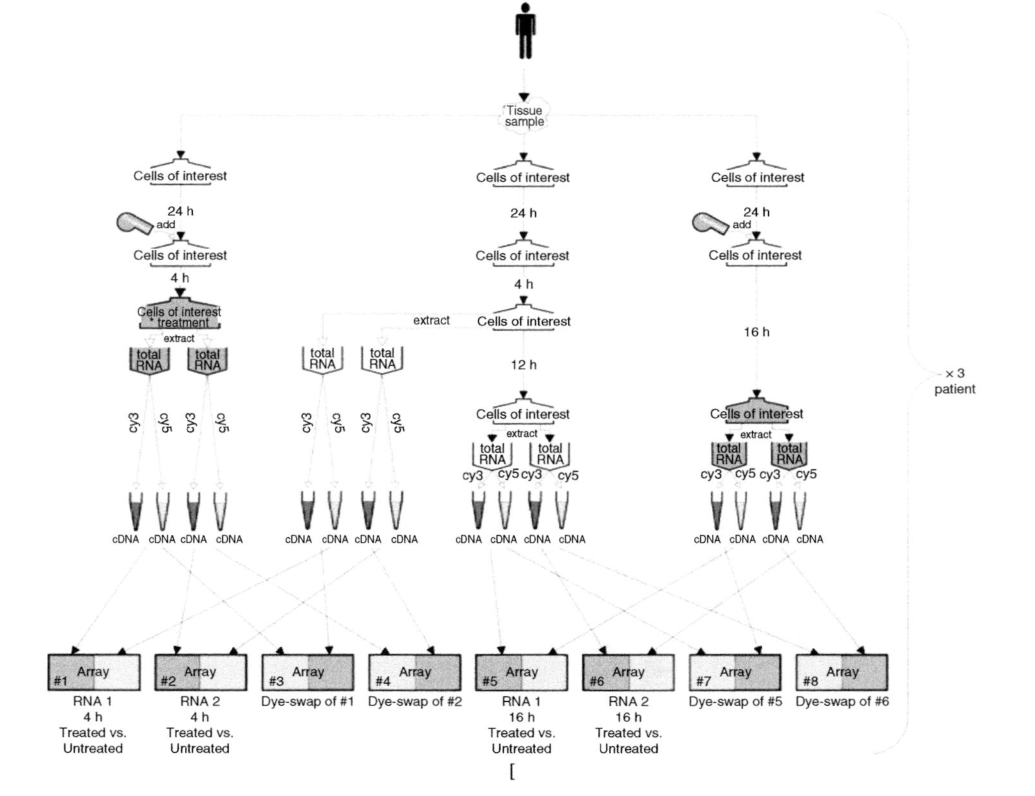

COLOR FIGURE 3.25
Example of an experimental plan.

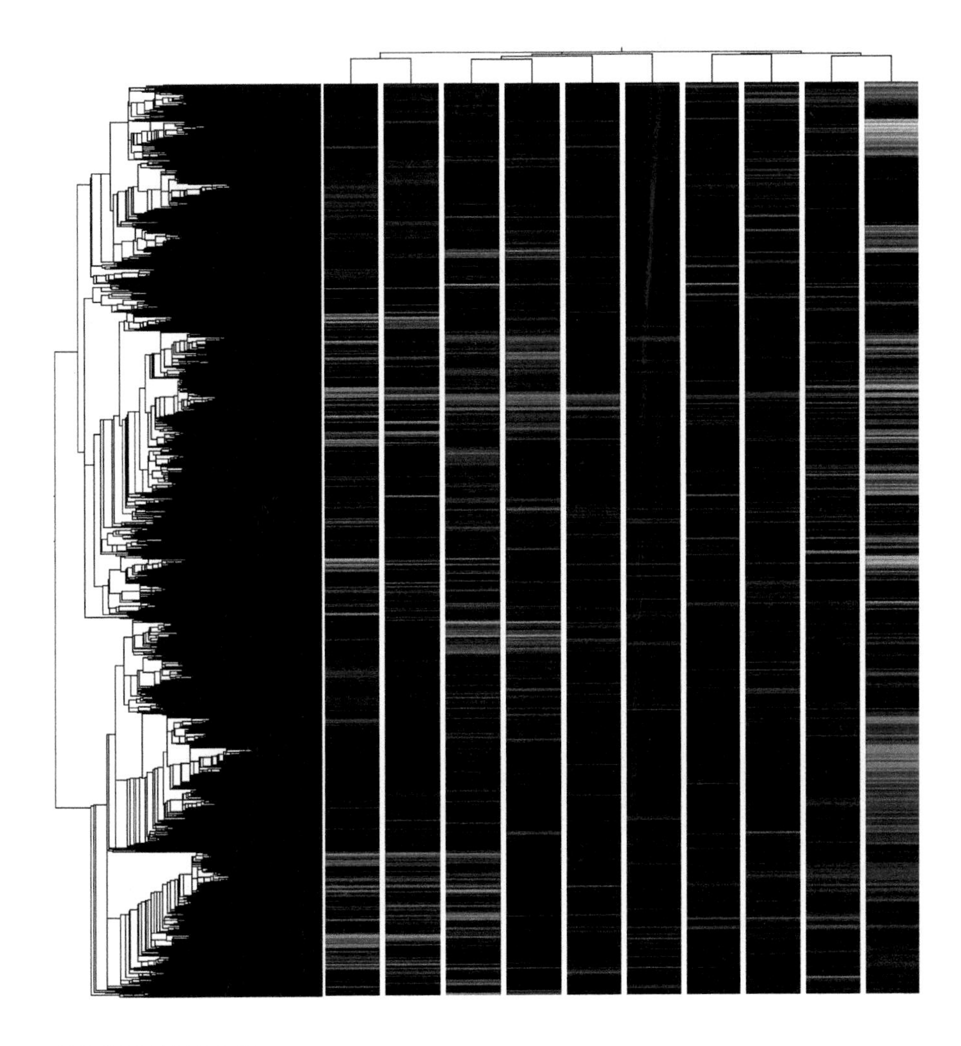

COLOR FIGURE 3.28
Graphical output of a hierarchical clustering. Columns represent arrays; each colored line within the columns is an individual gene. The branches along the top indicate which arrays (i.e. samples) are most similar to one another. The branches along the side indicate which genes are most similar to one another in terms of expression across the four arrays. Interesting genes may be the ones that have high expression (red) in one set of arrays, but low expression (green) in another set.

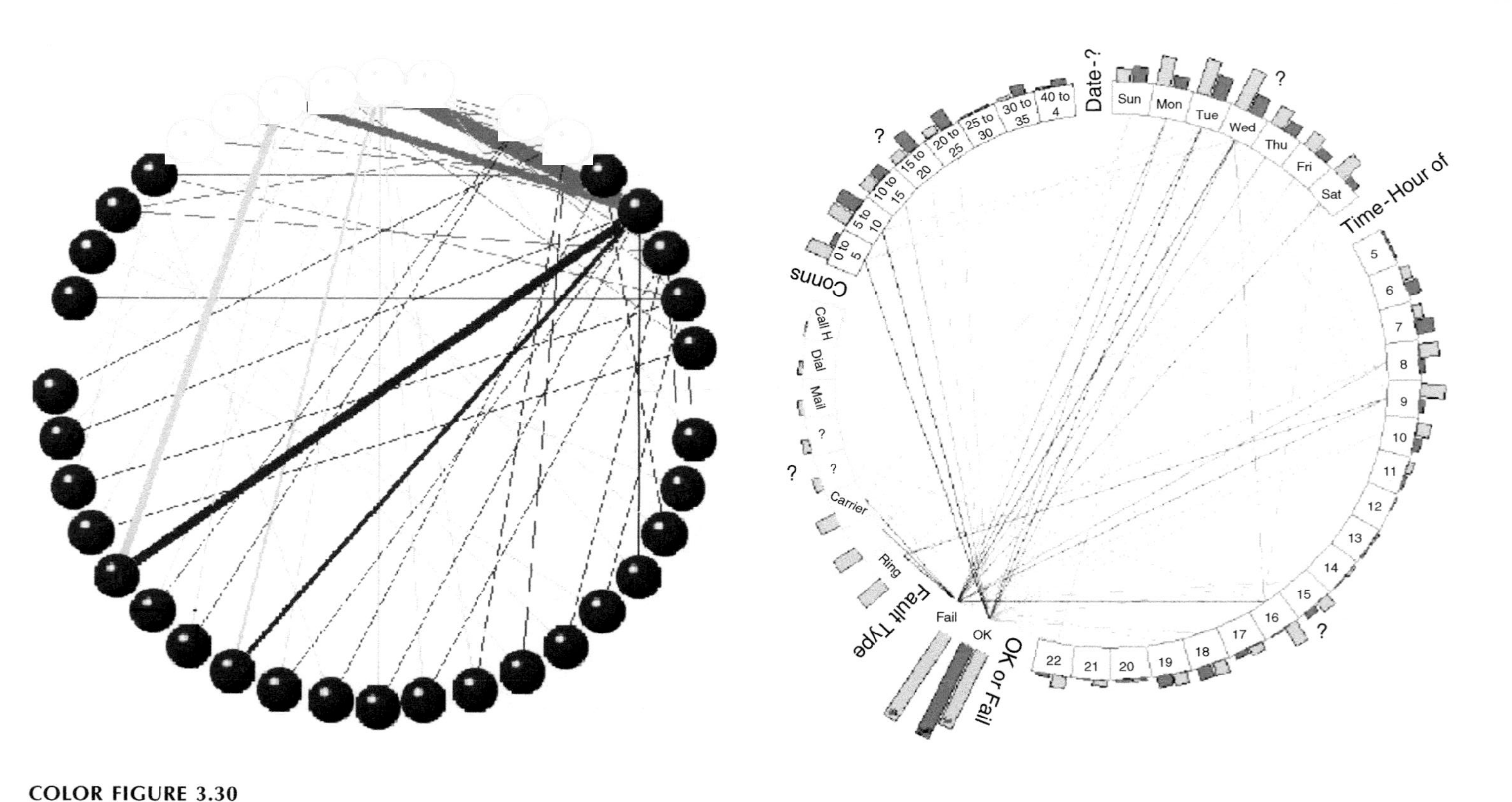

COLOR FIGURE 3.30
Data mining; link analysis (*left*) and Daisy chart (*right*). (Source: http://www.daisy.co.uk/daisy.html. With permission.)

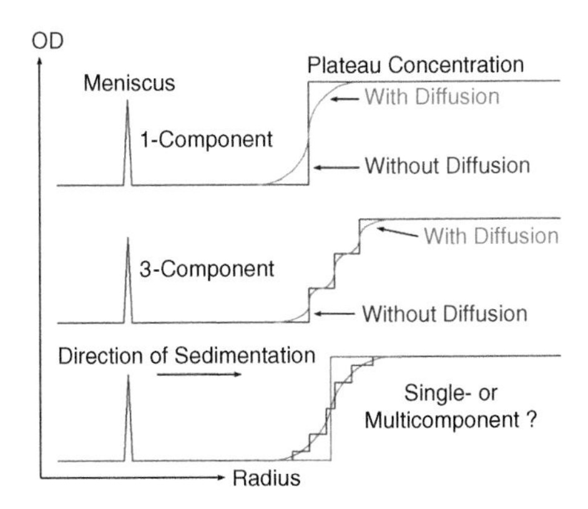

COLOR FIGURE 4.9
The effect of diffusion on the boundary shape in a sedimentation velocity experiment. Shown in blue is a scan trace for a hypothetical sample with $D = 0$. Shown in red is the actually observed trace with a non-nonzero diffusion coefficient. As multiple components sediment, the diffusion of each component obliterates the separation due to sedimentation and it becomes increasingly difficult to tell multiple components apart from each other (center panel). In the extreme case, a Gaussian sedimentation coefficient distribution of particles can be indistinguishable from a single component system with a large diffusion coefficient (bottom panel), unless a global method is employed that takes multiple time points (scans) into consideration.

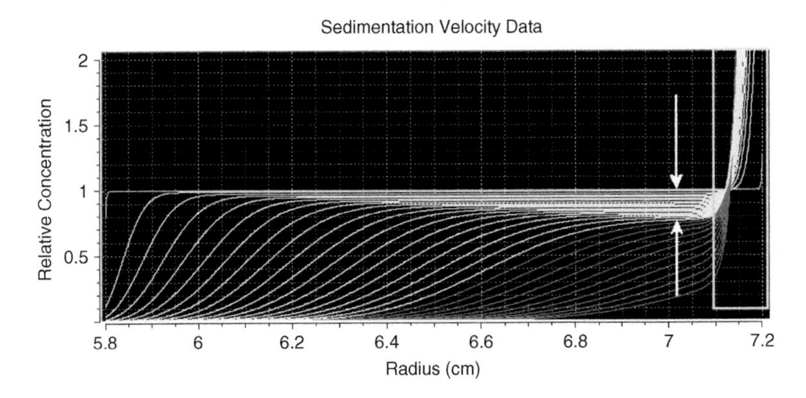

COLOR FIGURE 4.10
Simulated sedimentation velocity experiment showing 30 scans for a 50 kDa solute sedimenting for 8 hours at 35,000 rpm. Each scan represents a snapshot of the concentration profile at a different time during the sedimentation process. The scan depicted in green shows a mostly uniform concentration distribution at the beginning of the experiment, before sedimentation and diffusion have commenced. Later scans are shown in yellow. Note the radial dilution effect, marked as the plateau concentration difference between the two white arrows. The steep concentration gradient between the light blue lines results from sedimented material collecting at the bottom of the cell and back-diffusing into the cell. Scans shown in red are collected at the end of the experiment and are starting to lose their plateau concentration. Most of the material in the red scans has pelleted at the bottom of the cell at this point in the experiment. The longer the sample sediments, the more time the sample has to diffuse. Diffusion causes boundary spreading, which is more pronounced in later scans.

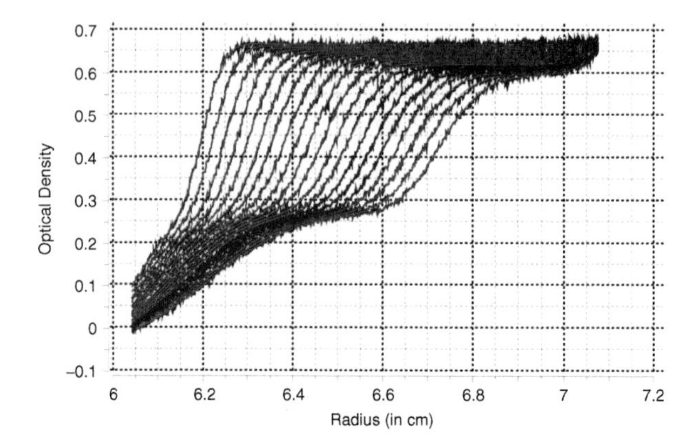

COLOR FIGURE 4.14

Finite element solution for a two-component, noninteracting mixture of lysozyme and a 208-basepair DNA fragment. Blue lines indicate experimental data, the red lines represent the finite element solution obtained in a nonlinear least squares fit. Note that the finite element solution closely traces the experimental data, including border effects such as back-diffusion near the bottom of the cell. Experimental parameters determined from this fit are: sLys, 1.66×10^{-13}; $D_{Lys} = 1.036 \times 10^{-6}$; $C_{Lys} = 0.324$ OD; $MW_{Lys} = 13.7$ kDa; $f_{Lys} = 3.81 \times 10^{-8}$; $s_{DNA} = 5.25 \times 10^{-13}$; $D_{DNA} = 2.341 \times 10^{-7}$; $C_{DNA} = 0.412$ OD; $MW_{DNA} = 123.0$ kDa; $f_{DNA} = 1.69 \times 10^{-7}$. These data illustrate the detail and accuracy of information that can be obtained from sedimentation velocity experiments. This plot has been generated with the UltraScan software. (From Demeler, B., UltraScan Data Analysis Software for the Analytical Ultracentrifuge. http://www.ultrascan.uthscsa.edu. With permission.)

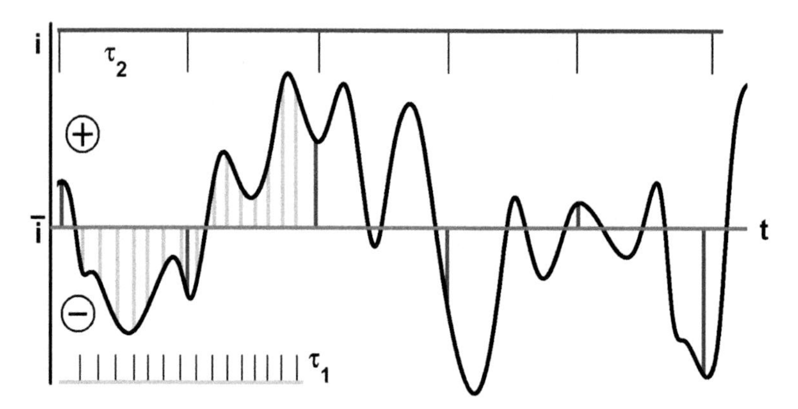

COLOR FIGURE 4.21

Measurement of the autocorrelation function. The intensity fluctuation, i, is plotted against time t. An average intensity $\bar{i}$ is determined and intensity values above this value are considered positive, values below the line are negative. Shown in green is a short sample time τ_1 leading to high correlation, since adjacent intensity values $i(\tau_1)$ tend to have the same sign and their products are positive, contributing to the autocorrelation sum. A longer sample τ_2 is shown in blue, producing little correlation, since adjacent values have random sign, and positive and negative contributions cancel out. Modern data acquisition hardware allows simultaneous collection of up to 256 different sample times, leading to 256 data points in the autocorrelation function. Sample times as short as a microsecond allow accurate diffusion measurements of even small particles.

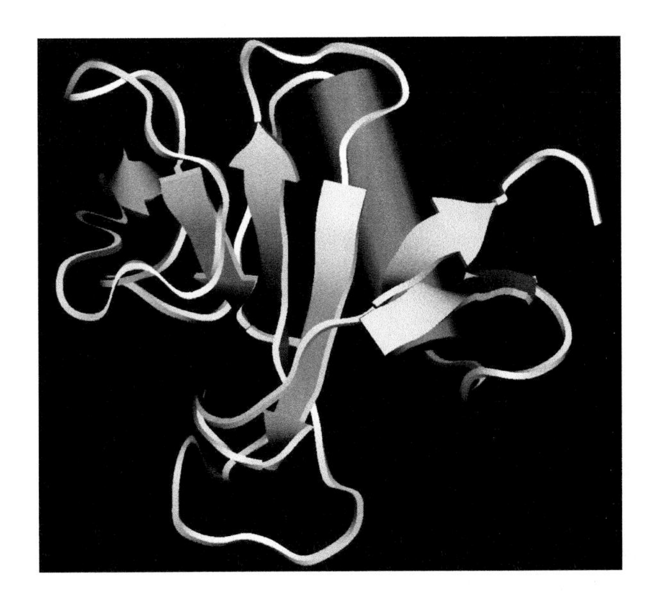

COLOR FIGURE 4.23
Secondary structure representation of a protein. Secondary structure cartoon of penicillin acylase: a small protein with a five-stranded antiparallel beta sheet (yellow ribbons) interacting with a single alpha helix (red extended cylindrical structure) on its "back" side. (From Molecular Simulations Inc. (MSI). With permission.) Protein motif with beta sheet (golden) and alpha helix (red).

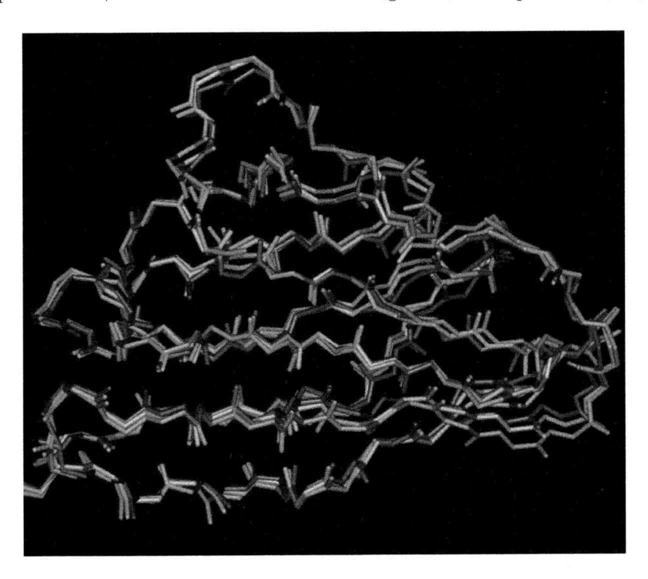

COLOR FIGURE 4.24
Energy minimization of homologous protein structures. The homology model as it undergoes minimization. The structures are colored red, orange, yellow and green. They represent the results from the starting model, fixing splices, fixing side chain clashes, minimizing whole structure respectively. (From MSI. With permission.)

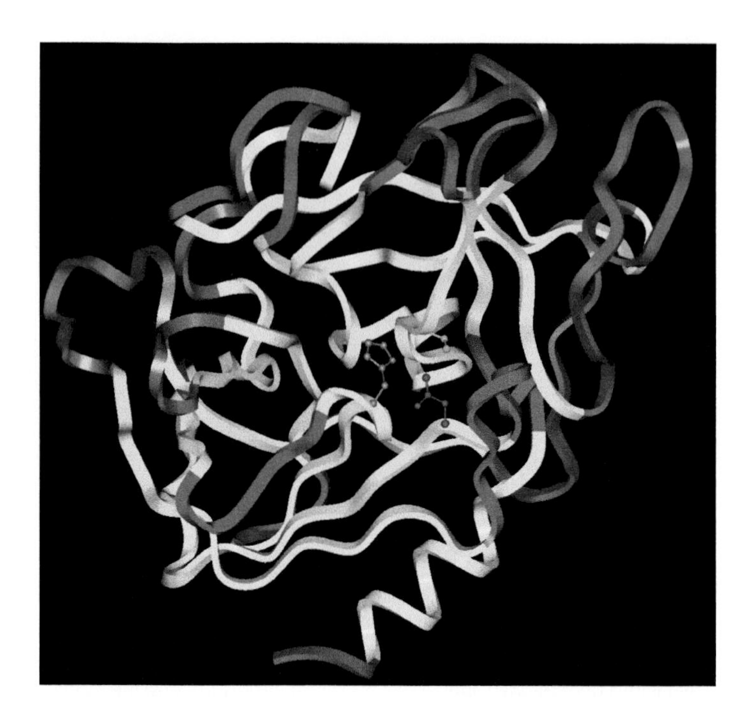

COLOR FIGURE 4.27
Final model structure of serine protease. Structurally conserved regions are shown in yellow, loop regions in blue and insertions in red. The catalytic triad is also shown. (From MSI. With permission.)

```
ACB (   E16) IVNGEEAVPGSWPWQVSLQDKT---GFHFCGGSLINENWVVTAAH(E57   )
TGD (    17)  VGGYTCGANTVPYQVSLNS------GYHFCGGSLINSQWVVSAAH(57    )
EST (    16) VVGGTEAQRNSWPSQISLQYRSGSSWAHTCGGTLIRQNWVMTAAH(57    )
RP2 (   A16) IIGGVESIPHSRPYMAHLDIVTEKGLRVICGGFLISRQFVLTAAH(A57   )

ACB (   E58) CGV-TTSDVVVAGEFDQGSSSEKIQKLKIAKVFKNSKYNSL--TI(E99   )
TGD (    58) CY--KSGIQVRLGEDNINVVEGNEQFISASKSIVHPSYNSN--TL(99    )
EST (    58) CVDRELTFRVVVGEHNLNQNNGTEQYVGVQKIVVHPYWNTDDVAA(99B   )
RP2 (   A58) CKGRE--ITVILGAHDVRKRESTQQKIKVEKQIIHESYNSV--PN(A99   )

ACB (  E100) NNDITLLKLSTAASFSQTVSAVCLPSASDDFAAGTTCVTTGWGLT(E144  )
TGD (   100) NNDIMLIKLKSAASLNSRVASISLPTS--CASAGTQCLISGWGNT(144   )
EST (   100) GYDIALLRLAQSVTLNSYVQLGVLPRAGTILANNSPCYITGWGLT(144   )
RP2 (  A100) LHDIMLLKLEKKVELTPAVNVVPLPSPSDFIHPGAMCWAAGWGKT(A144  )

ACB (  E145) RY-|ANTPDRLQQASLPLLSNTNCK--KYWGTKIKDAMICAGAS-(  gap )
TGD (   145) KSSGTSYPDVLKCLKAPILSDSSCK--SAYPGQITSNMFCAGYLE(186   )
EST (   145) R-TNGQLAQTLQQAYLPTVDYAICSSSSYWGSTVKNSMVCAGG-D(186   )
RP2 (  A145) GVR-DPTSYTLREVELRIMDEKACV---DYRYYEYKFQVCVGSPT(A186  )

ACB ( gap  ) -GVSSCMGDSGGPLVCKKNGAWTLVGIVSWGSSTCS--TSTPGVY(E228  )
TGD (   187) GGKDSCQGDSGGPVVC----SGKLQGIVSWGSG--CAQKNKPGVY(228   )
EST (   187) GVRSGCQGDSGGPLHCLVNGQYAVHGVTSFVSRLGCNVTRKPTVF(228   )
RP2 (  A187) TLRAAFMGDSGGPLLC----AGVAHGIVSYGHPD----AKPPAIF(A228  )
```

COLOR FIGURE 4.28
Multiple sequence alignment of four serine proteases. Legend: ACB; TGD; EST; RP2. (From MSI. With permission.)

COLOR FIGURE 4.29
Loop conformations located during a database search of the Protein Databank. Chosen conformation is shown in blue. (From MSI. With permission.)

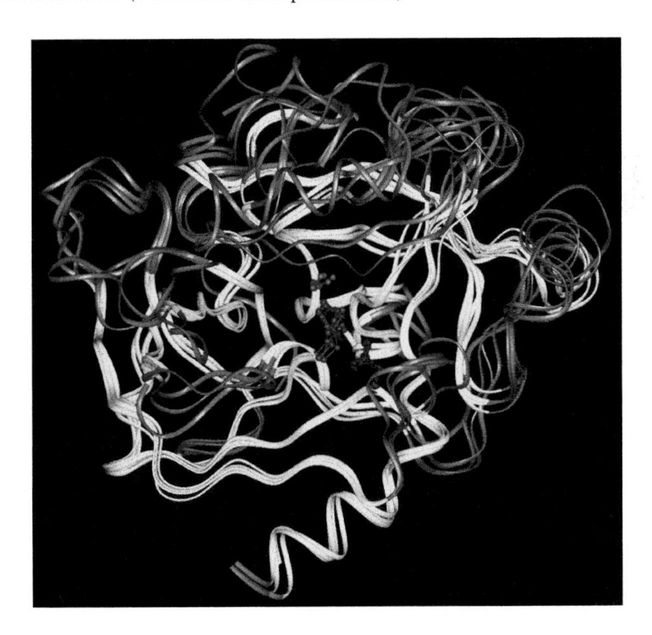

COLOR FIGURE 4.30
Four serine protease structures overlaid. Same proteases as shown in Color Figure 4.28 (sequence alignment). The structurally conserved regions are colored yellow. The catalytic triad is shown in ball and stick (serine: green, histidine: orange, aspartic acid: red.) (From MSI. With permission.)

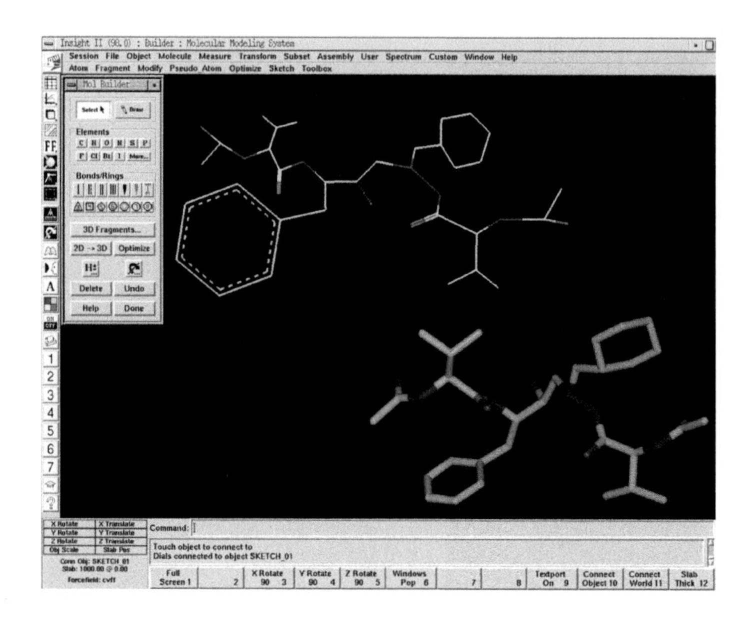

COLOR FIGURE 4.31

Molecular modeling desktop. Sketching a small molecule in 2-D (on the left) that can then be automatically converted into 3-D (on right) (MSI).

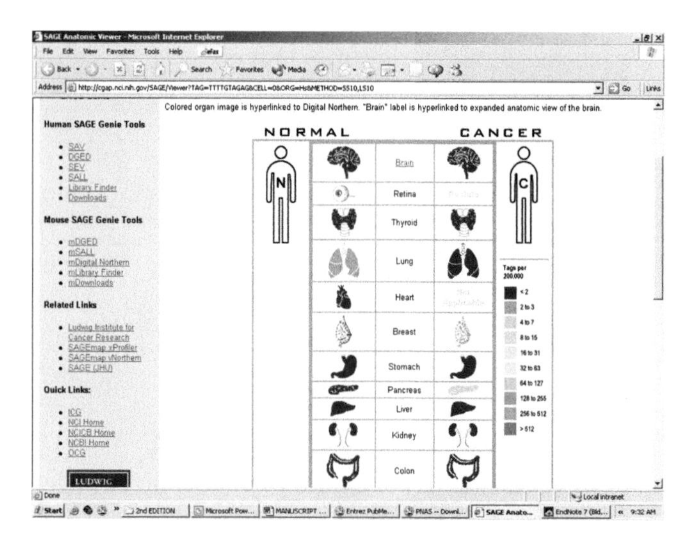

COLOR FIGURE 5.4

Human SAGE Genie Tools for p53 (TP53) Search query: TTTTGTAGAG. Search query: TTTTG-TAGAG, tissues only. Colored organ image is hyperlinked to Digital Northern. "Brain" label is hyperlinked to expanded anatomic view of the brain. SAGE is a gene expression tool Part of the National Institute of Health, U.S. Department of Health and Human Services; http://cgap.nci.nih.gov/.2 Differences in shading indicated differences in number of tags identified in normal and cancer tissues in respective organ tissue samples. (Shading scale on right: top [blue] less than 2 tags/200,000 bp, bottom (red) more than 512 tags/200,000 bp.)

1.6 DNA and RNA Structure

Nucleic acids were named after their cellular location and solubility properties when first discovered by Friederich Miescher in 1869. With Darwin's *The Origin of Species* published just ten years earlier, the search for the agents of change and the mechanism of heredity became an important task for chemists, physicians, and biologists. Miescher extracted a substance from the nuclei of cells whose solubility depended on the pH of the solution. He called the substance nucleic acid, but did not know that he had isolated the molecular entities of genes. The idea of genes, after all, was not conceived of until long after Gregor Mendel demonstrated as early as 1853 that specific traits could be inherited as independent entities, which today is known as the Mendelian inheritance of independent segregation and assortment of traits.

Today we know that nucleic acids come in two forms, the structurally abundant and functionally versatile ribonucleic acid, or RNA, and the chemically more stable, but less abundant deoxynucleic acid, or DNA. The latter is used for storage and inheritance of genetic material, while RNA contributes to energy metabolism, enzymatic activity in form of cofactors of protein enzymes and as RNA enzymes proper, as well as genetic information storage (some viral RNA) and retrieval (messenger RNA, transfer RNA), and peptide bond synthesis (ribosomal RNA). Newer research also shows the role of small interfering RNA structures regulating mRNA degradation and thus protein synthesis (RNA interference).

DNA's function as the sole bearer of genetic information in bacteria was demonstrated by Oswald Avery in 1944. Six years thereafter, Alfred Hershey and Martha Chase showed that not protein, but the DNA portion from bacteriophages alone was sufficient to infect bacteria.[1] Experimental evidence that DNA and not proteins are the evolutionary agent of change came some 75 years after Miescher first isolated DNA. In 1953, one year after Hershey and Chase published their results, the DNA double helix structure was discovered by James Watson, Francis Crick, and Maurice Wilkins who based their model on the x-ray crystallography data of Rosalind Franklin. The publication of this seminal structure[2] led to a mechanistic explanation of DNA replication[3] and ultimately the interpretation of the genetic code and its function in protein synthesis in 1963 by Robert Holley, Gobind Khorana, and Marshall Nierenberg.

The discovery of the genetic code marks the beginning of bioinformatics, because DNA, RNA, and protein sequences could now be identified and established as unique for every type of gene and its corresponding protein and that sequence differences for a specific protein found in different organisms correlate with the evolutionary history of these organisms. This insight forms the basis of molecular evolution. The initial experiments performed with bacteria in the 1940s and 50s suggested a one-to-one correspondence between a gene and its product, the protein. Molecular biologist in the 1970s, however, soon realized that the simple "one gene–one protein" hypothesis based on bacterial genetics did not hold true for higher organisms. Instead, careful comparison with recombinant DNA revealed that chromosomal gene structures are fragmented into coding and noncoding DNA segments. More complex relationships between gene and protein sequences emerged with the discovery of recombinatorial activity of chromosomal DNA and the splicing of RNA and proteins.[4] Molecular biologists learned how to make use of these processes in their laboratory settings, bringing life to recombinant DNA technology—the basis of gene technology and the proliferation of the biotechnology industry.

1.6.1 The DNA Double Helix

Nucleic acids are linear polymers that are composed of chemically distinct monomers called nucleotides. Nucleotides differ by their composition of aromatic base structure—purines and pyrimidines—linked to a phosphorylated ribose (sugar) unit. Their linear arrangement can be read as sequence of letters. It is the sequence pattern of these different building blocks read in triplets (codons) that constitutes the uniqueness of every gene in an organism and of genomes from one organism to another. RNA is composed of ribonucleotides with the four bases adenine (A), guanine (G), cytosine (C), and uracil (U), whereas DNA uses the deoxy form of the ribose and the four bases adenine (A), guanine (G), cytosine (C), and thymine (T), with the latter being the substitute for uracil. DNA, the carrier of genetic information, is found primarily in a dimer or double strand conformation giving rise to the double helix, as shown in Color Figure 1.13a, whereas RNA primarily forms monomeric or single-strand units, as shown in Color Figure 1.13b. However, RNA forms extended intramolecular double helical domains, but also hybridizes with DNA single strands as shown in Color Figure 1.13c.

The double helix structure of DNA is stabilized by hydrogen bonds within pairs of bases from opposing single strands. As a rule, only adenine and thymine (AT) and guanine and cytosine (GC) form base pairs in chromosomal DNA. The precision of this pairing lies in the thermodynamic and conformational limitation of hydrogen bond formation between purine (A,G) and pyrimidine (C,T,U) structures of nucleotides.

Single DNA strands are not stable, but associate (hybridize) with a second single strand with a complementary sequence (only AT or GC pairs) to form an *antiparallel double helix structure*, where both strands intertwine. At the center of the helix the bases H-bond with each other. The bases point toward the helix center and H-bond in the following manner only:

G with C 3 H-bonds
A with T 2 H-bonds

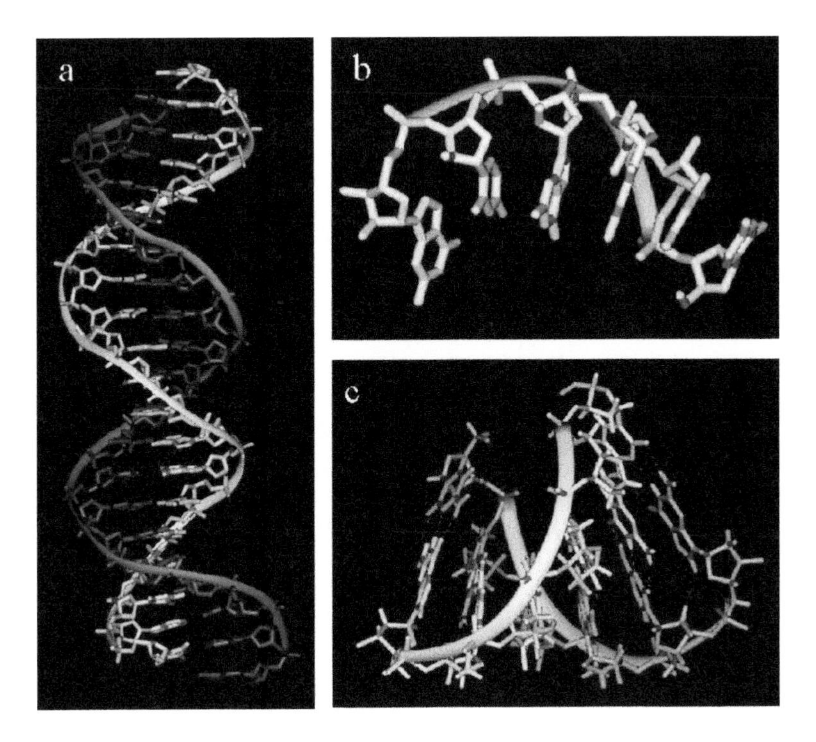

FIGURE 1.13 (See color insert)
DNA and RNA structures. (a) DNA double helix; each single strand is colored differently and the backbone indicated with an idealized solid line respectively. (b) RNA single strand; free base structures pointing downward. (c) RNA double helix with base pairs stabilized through hydrogen (dotted lines). RNA does not form extensive double helical structures, but the more typical stem and loop motifs, where a single RNA strand folds back on itself with the stem forming a short double helix.

The rigidity and linear geometry of the H-bonds restricts base pair (bp) formation. The plane of the base pair lies almost perpendicular to the helix axis (6° tilt). This double helix conformation is called the right-handed *B-DNA* form and represents the physiological form of chromosomal DNA. The B-DNA is a *right-handed, antiparallel* double helix with 2 nm in diameter. The physical parameters are:

10 bp per turn
A twist of 36° per bp
0.34 nm (3.4Å) per base
Pitch 3.4 nm (34Å)
Diameter 2.0 nm
Base plane tilt away from being perpendicular to helix axis 6°

The B-DNA as described by the above parameters represents an idealized helix. The conformation of "real" DNA, however, deviates slightly from the B form in a sequence-dependent manner as well as depending on the interaction with DNA-binding proteins. Two important features of the double helix are the *minor* and *major grooves* winding along the helix surface. In those grooves, parts of the aromatic ring structures of the purines and pyrimidines are exposed to the surface of the helix. The major groove is the site where proteins bind to DNA in a sequence-specific manner reading the genetic information.

The specific conformational constraints of base pairing made it clear that the sequence of the bases in the polymer encodes the genetic information for the synthesis of proteins and complementary hybridization suggested a mechanism for faithful sequence duplication during reproduction.[2]

A measurable quantity of DNA is its contents of G + C and A + T base pairs (besides the sequence per se) which differs from organism to organism and is characteristic of an organism's genome.[5] The G + C content in mammals varies from 39 to 46% and in bacteria from 25 to 75%. The G + C content exhibits an important correlation with the stability of the double helical conformation of DNA due to the difference in hydrogen bonding capacity of GC pairs with 3 H-bonds and AT pairs with two H-bonds. This has been shown through heat denaturation of the double helix into a single strand conformation. Measuring the double strand (ds) vs. single strand (ss) content of purified DNA is done by determining the UV absorbance at a wavelength of 260 nm (typical absorbance wavelength of aromatic ring structures) over a temperature range of 20 to 90°C. The resulting curve shows a sigmoidal behavior indicating a cooperative process during denaturation and renaturation. Normally a sharp increase in absorbance indicates the transition from double to single strand form with the half point of transition being defined as the *melting temperature* T_m of the DNA sample in the test solution. The process is strictly reversible. At the *melting temperature* T_m, 50% of the DNA is found in ds and 50% in ss form. The melting temperature is directly proportional to the GC content (Chargaff rule). This behavior of DNA hybridization is exploited in modern molecular biology techniques such as PCR and DNA microarray studies (see Section 3.4).

1.6.2 Genomic Size of DNA

In prokaryotes, the entire genome consists of one single large DNA molecule, a chromosome, in circular form. The small genome of *Haemophilus influenzae* contains more than 1.8 million base pairs. Larger bacterial genomes hold up to 9 million base pairs. Because of the large size of circular DNA, the chromosome is compacted in a *super-coiled* form. Eukaryotes with much larger genomes have their DNA fragmented into several chromosomes that contain noncircular DNA. Consider the size and structure of chromosomal DNA being a very long, linear polymer. Fragmentation of genomes into smaller units facilitates DNA duplication during reproduction and cell division. Often DNA is found to be methylated, the result of an enzymatic modification that serves as chemical protection. Methylated bases are often not recognized by

DNA binding proteins such as nucleases that hydrolyze the phosphate ester bond between nucleotides. In this way, bacteria protect their own DNA from degradation, whereas the DNA of intruders (like viruses) which is not methylated can be degraded. Methylation occurs at *N* 6-methyl-D-adenosine and 5-methyl-D-cytosine. In eukaryotes, methylation is found to be an important mechanism controlling chromosome packing and gene expression activity.

References

1. Hershey A.D. and Chase, M., Independent functions of viral protein and nucleic acid in growth of bacteriophage. *J. Gen. Physiol.*, 1952, 36, 39–56.
2. Watson, J.D. and Crick, F.H., Molecular structure of nucleic acids; a structure for deoxyribose nucleic acid. *Nature*, 1953, 171(4356), 737–738.
3. Watson, J.D. and Crick, F.H., Genetical implications of the structure of deoxyribonucleic acid. *Nature*, 1953, 171(4361), 964–967.
4. Paulus, H., Protein splicing and related forms of protein autoprocessing. *Annu. Rev. Biochem.*, 2000, 69, 447–496.
5. Chargaff, E. et al., The composition of the desoxyribonucleic acid of salmon sperm. *J. Biol. Chem.*, 1951, 192(1), 223–230.

1.7 DNA Cloning and Sequencing

Bioinformatics often comes down to database mining—the extraction, sorting, and analyzing of sequence information of genes, genomes, and proteins. Where does this information come from and how is it obtained? Simply put, genes have to be sequenced and to do so DNA or RNA has to be isolated and cloned into the appropriate form rendering them amenable to sequencing and manipulation in the laboratory. Cloning and sequencing are not, as such, part of bioinformatics, but are tools to generate the raw sequence information stored in databases such as GenBank. Bioinformaticians then analyze these sequences. Their results influence cloning and sequencing techniques. The increasing number of sequences improves the quality of statistical analysis, while the development of new bioinformatics software allows for the identification of biological function associated with sequence patterns, thus allowing faster detection and cloning of novel genes. In addition, sequencing and functional descriptions of genes in one organism facilitate the identification of their homologs in other organisms, while sequence similarities can be used to predict the function of newly discovered but biologically nondescriptive sequences.

Due to this crossfertilization and the predictive value associated to sequence similarities, analyzing the rapidly accumulating database information must be done with care if the underlying methods of obtaining this information are to be accurately evaluated in a timely manner. For any bioinformatician, it is thus important to understand the source of a DNA or protein sequence. Consequently, it is helpful to discuss some methodology of how genes are cloned and sequenced. As for amino acid sequences, they are typically predicted by translating the nucleotide sequence using the appropriate genetic code. Due to post-transcriptional and post-translational modifications, the actual amino acid sequence of a protein may differ from the genomic sequence. While protein sequencing is almost never used to obtain full sequences, sequencing of short peptides with five to fifteen amino acids has regained a central role in proteomics. The identification of proteins from tissue samples is possible by microsequencing short peptide fragments by mass spectrometry and matching the calculated sequences of several peptides to the nucleotide sequence of a gene in GenBank (see Proteomics Section 4.1 for details).

1.7.1 DNA Cloning

Cloning is commonly known as the process of asexually producing a group of cells (clones), all genetically identical, from a single ancestor. Here, it refers to the use of DNA manipulation procedures to produce identical copies of a single gene or segment of DNA through recombinant DNA technology. A desired gene or DNA fragment is cut out of its chromosomal location and inserted into vector DNA that is used for replication (amplification) in a host organism. Such cloning vectors are DNA molecules originating from viruses, bacteria, or yeast that contain proper strings of promoter sequences to control DNA duplication in a host cell. For instance, bacterial DNA polymerase will specifically control the *bacterial vector* DNA. Mammalian DNA fragments inserted into the expression vector are integrated without loss of the vector's capacity for self replication in its natural cellular environment, allowing foreign or recombinant DNA to be reproduced in large quantities in host cells. Examples of cloning vectors are plasmids (bacterial origin), cosmids (viral origin) and fosmids (F-plasmid or bacterial fertilization factor), and yeast and bacterial artificial chromosomes (YACs, eukaryotic origin; BACs, ~150 kbp inserts). Vectors can also function as *expression vectors* when they contain the promoter elements necessary for gene regulation. This feature is used to produce large quantities of mRNA and proteins in host organisms that normally do not contain or express these genes.

Large collections of DNA fragments inserted in expression vectors are stored in *clone libraries* from which scientists can readily buy the necessary material for their genetic research. In fact, behind every sequence stored electronically in a computer database (electronic sequencing) is a physical library of tissue samples and cloned DNA. These libraries are not restricted to functional DNA fragments. When originating from genome projects, they are often random collections of genomic DNA obtained through shotgun techniques such as mechanical shearing or *in vitro* radiation-induced chromosome fragmentation (radiation hybrid mapping). If a vector DNA contains genes it will be possible to functionally express or transform them into cell lines and whole organisms (transgenic animals and plants). For information about clones used in genome projects see NCBI's clone registry at www.ncbi.nlm.nih.gov/genome/clone. The clone registry provides information about genomic clones and libraries, including sequence data, genomic position, and distributor information.

1.7.2 Transcriptional Profiling

Before a gene can be cloned, its location on the chromosomes must be identified. The activity pattern of a gene (i.e., within which cells in a body and at what time of development or life stage a gene is expressed and a protein synthesized) is the first piece of information used to identify the chromosomal location of a gene. Specifically, the expression of certain genes may be restricted to specific cell types, tissues, or organs; their activity patterns may change from healthy to diseased (like tumors), and these may differ among young and old people.

In transcriptional profiling, the transcript levels or concentrations of messenger RNA are measured directly. Cytoplasmic levels (concentrations) of mRNA are good indicators for gene activity and form the bases of expressed sequence tags or ESTs, functional sequence information that has been crucial in whole genome projects. High levels of mRNA are in many, but not all, cases indicative of the presence of a protein. The presence of protein levels must be demonstrated independently (see proteomics), rather than relying on indirect measure based on mRNA levels.

Identifying mRNA is done by hybridizing (binding) radioactive labeled oligonucleotides in a sequence-specific manner to isolate target mRNA. Obviously, some sequence information must be obtained in advance. This information could have been derived from short amino acid sequences obtained from protein fragments or peptides, or by searching the DNA databases for sequences with desired properties such as a human homolog to a known gene from a mouse or rat, or simply a similar sequence representing a potential novel gene, and so forth. Comparing the hybridization pattern of different samples at varying times during the life cycle of an organism or cell, before or after differentiation during development or under varying conditions (resting vs. hormone-induced state), can be used to construct a time–space map of where in the body a specific gene or groups of genes are actively expressed.

It is imperative to understand the importance of the structure of coding sequences of eukaryotic genes. Eukaryotic genes are organized on the chromosomes in the cell nucleus, which differs from the sequence found on the mRNA. A cDNA sequence of a eukaryotic gene is normally shorter than the genomic version due to the organization of genes into coding (exons) and noncoding or intervening regions (introns). Although the entire gene (intron plus exon plus control sequences) is transcribed into a primary transcript located in the nucleus of the cell, the final mRNA product used in the cytoplasm is the result of catalytically modifying the primary transcript to eliminate the introns and stitch or splice together the exons into a contiguous coding sequence. This leaves a shortened mRNA—a combination of all exon fragments found at the genome level. This is why the use of mRNA for the synthesis of cDNA yields synthetic genes that differ from their genomic origin and can be cloned into vector DNA for easier use in the laboratory (such as *in vitro* biosynthesis of proteins, transformation of DNA into new cell lines, and transgenic animals). Splicing of exons from the primary transcript into a final mRNA can use different sets of exons from the same gene yielding multiple proteins called splice variants. Often, splice variants affect the cellular localization or regulatory mechanism of an enzyme.

1.7.3 Positional Cloning and Chromosome Mapping

An alternative strategy used for the detection of hereditary disease genes is positional cloning. Most commonly, a gene is identified by following a familial inheritance of its phenotype or through quantitative trait loci (QTL) analysis in animal models or cell culture. QTL is based on the statistical correlation between the appearance of a phenotype and the presence of genetic markers on a piece of chromosomal DNA. In general, genes are easiest to identify when associated to a disease or one of several traits associated to a phenotype. A gene associated to a trait that contributes to the development of a disease is known as an allele (a variant of a gene in a population). A gene or allele can be located on the chromosome using genetic markers. Markers are short, easily detectable sequences usually found on nearby noncoding parts of the genome. For this method, family histories must be available for analysis of the population genetics for those genes that contain mutations and appear at a specific frequency in the human population (alleles). An allele refers to a particular gene within the genome of every individual in a population. Known alleles linked to hereditary diseases are collected in the On-line Mendelian Inheritance in Man[1] (OMIM) database. This database is discussed in more detail in Section 5.1.2. The actual sequence of the gene and nearby noncoding segments, however, often vary from individual to individual due to the random occurrence of mutations; SNPs, and deletions and insertions of single nucleotides or short sequences. Patterns of these sequence variants are often stable within a population (e.g., an SNP must be present in 1% or more of a population to be considered a stable marker and not a random mutation) resulting in haplotypes that are cataloged in the HapMap Project. For information see the International HapMap Project at www.hapmap.org.

SNPs occur on average once every 300 to 500 bp, representing the lower end of length distribution of genes. This means that polymorphism is a fairly

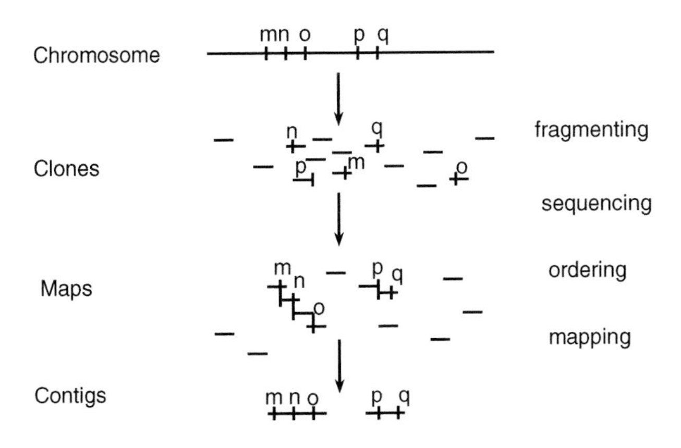

FIGURE 1.14
Mapping and sequencing of chromosomes.

common feature of our genome. Because of the exon intron architecture of our genes, these single nucleotide mutations occur mostly in noncoding regions of the genome, which puts them outside natural selection mechanisms affecting genes and thus makes them useful for mapping genomes as shown in Figure 1.14.

Once a chromosomal location of a gene of interest has been identified via a nearby genetic marker or markers, clones with large inserts are identified by physical mapping. Due to their large size, the genomic region may contain several genes that are now subject to sequencing. Finally, a mutation analysis compares the identified gene(s) in affected and unaffected members of a population.

A chromosome containing five markers (m, n, o, p, and q) is fragmentized and the fragments (clones) sequenced for markers. The identified clones are ordered and mapped by comparing overlapping strings of sequences. As long as fragments show such overlapping sequences, they can be linked into continuous segments (contig 1 *mno* and contig 2 *pq*). If some fragments are missing, the entire chromosome map cannot be reproduced, but is rather represented by several contigs. Since fragments have been sequenced in full, the corresponding contig sequences are known and can be stored in the database (e.g., GenBank). Depending on the extent of sequences covered in genome sequencing projects, so-called complete genomes cover at least 95% of the euchromatin structure of chromosomes. Now that the entire human genome sequence is known, identification of mutations related to human diseases proceeds at a faster pace. Identification of such genes will be facilitated with the help of full genomes from related organisms.

The human genome sequence is, in fact, the consensus sequence of a mix of some 40 to 50 individuals revealing little information about allelic variants of humankind. Mapping allelic variants and SNP distribution is the goal of the HapMap project with sequence information collected from individuals of related groups. Thus, mutation analysis of different members of the population is still a necessary step in the positional cloning approach, since no individual's genome is currently sequenced in its entirety, although technologies are being developed to do just that. With whole genome sequences available in databases, polymorphism databases are specifically useful to quickly find genetic markers and the nearby genes of interest. In addition, medical databases for every disease, susceptibility for infection, cancer, and possibly psychological traits can be envisioned for the future.

1.7.4 Polymerase Chain Reaction (PCR)

The technique that revolutionized DNA amplification is PCR.[2] It was developed in 1985 by Kary B. Mullis, who was then working at Cetus Corporation. In 1993, Mullis received the Nobel Prize in Chemistry for his contribution to molecular biology, and today his technique is used in virtually every biomedical laboratory in the world. The process has been automated, and machines that amplify DNA from small quantities into large ones are commercially available. The process, from template design of primers or oligonucleotides used to hybridize to a target gene sequence in the genome library to mapping an organ for gene expression, is done by computer programs. The success of PCR is based on the ability to amplify DNA fragments *in vitro* and independent of any cellular machinery.

The increasing numbers of sequences allow the search for functional units of unknown genes. Large-scale identification of gene expression by measuring messenger RNA levels allows researchers to keep pace with the sequencing results of genome projects (public and proprietary libraries and databases). DNA sequences can be used to generate short search sequences to screen for mRNA. In an effort to increase the efficiency of finding good drug targets, the pharmaceutical industry is developing multiarray plate and microchip assays[3] where hundreds, even thousands, of gene fragments or cell types can be screened in a single assay (see also Section 3.4). DNA chip technology developed by Affymetrix in Santa Clara, California (www.affymetrix.com) is leading the technology push to determine tissue distribution of expressed genes and expression sequence tags (ESTs).

The importance of PCR to bioinformatics—and genome projects in particular—is its ability to amplify DNA without any biological information attached. This means that both coding and noncoding regions can be analyzed as long as short stretches of sequence (between 10 and 20 nucleotides long) are known. The technique is extremely sensitive to initial sample quantity because of the enzymatically controlled DNA amplification process in noncellular test tube solutions.

Viral reverse transcriptase and heat-stable DNA polymerases from thermophilic bacteria form the cornerstone of today's molecular biology. PCR and cDNA formation allow for the rapid identification of novel genes even from tiny tissue fragments, a feature prominently exhibited in forensic science.

1.7.5 Sequencing Technologies

> A major focus of the Human Genome Project is the development of automated sequencing technology that can accurately sequence 100,000 or more bases per day at a cost of less than 50¢ per base. Specific goals include the development of sequencing and detection schemes that are faster and more sensitive, accurate, and economical. Many novel sequencing technologies are now being explored, and the most promising ones will eventually be developed for widespread use. Second-generation (interim) sequencing technologies will enable speed and accuracy to increase by an order of magnitude (i.e., ten times greater) while lowering the cost per base. Some important disease genes will be sequenced with technologies such as high-voltage capillary and ultra-thin electrophoresis to increase fragment separation rate, and the use of resonance ionization spectroscopy to detect stable isotope labels. Third-generation, gel-less sequencing technologies aiming to increase efficiency by several orders of magnitude are expected to be used for sequencing most of the human genome. These developing technologies include enhanced fluorescence detection of individually labeled bases in flow cytometry; direct reading of the base sequence on a DNA strand using scanning tunneling or atomic force microscopy; enhanced mass spectrometric analysis of DNA sequence; and sequencing by hybridization to short panels of nucleotides of known sequence. Large-scale pilot sequencing projects will provide opportunities to improve current technologies and will reveal challenges investigators may encounter in larger-scale efforts. (From: Primer on Molecular Genetics, Dennis Casey, Dept. of Energy, 1992.)

Fluorescent labeling instead of radioactive markers that require long exposures of electrophoresis gels on x-ray sensitive films of DNA fragments has radically improved the speed of sequencing DNA by using the Sanger dideoxy chain termination method[4] (Applied Biosystems, 1987; Taq cycle sequencing, 1990). This method is based on enzymes that are capable of synthesizing DNA. By adding nucleotide substrates that block the elongation process, fragments of different lengths are being produced. By carefully separating them, fragments differing in only one nucleotide can be distinguished, thereby allowing us to read the sequence of the cloned DNA in its entirety. As we know which fragment comes from the dG, dA, dC, or dT termination reaction, the parallel separation of the mixtures reveal the nucleotide sequence immediately.

Alternatively, the Maxam–Gilbert technique[5] is based on enzymes that cleave the DNA clone at specific bases resulting in a mix of fragments of different lengths. Gel electrophoresis allows the separation of these fragments at a resolution of one nucleotide difference. Information and experimental protocols on sequencing can be found in manuals such as those published by Sambrook et al.[6] and Glover.[7]

References

1. Schorderet, D.F., Using OMIM (On-line Mendelian Inheritance in Man) as an expert system in medical genetics. *Am. J. Med. Gene.*, 1991, 39(3), 278–284.
2. Mullis, K., et al., Specific enzymatic amplification of DNA in vitro: the polymerase chain reaction. *Cold Spring Harbor Symposium on Quantitative Biology,* 1986, 51(Pt 1) 263–273.
3. Yershov, G. et al., DNA analysis and diagnostics on oligonucleotide microchips. *Proc. Natl. Acad. Sci. U.S.A.*, 1996, 93(10), 4913–4918.
4. Sanger, F. et al., DNA sequencing with chain-terminating inhibitors. *Proc. Natl. Acad. Sci. U.S.A.*, 1977, 74(12), 5463–5467.
5. Maxam, A.M. and Gilbert, W., A new method for sequencing DNA. *Proc. Nat.l Acad. Sci. U.S.A.* 1977, 74(2), 560–564.
6. Sambrook, J., Fritsch, E.F., and Maniatis, T., in *Molecular Cloning: A Laboratory Manual.* Cold Spring Harbor Laboratory Press, NY, Vol. 1, 2, 3, 1989.
7. Glover, D.M., *DNA Cloning Volume I: A Practical Approach.* IRL Press, Oxford, 1985.

1.8 Genes, Taxonomy, and Evolution

Genes are the hereditary units of all forms of life and are made of deoxyribonucleic acid, or DNA, with the exception of select viral RNA genomes such as the human immunodeficiency virus (HIV, a retrovirus). Based on a morphological criterion pertinent to their genome, namely the presence or absence of a cell nucleus, all organisms can be grouped into two major morphological groups: the eukaryotes and prokaryotes. The latter are single-cell organisms lacking internal membrane organelles and have been subdivided into the two domains bacteria and archaea, respectively, based on molecular analysis of ribosomal RNA.

Being prokaryotes and having similar morphological characteristics, bacteria and archaea differ in their basic genomic structure and metabolism, importantly at the level of translation and protein biosynthesis, but also lipid content and cell wall structures. The grouping of all living organisms into three domains (eukarya, bacteria, and archaea) has been based on analysis of ribosomal RNA. With dozens of complete genomes from members of all three urkingdoms, the archaea branch appears to have eukaryotic, as well as eubacterial characteristics, depending on which set of proteins and metabolic pathways are being studied (Figure 1.15).[1] Taxonomic classification of organisms into various branches and kingdoms has changed over the last

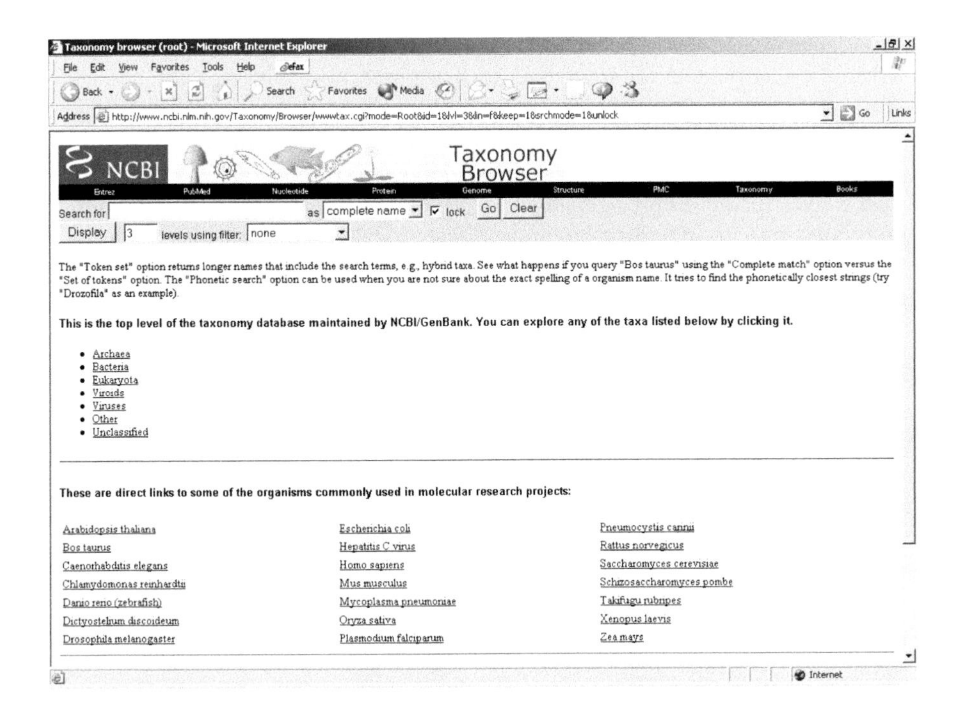

FIGURE 1.15
Taxonomy Browser at NCBI. Top level: the main menu of NCBI's taxonomy browser offers seven links to distinct "classes" of organisms. They include three urkingdoms: archaea (single cellular), eubacteria (single cellular), and eukaryota (single and multicellular; contain nucleus and other internal organelles). The other links refer to viruses and unclassified organisms. Viruses are not considered complete life forms and need the cellular machinery of the organisms of the three urkingdoms to proliferate. The taxonomy browser contains only organisms for which an entry exists in one of the databases: nucleic acid sequence, amino acid sequence, or protein structure.

century. The question of the correct "tree of life" is as old as taxonomy itself. Molecular evolution based on bioinformatics analysis rather than morphological structures adds new and powerful tools for finding answers to biological classification. While neither genotypical nor phenotypical taxonomy alone will provide the final answer, molecular biology already replaces "older" morphological branches of evolutionary biology. The current three domain structures, for instance, divide the previous single kingdom Monera into two radically different branches of life that are equally different from each other as each is different from the domain eukarya. Questions regarding the origin and coexistence of three basic organismal structures have fueled an ongoing debate among evolutionary biologists regarding which one occurred first and how they came into existence. The genome projects involving organisms from all three domains will certainly contribute to our understanding of this problem.

Bacteria are prokaryotic single-celled organisms that contain genomes with a highly compact gene structure and organization. Simply put, all genes contain a single coding region that corresponds to the amino acid sequence of the protein it is coding for. These coding regions are flanked by control regions that define the way proteins have access to the DNA for replication and transcription. Often, genes are grouped within functional units that are regulated in a coordinated manner. These genes code for multiple enzymes that constitute metabolic pathways. The corresponding multi-gene structures are called operons. The operon reflects the functional unit of a group of genes that are expressed in concert, e.g., their expression is coordinated by a single transcriptional unit.

The genome structure in eukaryotic organisms is more complex than in eubacteria and archaea. Most of their genes are not simple, single-coding frames, but are fragmented into exons and introns, which are coding and noncoding regions, respectively (note that prokaryotes have group I and II introns mostly in structural RNA genes). Eukaryotic genomes contain only about 5 to 15% coding regions, or genes, meaning that the vast majority of DNA is either not coding for any proteins or is not known to be doing so. In the human genome, only 1.2% are actually coding for proteins. Noncoding DNA, however, may be important in the regulation or "behavior" of the genome as a whole, specifically during meiosis, an important process in the reorganization of chromosomal DNA during sexual reproduction, and other stages in the life cycle of an organism. Archaea, which are morphologically indistinguishable from bacteria, are more closely related to eukaryotes in their genome structure and certain—but not all—metabolic pathways. With access to several completed genomes, it becomes evident, however, that the classification into urkingdoms is strongly dependent on the metabolic group of enzymes studied. The archaea may be the oldest group of organisms—in other words, the modern group of organisms that most closely resembles a potential common ancestor of all life on earth. Again, it is the universal existence of the genetic code that strongly suggests such a single, common ancestor organism.

Phylogenetic trees are visual ways of understanding evolutionary relationships. These trees contain information about the genes of thousands of species, their taxonomy, and evolutionary relationship sometimes referred to as "the tree of life." The tree of life is a metaphorical representation of the diversifying life forms on earth originating from a common ancestor. It is believed that there is only one such tree (e.g., a single progenitor "cell") and that life is not of multiple origin. The true origin of life on Earth, however, is obscured by the fact that extensive horizontal gene transfer between organisms occurs, effectively challenging the usefulness of the tree as a representation of the actual evolutionary history.

The Tree of Life Project at the University of Arizona (http://tolweb.org/tree) provides a visual overview of the phylogeny and biodiversity of all life on Earth. This is a very useful reference website for molecular biologists who often have no formal training in evolution, zoology, botany, and ecology.

The project contains information about the diversity of organisms on Earth, their history and characteristics. It is a multi-authored website coordinated and created by David R. Maddison at the University of Arizona. Clearly, we come to a point in biology where merging molecular with taxonomic data is not only feasible, but necessary to advance our understanding of evolution of life on earth.

Why would such two radically different gene and genome structures exist today? Both the prokaryotic and eukaryotic genome architectures have their advantages. The tight organization of genes on small chromosomes, where each gene comes in mostly one copy allows for fast growth. However, it makes the prokaryotic genome vulnerable to mutations since almost every mutation will hit a gene. Assuming that most mutations are detrimental, these "economic" genomes (they can be copied very rapidly) make individual cells vulnerable to disease and death. Two evolutionary strategies exist to cope with this vulnerability and to turn it into an advantage. First, simple genomes of bacteria and viruses allow a microbial population to grow rapidly and in large numbers with a capacity of adapting environmental stress by generating large numbers of mutations. Some individuals will have beneficial mutations that will carry the population on, even as a large percentage of a bacterial colony may die.

Clearly, while this strategy of genomic simplicity and economic efficiency works well for microbial organisms, they have not changed their overall appearance over some three billion years of existence. Bacteria were and remain small, single-celled prokaryotes. A different picture emerges with the inclusion of large amounts of noncoding sequences to the extent that genomes must be physically partitioned into chromosomes in order to manage replication and expression efficiently. The implications of noncoding regions in DNA on evolution are tremendous and largely explain the evolution of morphological complexity of eukaryotic organisms, from single-celled yeast to multicellular fungi, plants, and animals.

How is evolution of complexity thought to occur? Because mutations are random events, the noncoding parts of eukaryotic chromosomes absorb most of these changes in base composition and serve as a "playground" for chromosomal recombination and accumulation of silent mutations. Polymorphic markers are found in this portion of the DNA. A surprising finding of genetic analysis of clusters of genes from different individuals reflects the high frequency of nucleotide sequence differences between individuals (restriction fragment length polymorphism, or RFLP): reflects sequence variations in DNA sites that can be cleaved by DNA restriction enzymes. This polymorphism is used in forensic science. DNA "fingerprinting" yields information unique to one individual in several billion. The sensitivity of PCR is enough to amplify DNA for fingerprinting analysis from microscopic tissue samples, bloodstains, dead skin, or a single hair found at a crime scene.

In addition to the accumulation of noncoding sequences, eukaryotic organisms acquire complexity through polyploidy, i.e., they contain at least two sets of the genome. For instance, humans are diploid, having two chromosomal sets (23 chromosomes per set)—one from the mother and one from the father. Polyploidy possibly is a prerequisite for sexual reproduction, the random mixing of maternal and paternal genome variants during fertilization. In diploid organisms, each individual contains pairs of chromosomes with sister chromosomes inherited from each parent and randomly sorted during meiosis. The large amount of noncoding sequence allows the recombination (crossing over) of maternal and paternal copies of a gene (alleles) among homologous chromosomes before passing it on to the next generation. For a majority of genes, paternal and maternal origins are obscured. A small set of genes found on X,Y, and autosomal chromosomes are marked during meiosis (genetic imprinting). These marked genes control the development of the fetus, and their function is to maintain the male-to-female ratio at about 50%. Parthenogenesis, i.e., cloning or asexual reproduction, is prevented. An exception is found for mitochondrial and plastid DNA in higher organisms. Organelles are always inherited from the egg, i.e., the maternal side.

Consequently, sexual reproduction, unlike prokaryotic cell division, has evolved a mechanism that promotes allelic variability among individuals in a population. The result of the extended genomic content, genome duplicates, and recombination are individuals (except identical twins) that are genetically distinct from parents, siblings, and offspring. Cloning of animals from stem cells will bypass this random variation of alleles. The importance of genome structures in sexually reproducing organisms is evident from disease-causing translocations. A piece of one chromosome is transferred to a nonhomolog chromosome. Translocations are often reciprocal; that is, the two nonhomologs (e.g., chromosome 7 and 21) swap segments, causing a break within a gene destroying its function or creating a hybrid gene, or moving a gene under the influence of different promoters altering gene expression (causing cancer or developmental defects). Other diseases are the result of duplications, deletions, inversions, or ring formation causing aneuploidy where a particular chromosome or chromosome segment is under- or over-represented.

The importance of genome structures and chromosomal stability is also evident from the molecular evolution of histone proteins—the proteins responsible for packing DNA into condensed chromosomes during cell division. There, chromosomes form the tightly packed coils visible under the light microscope as X-shaped structures (pairs of chromosomes; see karyotype). Analyzing the amino acid sequence of histone proteins has revealed a highly conserved protein family in all eukaryotic organisms. They are a key feature of the genetics of animals, plants, and fungi. Their conserved sequences indicate a single ancestor cell or organism from which all modern eukaryotes are derived. Indeed, histone proteins are used as molecular clocks, molecular rulers to measure the phylogenetic distance (time since separation of two species) between distantly related organisms. The rate of mutations accumulating in a gene (nucleotide sequences which are not rejected) over time is a direct measure of the importance of the phenotype (e.g., a protein, cellular structure, physiological process) to the viability of the individual. Histone genes show an extremely low mutation rate over hundreds of millions of years, indicating that the function of these proteins is very sensitive to amino acid changes (mutations) and thus their structure must be essential for a eukaryotic organism to survive.

At first sight, this conserved feature of histone proteins is surprising, because they form millions of repetitive protein complexes like pearls on a string, binding the DNA in regular intervals irrespective of sequence. How could this be, that a highly conserved protein could bind DNA independent of its nucleotide sequence along chromosomes? The answer comes from the role histone proteins play in gene expression regulation, the epigenetic control of the genome. Specific amino acids in histone proteins can be modified post-translationally. This modification is a regulator component of chromosome packing and gene expression activity. Entire sections of chromosomes can be made accessible to transcription factors, while other sections are permanently inactivated through chromosome compaction. Thus, cell type-specific patterns of posttranslational modifications of histone proteins explains the precise unfolding of developmental programs, cell differentiation, and organ growth.

That chromosome structures are sensitive to disturbances indicates the importance of the large portion of noncoding sequences. They control how genes are inherited, and what happens to genes and when it happens. Understanding the latter is called functional genomics. The attempt to catalog the gene content, organization, and temporal expression patterns of a genome, will allow us to better understand the evolution of cellular function.

References

1. Woese, C.R., et al., Archaebacteria. 1978, *J. Mol. Evol.*, 11(3), 245–251.

2

Biological Databases

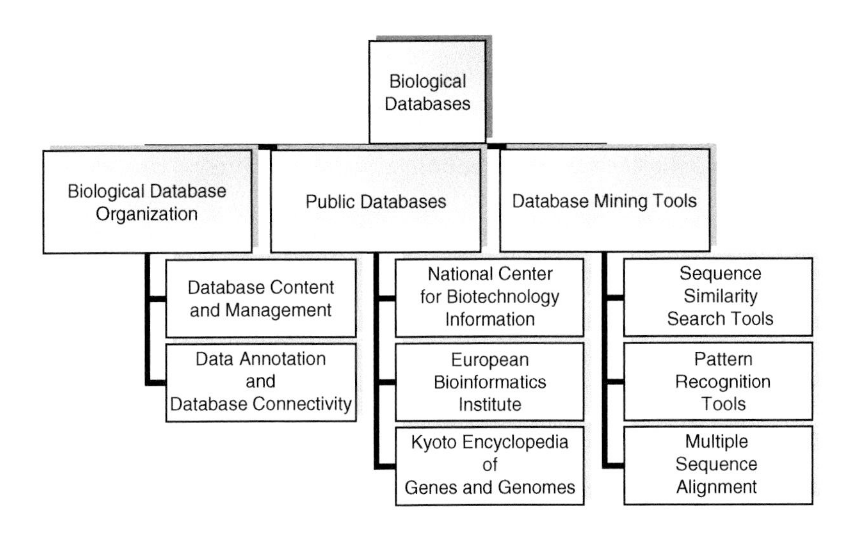

FIGURE 2.1
Chapter overview.

2.1 Biological Database Organization

2.1.1 Database Content and Management

The exponential growth of biological data over the past decade has created an enormous challenge to make effective use of the accumulated information. Correctly cataloging, labeling, and connecting sequence, structural, and functional information of genes and proteins of various organisms will facilitate the discovery of new biological trends and laws crucial to our understanding of life on Earth as complex systems. Information stored must be correct, complete, and internal relationships among elements easy to navigate.

Computational tools and databases are essential to the management and identification of patterns among database elements that reflect biological systems. The National Center for Biotechnology Information (NCBI) in the United States[1] and the European Bioinformatics Institute (EBI) in England[2] are two main life science servers responsible for dealing with this staggering volume of data. They both maintain reliable databases and analytical software that serve as valuable tools for today's scientific community. New entries to their databases are submitted every day, and their busy scientific staff adds the new data to the appropriate database. This allows the scientific community that subscribes to their databases to stay well informed while facilitating the progress of a variety of projects. The services offered by the servers (e.g., NCBI and EBI) are made possible by fast, elaborate computers that can perform the necessary analytical tasks, and the Internet interface that facilitates the electronic communication efforts. Typical uses of software tools in the life sciences are analyzing sequences, designing small molecules in drug development, predicting the function of proteins, and simulating the role of gene networks in cellular mechanisms and pathogenesis.

The usefulness of these and other centralized databases lies in their utility as free repositories of the collective knowledge of the structure and function of genes and proteins in biology and medicine. They are electronic libraries to be read using bioinformatics tools. These tools have become indispensable for any biological or medical research. Since the data pool is so enormous, particular care needs to be spent on database information storage, organization, and retrieval. The organization of biological databases starts with raw data; DNA sequences, protein structures, and more recently functional data (bioarray data; see Section 3.4).

A functional database must provide for accurate submission, annotation, and retrieval of raw data. It also must allow the construction of secondary and tertiary databases including bibliographic, medical annotations, and interactivity to find information to biologically relevant questions (queries) such as sequence similarity searches and alignments, the identification of genes, comparison of genomes, the translation of DNA sequences into amino acid sequences and back, and the identification of homologs and paralogs (evolutionary related sequences) along taxonomic groups.

2.1.2 Data Submissions

The main sources of sequence information in central databases are through direct contributions from scientists. Twenty years ago, it was not uncommon to find people reading DNA or amino acid sequences to each other over the telephone, thereby causing an "error rate" that might have exceeded the mutation rate of natural DNA replication or transcription processes. Today, sending files by e-mail and transferring them to and from GenBank (Figure 2.2) or SWISS-PROT is extremely easy, fast, and virtually error free. The current development of the Internet has made the process of reading and submitting information to NCBI, EBI, or the DNA Data Bank of Japan (DDBJ) indeed

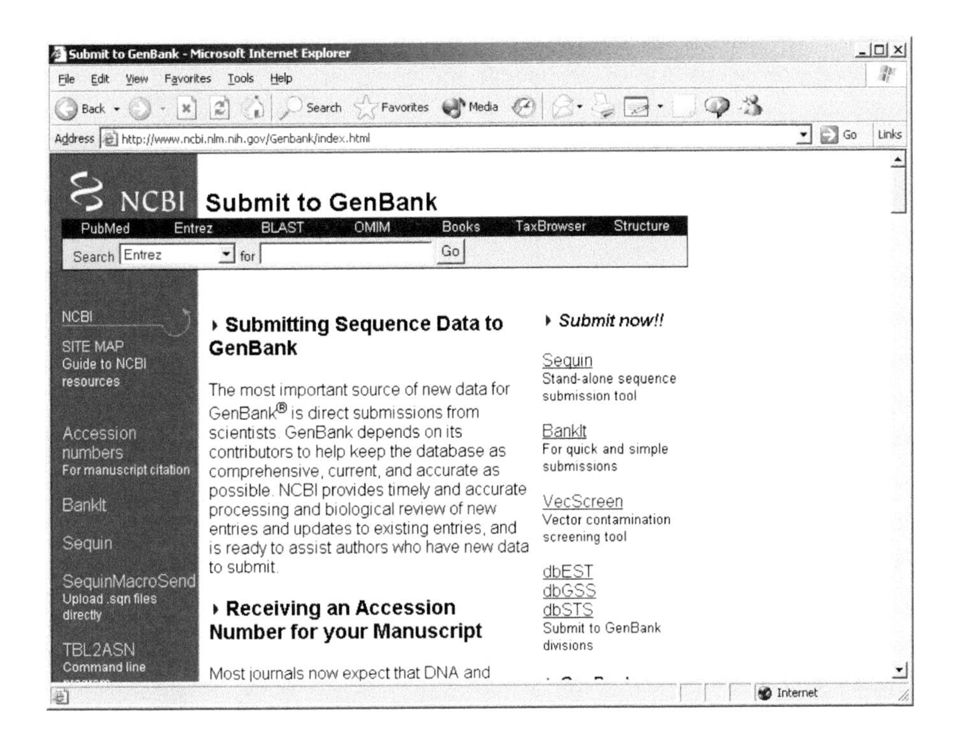

FIGURE 2.2
Submitting sequence data to GenBank.

very easy. BankIt (WorldWide Web direct deposit) or Sequin (a standalone program) are provided by NCBI to send sequence information and biological annotation to GenBank's staff scientists who assign them accession numbers for immediate release to the public (usually within 48 hours). Daily exchanges of new submission data between GenBank, EBI, and DDBJ ensure that the information submitted by the researcher is nonredundant (submitted only once) making them virtual mirror sites.

Authors are able to update their original information. Normally, scientists have a single sequence of a gene they have discovered and some relevant biological information. The genome projects, however, require specialized submission procedures for sequence information originating from ESTs (expressed sequence tags), STSs (sequence tagged sites), and GSSs (genome survey sequences). These sequences differ from traditional sequences of functional genes or proteins in their relative short length and large number.

ESTs are short sequences of 300 to 500 bp and represent actually expressed genes because they are obtained through polymerase chain reaction (PCR) of messenger RNA extracted from tissues and cells. In addition to their proper sequence, these short sequence tags are markers that are helpful in locating (map) genes on chromosomes. EST submissions, therefore, include both sequence and genome map information (Figure 2.3). They are often submitted in batches from dozens to thousands and contain redundant information with

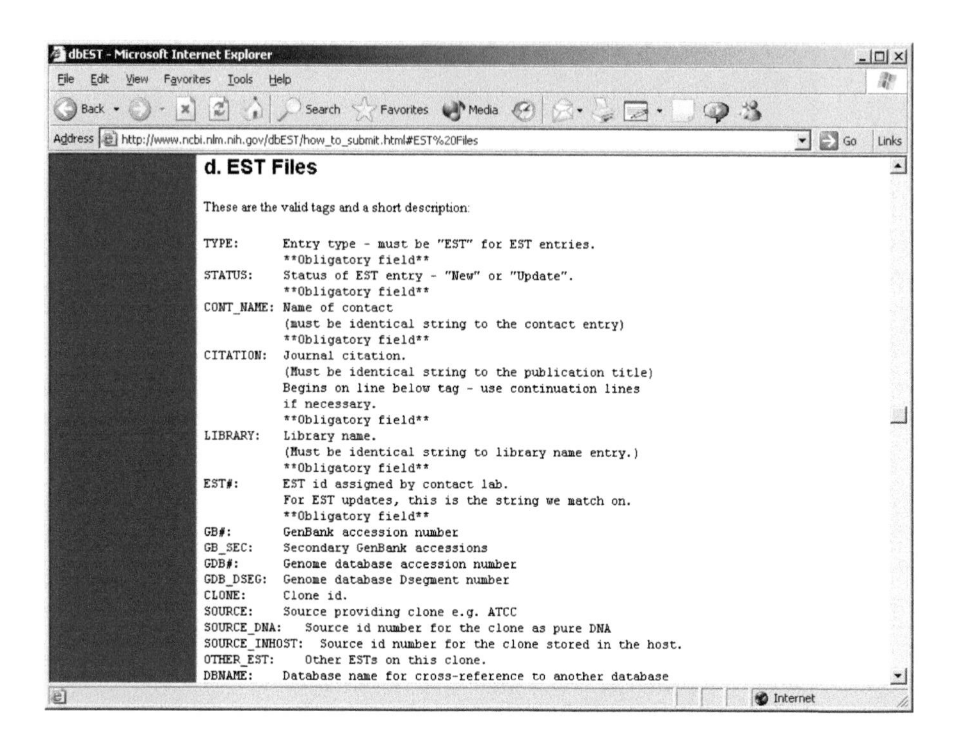

FIGURE 2.3
Submission format for EST files at GenBank. EST expressed sequence tag.

regards to citation, submission data, and library information. GenBank provides online information about all submission requirements.

STSs are similar to ESTs in length and number of submitted sequences per batch. However, they do not represent gene expression patterns but provide unique identifiers within a given genome identifiable by PCR. Although ESTs are the fastest growing segment of public databases, STS and single nucleotide polymorphism (SNP) sequences will soon outnumber them because of the large percentage of noncoding regions of these genomes and the increasing input of sequences from genome projects.

Because of their value to the scientific community, genomic sequences submitted to NCBI are processed (genome center, clone name, accession number) on a daily basis, and can be submitted in various draft stages long before they are completed. The high throughput genomic (HTG) sequences division at NCBI (see Figure 2.4) distinguishes three phases: (1) unfinished, unordered, (2) unfinished, ordered, and (3) high-quality finished sequences that do not contain any sequence gaps. Because of the accelerated pace of high-throughput sequencing and submission, the early identification of an error is important.

To expedite this process, NCBI has set up streamlined submission procedures and deadlines to ensure the prompt and error-free release of new sequences into its Entrez[3] system. Without speed and accuracy, any data analysis will suffer

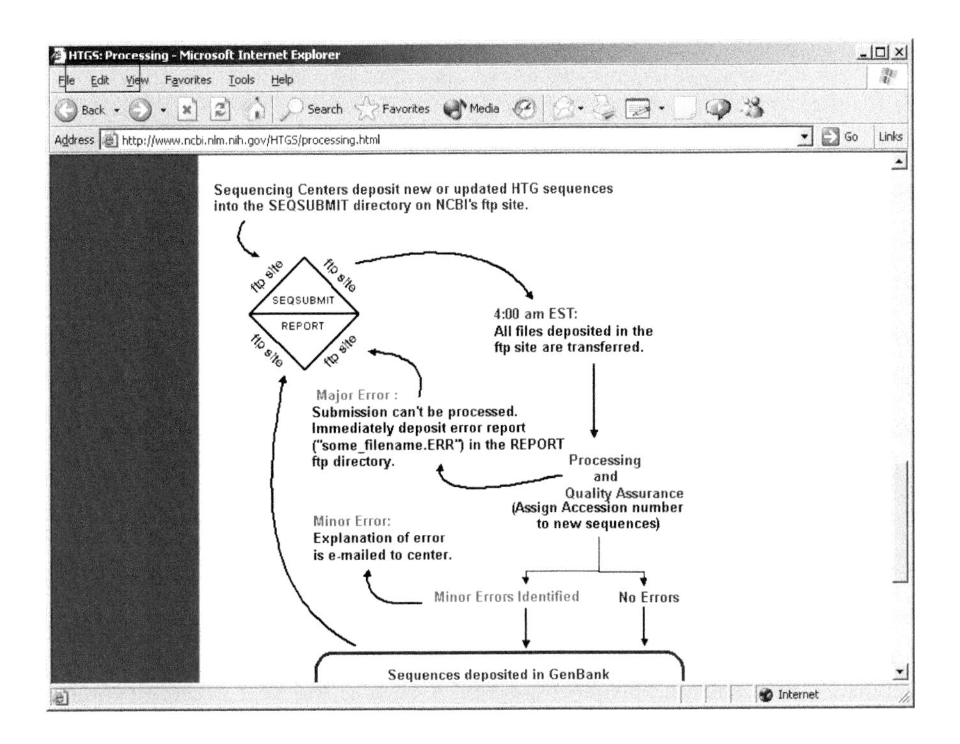

FIGURE 2.4
HTGS, processing: submission error communication procedure using FTP protocol transmission.

greatly. Errors propagate very quickly through electronic media and because much of bioinformatics is used for predictive purposes (identifying novel genes, novel functions, drug discovery, predicting structures, and phylogenetic relationship), errors at the sequence level will result in erroneous interpretation and conclusions. The greater concern at NCBI, however, is with erroneous annotation of deposited sequences. To solve this problem, trained scientists must comb the databases and correct any mistakes. Again, the propagation of faulty annotation reduces the reliability of data interpretation.

Although GenBank relies on direct submission of individual scientists and high throughput sequencing centers (such as Sanger Center, TIGR) the staff at NCBI continuously screens biomedical journals for published sequence and structure data and use them for annotation purposes. According to GenBank's release 138.0 (October 2003):

> GenBank Release 138.0 is a release of sequence data by NCBI in the GenBank flat file format. GenBank is a component of a tri-partite, international collaboration of sequence databases in the U.S., Europe, and Japan. The collaborating databases in Europe are the European Molecular Biology Laboratory (EMBL) at Hinxton Hall, U.K., and the DNA Database of Japan (DDBJ) in Mishima, Japan. Patent sequences are incorporated through arrangements with the U.S. Patent and Trademark Office, and via the

collaborating international databases from other international patent offices. The database is converted to various output formats, including the Flat File and Abstract Syntax Notation 1 (ASN.1) versions. The ASN.1 and Flat File forms of the data are available at NCBI's anonymous FTP server: ftp.ncbi.nih.gov (from ftp://ftp.ncbi.nih.gov/genbank/gbrel.txt).

2.1.3 The Growth of Public Databases

To grasp the extent of today's vast computer network and the tremendous flow of genetic data, we must go back some 40 years, when molecular biology was in its infancy and none of today's techniques existed. The genetic code had just been solved by Marshall Nierenberg in 1965 and the discovery of restriction enzymes — the tools used to cut DNA — occurred soon thereafter in 1968 by Werner Arber. In the 1960s, sequencing of proteins was faster than sequencing nucleic acids; biochemists spent months and years establishing amino acid sequences by sequentially splitting amino acids off large amounts of purified proteins. Pioneers such as Margaret Dayhoff, one of the first biologists to make use of comparing amino acid sequences for evolutionary information,[4] recognized the need to establish public sequence databases and her input was instrumental in developing computer-based analysis tools. As a result of Margaret Dayhoff's efforts, the first protein sequence database was established in the early 1960s.

Today, amino acid sequences are routinely obtained by means of molecular biology; i.e., by sequencing the gene first and then inferring the amino acid sequence from the DNA sequence according to the appropriate codon usage. Amino acid sequencing, however, is still used to analyze short peptides. Its use has been boosted recently by the increased interest in proteomics. Peptide fragments obtained from protein expression profiles are micro-sequenced to determine molecular mass and charge (isoelectric focusing). Based on these short amino acid sequences obtained from protein extracts, protein expression profiles and post-translational modifications can be quickly analyzed.

It was not until the 1980s that DNA sequence databases were established. The fledgling Internet, particularly e-mail, facilitated the scientific community's ability to transfer information quickly and reliably. Before the use of web browsers, the standard methods for downloading files from remote computers (locations) were the file transfer protocols, FTP and Kermit, using public domain software packages. These are still used for communication and uploading and downloading files between supercomputer centers (see Pittsburgh Supercomputing Center; http://www.psc.edu). Until 1989, the most common forms of sequence submission and retrieval were surface mail (hard copies, floppy disks, and magnetic tapes), telex, and dial-up online networks. The Human Gene Mapping Library (HGML, Cold Spring Harbor Laboratories) manually updated their database annotations, and staff scientists at GenBank screened journals for published sequences. During that same period, only 50% of all entries were submitted directly by the scientists involved. Of these entries, 70% were in computer-readable form for UNIX-based Sun workstations. UNIX, of course, still has not been replaced. Internet browsers are user-friendly interface programs for Windows and Apple operating systems that transform PCs into terminals for UNIX-operated supercomputer centers and workstations.

In a period of just five years, from 1983 to 1988, the average lag from time of publication and submission of a DNA sequence to its availability, including annotation dropped from one year to five months. By 1988, the year of the Human Genome Initiative,[5] gene sequences representing some 1200 organisms were deposited in the databases of GenBank, EMBL, and Japan's DDBJ. Subscribers to GenBank received information every three months on a magnetic tape. By 1994 CD-ROM technology was adopted, but was prohibitively costly for academic users. A one-year subscription to EMBL's database[6] cost about $200 on tape and $400 on CD-ROM for noncommercial users in the U.S. Today, website entry and retrieval forms provide instant access to millions of sequences (Figure 2.5).

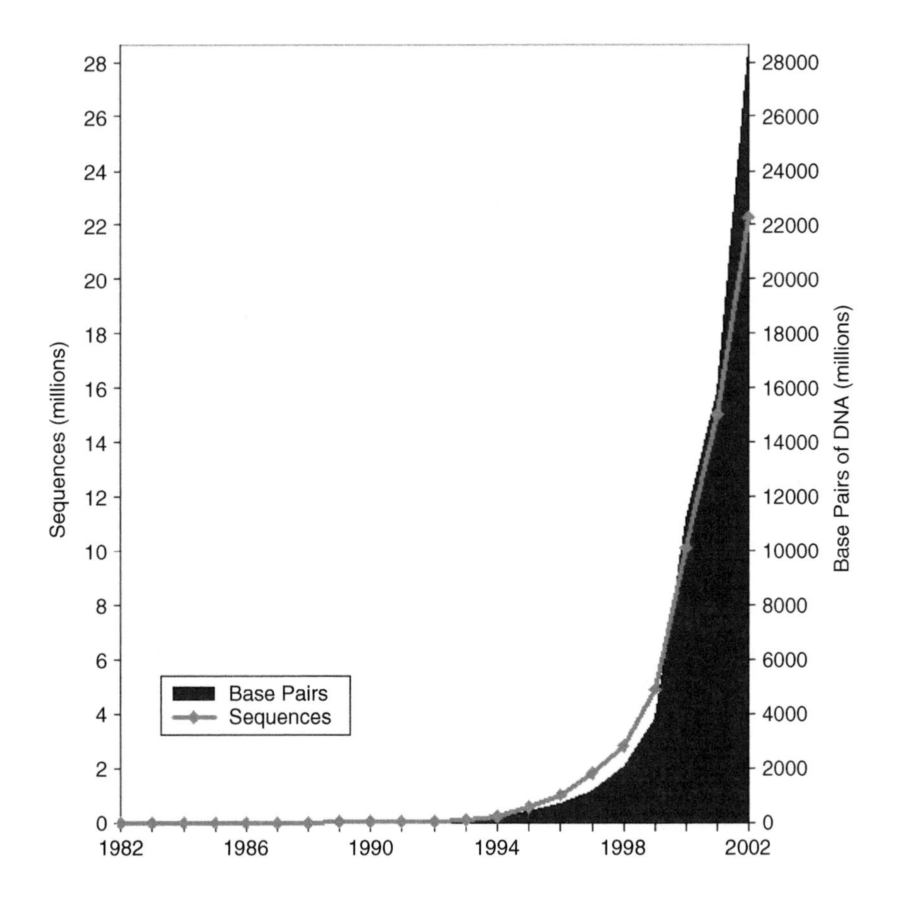

FIGURE 2.5
Growth of sequence and base pair content of GenBank.

2.1.4 Data Retrieval

The real power of a database comes with the software used to retrieve and analyze sequence information. The results of sequence analyses fall into four major categories of biological information; (1) comparing gene sequences for *similarities* for phylogenetic analysis, (2) identification of the gene structure including *reading frames, exon-intron* distribution, and *regulatory elements*, (3) prediction of *protein structural elements*, and (4) navigating *genome maps*, the distribution of genes and noncoding elements on chromosomes and their assessment within the context of metabolic pathways.

The currently available data for DNA and protein sequences is so enormous that searching for information is dubbed, "biological data mining," similar to a real gold mine with hidden treasures where common rocks have to be separated from gold nuggets. Search engines perform two basic tasks: (1) simple string searches for information retrieval of stored data [GenBank (nucleotides and proteins), PubMed (MEDLINE), 3-D structures, genomes, and taxonomy databases] and (2) similarity searches (e.g., Blast) to retrieve, align, and compare sequences or structures.

The first step in sequence analysis includes retrieving sequences based on specific criteria, one of which is similarity or identity between sequences that could be obtained through a search tool such as basic local alignment search tool (BLAST).[7] But if no sequence is known or available the NCBI's search engine at either the nucleotide or protein level can be screened by typing in a keyword referring to the name of a protein, the name of authors doing research on a protein of interest, or the proper accession number. These searches retrieve reports indicating the number of entries in the selected database containing the keyword anywhere in the data file.

As an example we are looking for bacterial proteins called porins. On the Entrez interactive search site (Figure 2.6) across multiple databases we type in the keyword "maltoporin."

The query details the existence of 50 entries found in the "nucleotide" section (GenBank sequence database). There are additional entries on protein sequences, protein structure, genomes, and research articles. Clicking on "nucleotide" link brings us to the "Entrez Nucleotide" site where we have a summary of all 50 individual entries. Depending on the interest in specific genes, proteins, and organisms, selecting the appropriate links will download the sequence information (FASTA report), annotated information (GenBank report), literature links (Medline link), a graphical viewer (Applet Java based; Figure 2.7), linking to related protein or nucleic acid sequences selected for significant alignments (Figure 2.8) or taxonomic report listing (Figure 2.9). The last entry on the list with accession number M16643 links to the *E. coli* lamb gene encoding maltoporin, an outer membrane protein of this bacterial species that selectively facilitates transport of maltodextrins (a form of oligoglucose) across the outer membrane.

```
M16643
E. coli lamb gene encoding malto-oligosaccharide-selective
pore protein (maltoporin), 5 end
gi|146589|gb|M16643.1|ECOLAMB[146589]
```

Clicking on the M16643 link and selecting the FASTA display option shows the 5 end of the gene.

```
>gi|146589|gb|M16643.1|ECOLAMB E.coli lamb gene encoding
malto-oligosaccharide-selective pore protein (maltoporin),
5 end
TCGACTGCATAAGGAGCCGGGCGTTTAAGCACCCCACAAAACACACAAAG
CCTGTCACAGGTGATGTGAAAAAAGAAAAGCAATGACTCAGGAGATAGAATGATG
ATTACTCTGCGCAAACTTCCTCTGGCGGTTGCCGTCGCAGCGGGCGTAATGTCT
GCTCAGGCAATGGCTGTTGATTTCCACGGCTATGCACGTTCCGGTATTGGTTGGA
CAGGTAGCGGCGGTGAACAACAGTGTTTCCAGACTACCGGTGCTCAAAG
TAAATACCGTCTTGGCAACGAATGTGAAACTTATGCTGAATTAAAATTGGGTCAG
GAAGTGTGGAAAGAGGGCGATAAGAGCTTCTATTTCGACACTAACGTGGCCTAT
TCCGTCGCACAACAGAATGACTGGGAAGCTACCGATCCGGCCTTCCG
TGAAGCAAACGTGCAGGGTAAAAACCTGATCGAATGGCTGCCAGGCTCCACCATC
TGGGCAGGTAAGCGCTTCTACCAACGTCATGACGTTCATATGATCGACTTCTAC
TACTGGGATATTTCTGGTCCTGGTGCCGGTCTGGAAAACAT
```

Selecting the "proteins" under the links menu shows one entry for the maltoporin precursor protein of *Escherichia coli* (accession number AAA24059).

```
AAA24059
maltoporin precursor
gi|551815|gb|AAA24059.1|[551815]
```

The FASTA display shows the amino acid sequence of the N-terminal part of the protein, according to the paper published by Heine, Kyngdon, and Ferenci (Medline link 87277420). According to the authors the deposited partial sequence of *E. coli* maltoporin contains determinants in the LamB (maltoporin) gene of *E. coli*, which influence the binding and pore selectivity of the protein.

```
>gi|551815|gb|AAA24059.1| maltoporin precursor
MMITLRKLPLAVAVAAGVMSAQAMAVDFHGYARSGIGTGSGGEQQCF
QTTGAQSKYRLGNECETYAELKLGQEVWKEGDKSFYFDTNVAYSVAQQNDWEATD
PAFREANVQGKNLIEWLPGSTIWAGKRFYQRHDVHMIDFYYWDISGPGAGLEN
```

The N-terminal amino acid sequence can now be used (FASTA format) to Blast (Blastp search) the nonredundant database (GenBank CDS translations

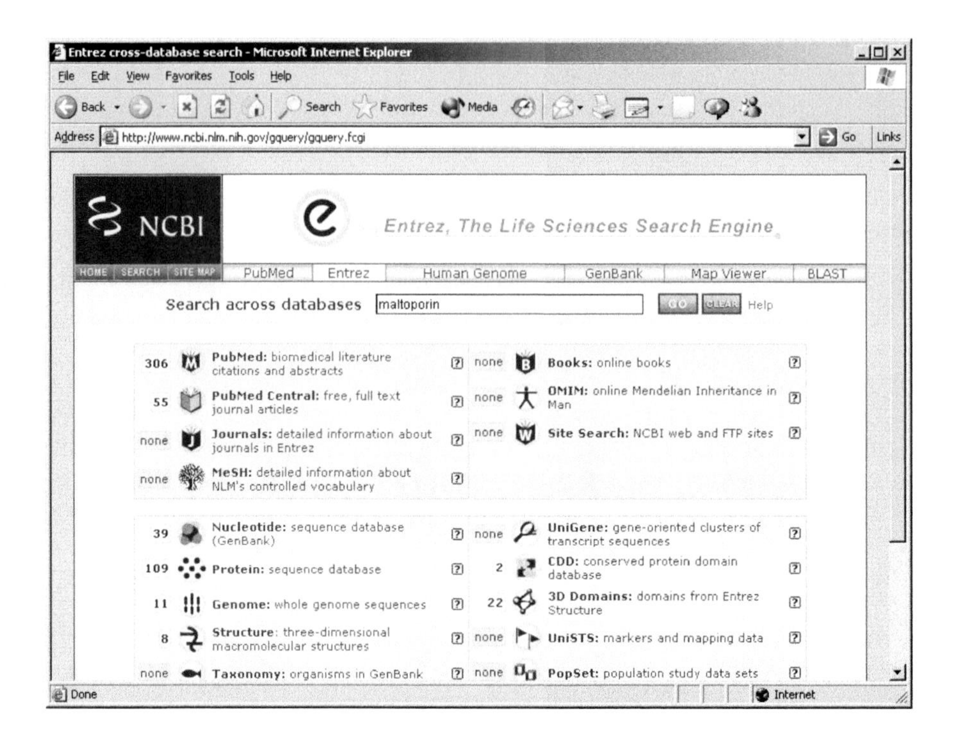

FIGURE 2.6
Entrez search "across database": result for "maltoporin"(www.ncbi.nlm.nih.gov/gquery/gquery.fcgi).

+ PDB + SwissProt + Spupdate + PIR; 1,543,518 sequences; 504,497,942 total letters) to find related sequences. This *protein–protein* Blast search is executed by selecting and copying the sequence information from the Fasta format (amino acid sequence) window into the Blast search window and selecting protein–protein BLAST (blastp) form the main menu page of BLAST (www.ncbi.nlm.nih.gov/BLAST). Improving user friendliness and interactivity the "Blink" link provides precomputed Blast results for each Entrez protein sequence offering visual taxonomic information (phylogenetic relatedness of near neighbors and the structural database, including the conserved domain database, CDD).[8]

The Blast search results currently lists 64 hits, all of which are maltoporins of *E. coli* and related Gram-negative bacteria (*Salmonella typhimurium*; *Yersinia enterocolitica*; *Aeromonas salmonicida*; *Vibrio cholerae*; *V. parahaemolyticus*; *Klebsiella pneumoniae*). The results page displays a Java applet based interactive graphic (Figure 2.7) that links to a list of sequences with significant alignment scores (Figure 2.8) or sorted by taxonomic report according to species that have porin orthologs (Figure 2.9).

The level of a reported similarity indicates potential biological relationships across species and taxonomic divisions. Identities between sequences

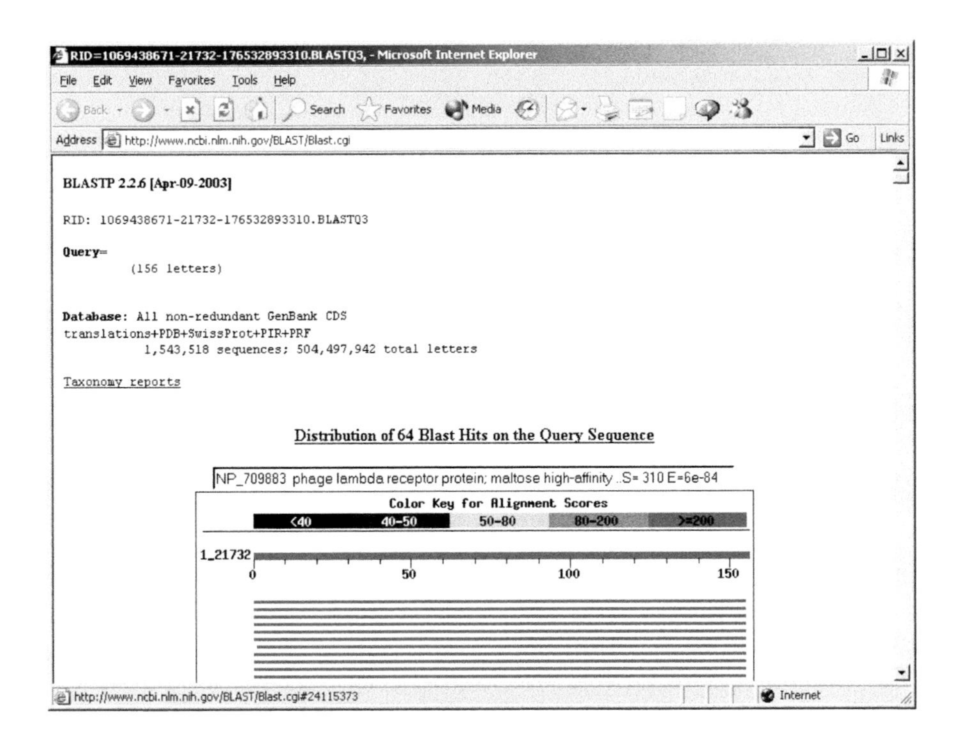

FIGURE 2.7
Graphical JAVA based browser interface. Blast search result for maltoporin precursor sequence >gi551815.

are measured as E values (expert value) and ranges from 0 (100% identity) or close to 0 as 3e-91 for entry number M16643 to larger values indicating lower identity. Larger values (>0.1 threshold) mean that the similarity is most likely due to chance and thus is a random hit. Hits with E values larger than the threshold indicates a relationship that can neither be established nor excluded *a priori*. The E-value estimates the expected number of random hits between sequences in the database. Thus, the E value depends both on the number of distinct sequences in the database and the length of the query sequence. Short sequences will naturally produce more random hits in a very large database.

So what does a "high" E value mean? It means that even though the DNA sequence is not sufficient to establish a relationship between two genes, additional information at the level of the corresponding amino acid sequence, motifs, and protein structure information may nevertheless establish such a relationship. With an increasing number of high-resolution structures available it can be firmly established that (super-secondary and tertiary) structures of related proteins show higher similarity than can be found by looking at the sequence alone. The reason is the redundancy in amino acid (sequence) combinations with respect to structural motifs.

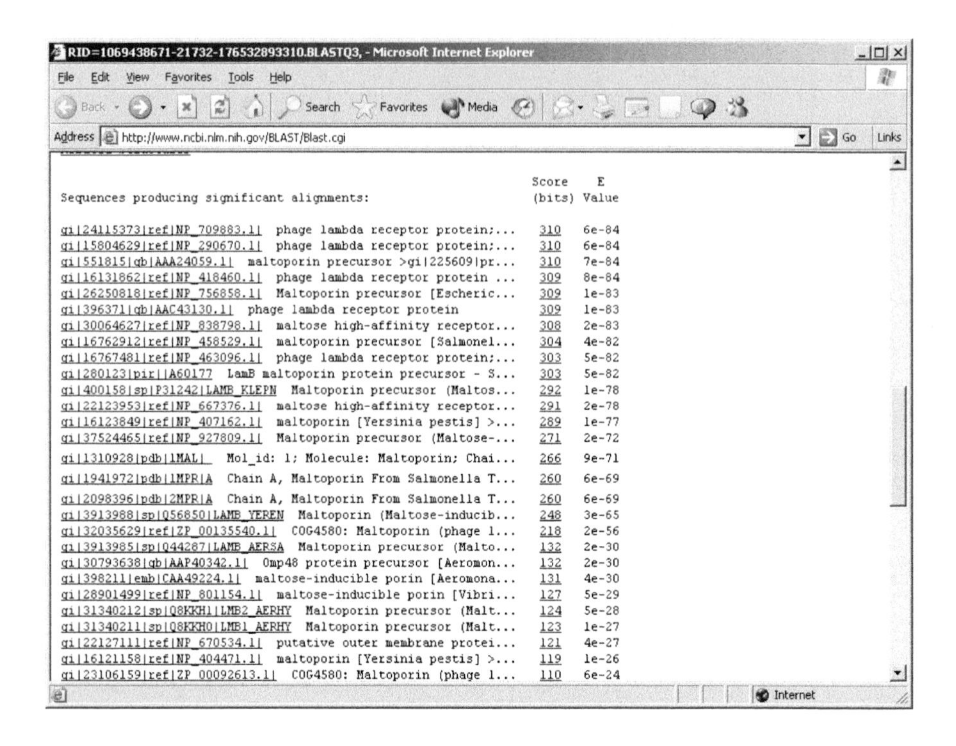

FIGURE 2.8
List of sequence producing significant alignments with score values. Blast search result for maltoporin precursor sequence >gi551815.

References

1. Woodsmall, R.M. and Benson, D.A., Information resources at the National Center for Biotechnology Information. *Bull. Med. Libr. Assoc.*, 1993, 81(3), 282–284.
2. Emmert, D.B. et al., The European Bioinformatics Institute (EBI) databases. *Nucleic Acids Res.*, 1994, 22(17), 3445–3449.
3. McEntyre, J., Linking up with Entrez. *Trends Genet.*, 1998, 14(1), 39–40.
4. Schwartz, R.M. and Dayhoff, M.O., Origins of prokaryotes, eukaryotes, mitochondria, and chloroplasts. *Science*, 1978, 199(4327), 395–403.
5. Roundtable forum, The human genome initiative: issues and impacts. *Basic Life Sci.*, 1988, 46, 93–109.
6. Kahn, P. and Cameron, G., EMBL Data Library. *Methods Enzymol.*, 1990, 183, 23–31.
7. Altschul, S.F. et al., Basic local alignment search tool. *J. Mol. Biol.*, 1990, 215(3), 403–410.
8. Marchler-Bauer, A. et al., CDD: a curated Entrez database of conserved domain alignments. *Nucl. Acids. Res.*, 2003, 31(1), 383–387.

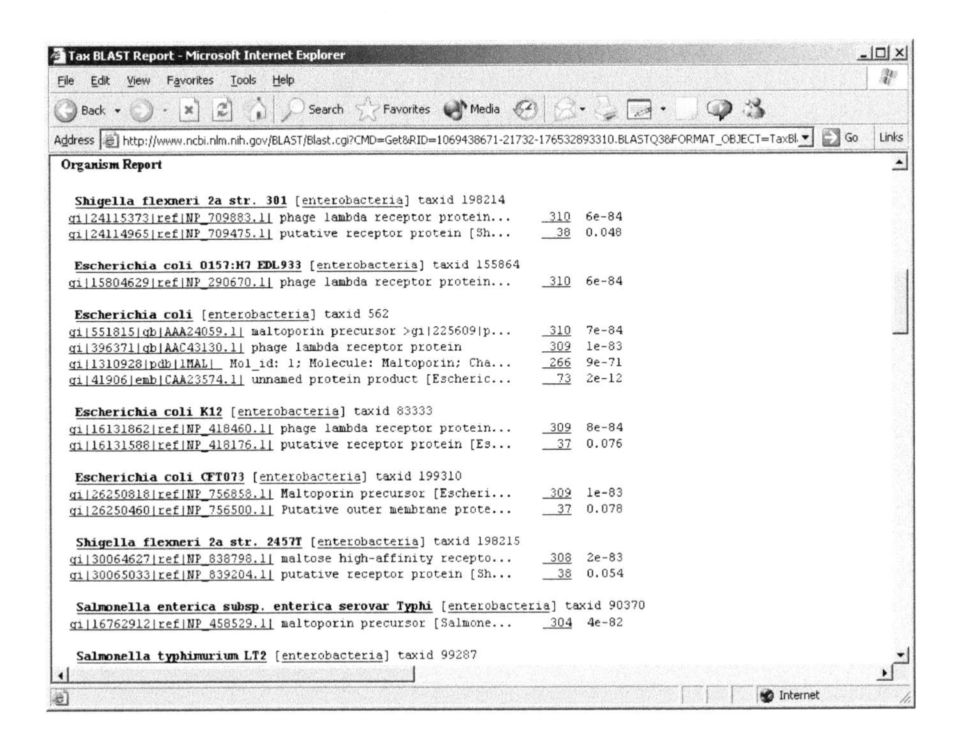

FIGURE 2.9
Taxonomy report; list of best matches by organism. Blast search result for maltoporin precursor sequence >gi551815.

2.1.5 Data Annotation and Database Connectivity

Scientists work in small independent groups, which often results in repetitive naming of identical genes and proteins. This is typically reflected in the submission of data for new, emerging fields; in this case, DNA sequences are submitted more than once and often under different names and with inconsistent annotations. Only specialists in the field would be able to recognize that seemingly different entries refer to one subject. It is similar to having your name listed as three entries in a telephone book—last name, first name, and nickname—each of which refers to the same person. For biological databases, these multiple entries with different names and overlapping sequences bring forth two important issues. First, the need to develop a clear terminology or controlled vocabulary across all biological faculties to completely and unambiguously annotate sequences and structures of genes and proteins in the databases. Such consistency across platforms and disciplines will facilitate automated database connectivity and thus improve biological knowledge. Second, to exploit sequence redundancy as quality control of the submitted data.

2.1.5.1 Annotation

To understand a gene or its sequence, one has to know its context. The context for a gene means all associated biological information that defines its *function*, its structure, its cellular (chromosomal) location, the structure and function of its product, protein or RNA, and its taxonomic ranking. The following is a list of the minimal annotations that may come with a DNA sequence:

Related sequences in database
Structure prediction/comparison with x-ray structure
ORF (open reading frame) if function is unknown
Domain structure
Transmembrane segments
Signal sequence
Consensus site for glycosylation, phosphorylation, and lipid anchors
Alternative nomenclature
Genetic information like regulatory sequences
Translation
Molecular weight, charge (2-D gels)
Bibliography

This list contains *intrinsic elements* of a gene record that are (except for bibliography) the result of applying bioinformatics tools. Additional annotation terms, however, must be obtained from experimental and contextual information. Does the gene code for an enzyme or a receptor? Does it code for a cytoplasmic or a membrane protein? Is it part of cell cycle control, energy metabolism, or transport? A complete annotation list should include all the *biological* information associated with it.

In which way should this information be listed and what terms should be chosen for consistent annotation? In order to be useful for retrieving, linking, and analyzing biological databases, data annotation depends on well-structured and controlled vocabularies (ontologies) for shared use across different biological disciplines.

While *systematics* has a long tradition in science, genome projects require a renewed effort to develop a common language for the description of genes and proteins, cells and organisms. Taxonomy developed the well-known terms ranking organisms as species (e.g., *Homo sapiens*), genus, family, order, class, phylum, kingdom, and domain. When it comes to naming genes and proteins, however, biological subdisciplines have their own terminology, hampering interdisciplinary research. Novel open biological ontology projects (Table 2.1) are beginning to address this question of extending systematics to the molecular level.

Today, information on a single gene can be found in various databases under various names and different annotation terms, making it appear as if these entries refer to entirely different genes and gene products, because of lack of

TABLE 2.1

Open Biological Ontologies

Project	Controlled Vocabulary	URL
Gene Ontology Consortium	Provides three structured networks of defined terms to describe gene product attributes: Biological process/Cellular component/ Molecular function	www.geneontology.org
Sequence Ontology	Set of terms used to describe features on a nucleotide or protein sequence and is intended to be used in concert with a heavily curated genome annotation project	song.sourceforge.net
Generic Model Organism Databases	Genome visualization and editing tools, literature curation tools, a robust database schema, biological ontology tools, and a set of standard operating procedures	gmod.sourceforge.net
Standards and Ontologies for Functional Genomics	Standards and ontologies with an emphasis on describing high-throughput functional genomics experiments	www.sofg.org
The Microarray Gene Expression Data	Ontology for describing samples used in microarray experiments	mged.sourceforge.net
Plant Ontology Consortium	Vocabulary for plant structures, and growth and developmental stages	www.plantontology.org

Source: From open biological ontologies (OBO) obo.sourceforge.net. With permission

common terminology. Only recently has the biological community started to use vocabulary by the *Gene Ontology Consortium* referred to as gene ontology (GO). It groups *proteins* within three structured networks: biological process, cellular component, and molecular function. As of September 2004 the gene ontology vocabulary contains 7400 terms for Molecular Function, 8888 terms for Biological Process and 1448 terms for Cellular Component. A search for GO terms for the water channel protein "aquaporin" shows the following result tree:

* Accession: GO:0015250

* Aspect: molecular_function

* Synonyms: aquaporin

* Definition:None

Term Lineage

* GO:0003673: Gene_Ontology (149784)

o GO:0003674: molecular_function (101079)

+ GO:0005215: transporter activity (9113)

GO:0015267: channel/pore class transporter activity (1125)

* GO:0015268: alpha-type channel activity (1099)

o GO:0015250: water channel activity (87)

Only information regarding molecular function is available. The cascade illustrates that there are currently 87 aquaporin sequences/structures deposited in public databases. Aquaporins belong to a class of transporters with "alpha-type channel activity" (containing alpha helical channel structure). The alpha-type channels are part of "channel/pore class of transporter activity." Channels form a part of 9113 described "transporter activity" terms.

2.1.5.2 Redundancy

Obviously, redundancy of sequence information in databases can be useful as quality control. It sometimes happens that two competing laboratories publish the sequence of the same gene, but with one or more base differences. Does this represent a true mutant coming from different strains of mice or a sequencing error? If the source of the gene is the same organism with the same traits, sequencing errors usually explains the differences.

As the genome projects develop into an organized enterprise, the elimination of redundancy is a major concern in streamlining and optimizing the databases. Redundancy not only comes from the fact that different researchers are interested in the same protein or gene, but that different techniques of cloning and sequencing random genomes creates fragments with little biologically relevant annotation.

> One gene, many sequences. GenBank is a comprehensive source of sequence data, but selecting candidates for physical mapping can be difficult. This is in a large part due to the presence of multiple sequence records that, while not identical to one another, are derived from the same gene… . Gene sequence entries possess differing amounts of flanking and intron sequence. Sequences of mRNAs can be incomplete or contain variation because of alternative splicing. Finally, ESTs are both fragmentary and have a higher error rate. For the UniGene set, all of these sequences are drawn together into a cluster if they are found to share statistically significant DNA sequence similarity in the 3 UTR. For the Washington University–Merck & Co. ESTs, these sequences are derived from oligo (dT)-primed mRNA, with directional cloning and sequencing from both the 5 and 3 ends. Clustering uses only 3 ends, but use of common clone identifiers places corresponding 5 ends into the clusters. (Source: www.ncbi.nlm.nih.gov.)

Redundancy, of course, is beneficial for certain aspects of genome mapping. Redundancy and homology are closely related concepts with homology actually referring to two (or more) different genes with great similarity. Those are likely paralogs in a population or orthologs in different species or taxonomic groups.

Redundancy also comes in a slightly different form that relates to pieces of individually submitted sequences that belong to the same gene or transcript cluster. Particularly for short sequence tags, be that expressed or genomic (site specific) tags, the UniGene feature provides information pages for known genes for which multiple sequence entries can be found in Entrez genes. It is important to note that NCBI's GenBank does not eliminate original database entries, but provides gene oriented sites from which all related information can be accessed, be that sequence information and its homologs, genome map location and nearby markers/genes, protein structures, and functional and medical information about diseases.

To provide concise view of the database, RefSeq is a molecular database at NCBI that contains a nonredundant set of GenBank sequences including genomic DNA contigs, mRNAs, and proteins for known genes, mRNAs and proteins for gene models, and entire chromosomes.

With the growing list of stored and annotated genes from various organisms, database interactivity is essential. At NCBI, a gene oriented view is implemented that allows biologists to search databases, and not just NCBI's site, for all available information related to a known gene. At NCBI, Entrez is the integrated, text-based search and retrieval system used for literature (PubMed), nucleotide and protein sequence, protein structure, genome, taxonomy, and expression databases.

2.2 Public Databases

The following section will predominantly focus on an overview of the U.S. NCBI, the EBI database, and the KEGG in Japan (Figure 2.10).

2.2.1 National Center for Biotechnology Information (NCBI)

In November 1988, the U.S. Senate recognized the need for computerized data processing in the biomedical and biochemical fields and passed legislation that helped to establish NCBI (Figure 2.11) at the NLM. NLM's focus is on maintaining biomedical databases, while NCBI is specifically involved in the development of new analytical tools to aid in understanding the

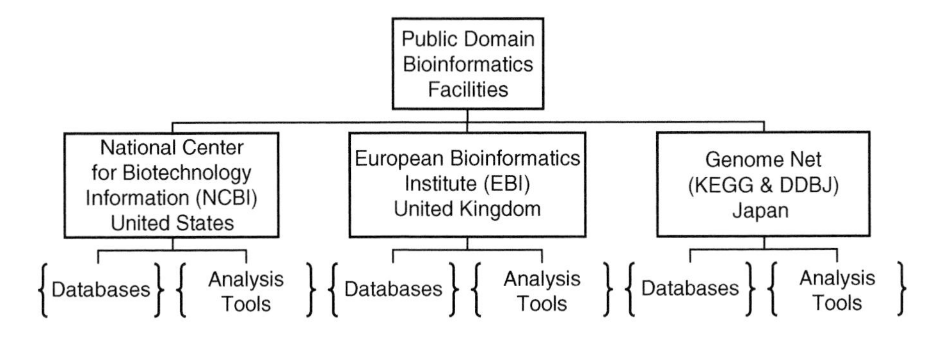

FIGURE 2.10
Primary public domain bioinformatics servers.

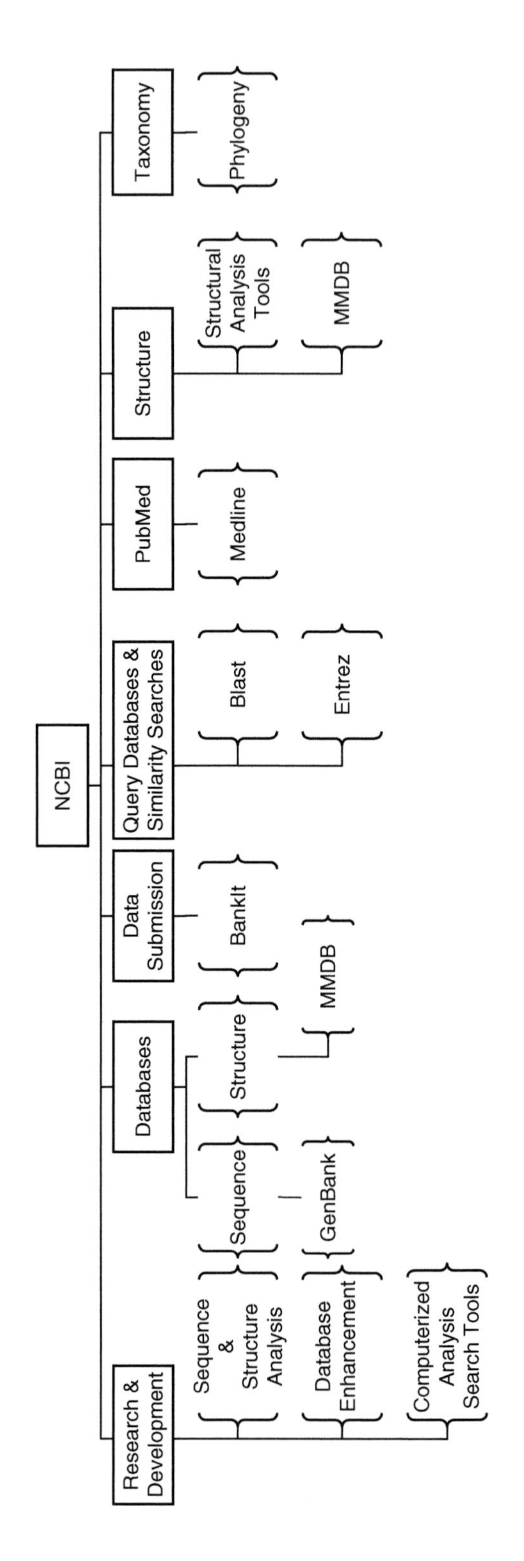

FIGURE 2.11
National Center for Biotechnology Information (NCBI).

molecular and genetic processes that are key players in pathogenic events. NCBI's four main tasks are:

1. To create automated machines that can analyze and store data pertaining to molecular biology, genetics, and biochemistry
2. To facilitate usage of the database and analytical software available to the scientific community (e.g., researchers, medical staff, etc.)
3. To coordinate worldwide efforts to gather biological data
4. To conduct research in computerized analysis of structure–function relationships for key biological molecules

2.2.1.1 *Who is Employed by NCBI?*

NCBI's scientific staff consists of computer scientists, molecular biologists, mathematicians, biochemists, research physicians, and structural biologists.

2.2.1.2 *What Kind of Research is Conducted at NCBI?*

The collaborative effort of NCBI's staff allows them to study the molecular basis of disease by using mathematical and computational tools. The three main facets of their studies are:

1. To analyze the sequence of the gene or gene product of interest
2. To gain a better understanding of the organization of the genes analyzed
3. To predict the structures of the molecules analyzed (e.g., proteins)

The analysis step could include the comparison of novel sequences to known homologs by comparing the sequence of the unknown protein or polynucleotide to proteins or polynucleotides whose sequence is known. Understanding the organization of the genes with respect to the whole genome could be a powerful tool for analyzing future novel genes whose functions are unclear or those that lack homology to any of the known sequences in the database. The step involving structure prediction of structurally unknown molecules, using homologs whose structures are known, would enable us to predict potential functional characteristics associated with molecules whose structure and function are not yet known.

2.2.1.3 *What Types of Databases are Supported by NCBI?*

- Protein sequence: these are experimentally sequenced proteins and translated nucleotide sequences from nucleotide libraries.
  a. Redundant protein sequence databases (e.g., PIR's complete database,[1] which consists of PIR1 + PIR2 + PIR3)
  b. Nonredundant or less redundant protein sequence databases (e.g., NR, SWISS-PROT,[2] and PDB[3])

- Nucleotide sequences (DNA and RNA): these are DNA and RNA sequences derived from less-automated sequencing projects (e.g., GenBank[4]) or automated sequencers (e.g., dbEST[5]).
  a. Redundant nucleotide sequence databases (e.g., dbEST)
  b. Nonredundant or less-redundant nucleotide sequence databases (e.g., GenBank)

2.2.1.4 What Do We Mean by Redundancy?

Redundancy in biological databases is a rather complicated issue. Should two alleles from the same locus be considered as one? What about the functionally identical enzymes (isozymes) in the same organism? How about tissue specificity in proteins and their relationship to their respective homologs in other tissues? These are all valid issues and therefore require each database to have its own definition of a redundant sequence. Most databases use automated measures to account for redundancy, especially in large projects. This method is qualitatively less sensitive than manual intervention, but makes up for it in speed. On the other hand, nonredundant databases allow access to redundant sequences for the sake of completeness.

The following is a list of the most frequently used protein sequence databases at NCBI:

- Alu: This is a selected set of translated Alu repeats.[6] This allows the masking of the potential Alu repeats in the query sequence. This database can also be retrieved through NCBI by anonymous FTP (under the/pub/jmc/alu directory).
- *E. coli*: This database specifically carries *E. coli* genomic CDS translations.[7]
- Kabat: This database deals with sequences that have immunological interest.[8]
- Month: This is a database of new or recently revised (within the last 30 days) CDS translations from GenBank and other protein sequence entries at the PDB, SWISS-PROT, and PIR libraries, combined.
- NR: This is a database of all nonredundant CDS translations from GenBank and other protein sequence entries at the PDB, SWISS-PROT, and PIR libraries, combined. In this database, the proteins with identical sequences are merged into a single file.
- PDB: These are protein sequences whose 3-D structures are known. This information is also found in NCBI's PDB mirror site in MMDB. The entries at PDB are predominantly nonredundant. In the case of identical sequence entries with multiple structures, the entry with the highest quality structure is kept. For crystal structures, this is generally the entry with the smallest resolution value (e.g., 1.8 Å preferred over 2.2 Å). However, other variables, such as complexed

structures with bound metals or ligands, will allow multiple structures for the same sequence in a given organism.

- SWISS-PROT: This is a database of the most recent release of protein sequence entries from the SWISS-PROT database.[2] It is now supported by EBI, an outstation of EMBL. This is one of the most informative cross-referenced, protein-sequence libraries available through the Internet. SWISS-PROT is a nonredundant database maintained by Amos Bairoch at the University of Geneva.
- Yeast: The yeast (*S. cerevisiae*) protein sequence database[9] stores the sequences generated from the yeast protein sequence projects.

The following is the list of some of the most commonly used nucleotide databases available through NCBI:

- Alu: This allows the masking of the potential alu repeats in the query sequence. This database can also be retrieved through NCBI by anonymous FTP (under the/pub/jmc/alu directory).
- dbEST: this is a questionably nonredundant database of GenBank, EMBL, and DDBJ EST entries. ESTs are single-pass cDNA sequences generated through automated sequencers with little or no human intervention. This will therefore increase the error frequency observed in these sequences relative to the rest of the sequence libraries. The most common errors observed in these entries are sequencing errors, heterologous sequence contaminations, and the presence of transcribed repetitive elements.
- dbSTS: This is a nonredundant database of GenBank, EMBL, and DDBJ STS entries.
- *E. coli*: This database specifically carries *E. coli* genomic nucleotide sequences.
- EPD: Eukaryotic promotor database[10] contains a list of all known eukaryotic promoter sequences in public domain libraries.
- GSS: The genome survey sequence contains single-pass genomic data, exon-trapped sequences, and alu PCR sequences.
- HTGS: This is the high-throughput genomic sequences database.
- Kabat: This database deals with sequences that have immunological interest.
- Mito: This database specifically deals with mitochondrial sequences.
- Month: This is a database of new or recently revised (within the last 30 days) entries that are found in GenBank + EMBL + DDBJ + PDB sequence libraries.
- NR: This is a database of all nonredundant GenBank + EMBL + DDBJ + PDB sequence entries. This database excludes EST, STS,

GSS, or HTGS sequence entries. Entries with a 100% sequence identity are merged as one.

- PDB: These sequences are derived from the 3-D structure of the molecule.
- Vector: This is the vector subset of GenBank (NCBI's nucleotide sequence database).
- Yeast: The yeast (*S. cerevisiae*) genomic nucleotide sequence database stores the sequences generated from the yeast genome project and other relevant yeast sequencing projects.

2.2.1.5 What are Some of the Services Offered by NCBI?

The following are the seven main databases and analysis tools supported by the NCBI server at their web site:

1. PubMed (Public MEDLINE)
2. BLAST: Basic Local Alignment Search Tool[11]
3. Entrez[12]
4. BankIt (WorldWideWeb submission)
5. OMIM (Online Mendelian Inheritance in Man)[13]
6. Taxonomy
7. Structure

2.2.1.5.1 PubMed

PubMed is the search service of the National Library of Medicine (NLM). It allows the user to gain access to over 9 million citations in MEDLINE and preMEDLINE, and is linked to participating online journals and related databases enabling the user to retrieve pertinent information in a speedy and efficient manner. Keywords may be used to retrieve journal articles that contain relevant topics. Multiple keywords may be used to increase the specificity of the search. Other search criteria such as author names and journal titles are also available for the user's convenience.

2.2.1.5.2 BLAST: Basic Local Alignment Search Tool

BLAST[11] is a set of similarity search programs that are designed to identify the classification and potential homologs for a given sequence. These programs are robust and capable of analyzing both DNA and protein sequences. BLAST programs are explained in further detail in Section 2.3.

2.2.1.5.3 Entrez

The scientific researcher is obligated to produce original nonredundant data that will serve to enhance the understanding of a particular principle. To prevent or minimize redundancy in published material, scientists must ensure the originality of their findings. This is not an easy task, but elaborate search tools with accessibility to relevant databases can facilitate the process. For instance, if a researcher has identified a particular trend in a family of proteins, the obvious next step would be to ensure the originality of his work. In other words, is this a new finding? To answer this not-so-trivial question, all possible citations with similar keywords would need to be searched. The results of this search could follow three main paths: the first path could lead you to a redundant dead end. In this situation, the data would be identical or very similar to previously answered questions. At this point, wise researchers would stop working on the redundant data and try to refocus their energy on other findings. The second path could lead to an original end point with no similar findings. In this case, the findings would be completely unrelated to anything in known citations. This could be good or bad; it could either signify an original finding or a mistake. At this point, the researcher would need to further investigate the findings and verify the steps followed in the protocol. This could further support the findings or allow the investigator to find the potential errors in either the protocol or the data analysis steps. The third path could lead to a relevant end point where citations would be found supporting the recent data without restating the same finding. This is the ideal situation for an investigator. The relevant citations could then be used as supporting references for the new findings. In any case, to conduct a reliable search a researcher must utilize a search engine that is not only efficient, but has access to all relevant, regularly updated databases. The government-supported search tools are generally the most reliable software available in the public domain. They are readily accessible through the Web and are very user friendly.

One of the most popular search engines is the Entrez at NCBI.[12] The Entrez Web interface (www.ncbi.nlm.nih.gov) allows the subscriber to gain access to bibliographic citations and biological data from a variety of reliable databases. Protein sequence information is retrieved from SWISS-PROT, PDB, PIR, and PRF. Proteins whose structures are known are retrieved from PDB. These proteins are incorporated into NCBI's Molecular Modeling Database, also known as MMDB.[14] The translated proteins and DNA sequences are retrieved from their parent DNA sequence databases (e.g., GenBank, EMBL, and DDBJ). For a bibliographic or citation search, Entrez uses PubMed's bibliographic database, which has access to over nine million biomedical articles from MEDLINE and pre-MEDLINE databases. Entrez also has access to chromosome mapping and genomic data. Entrez offers a variety of criteria for a particular search. For example, one could search a relevant database to find all possible terms that begin with a given word. Placing an asterisk at the end of the term allows Entrez to search for all possible words that begin with that particular term. For example a search for "inter*" will find all terms beginning with "inter," such as interstitium, intermolecular, etc. Entrez can also be used to conduct a smart search during which Entrez will search for phrases or groups of words. Entrez will automatically group the relevant terms together and exclude the unrelated terms from the grouping.

For instance, to locate all the possible citations from a particular author (e.g., Rashidi HR) that deal with a given subject (e.g., energetics), the user can enter the individual terms known about the author (e.g., Rashidi HR) and the subject of interest (e.g., energetics). Entrez will automatically recognize and group the relevant terms (e.g., the author's last name and initials), allowing the search engine to seek all relevant material from Rashidi HR that deals with energetics ("Rashidi HR" AND energetics). Entrez can also be made to group words that otherwise would be considered separate terms by using the automated grouping task. Inserting quotes would cause Entrez to group seemingly irrelevant terms into one (e.g., "brca 1"). Nevertheless, NCBI recommends that users allow Entrez to group the specified terms to minimize inaccurate retrievals. If the retrieval list from the search result is too long, Entrez will halt the search operation and inform the user.

One of the most accurate ways to retrieve a particular citation or sequence is through its identifier. An identifier is an index number assigned to a particular sequence or article in its relevant database. For instance, the identifier for MEDLINE citations is referred to as an UID number, while identifiers that pertain to a sequence are called GI numbers. To retrieve a MEDLINE citation with the UID 88067898, the user would simply input "UID 88067898" in the Entrez search engine to find the MEDLINE citation that is assigned this UID.

There are numerous search fields on Entrez, and experienced users find it to be useful and time efficient due to its adaptable nature. The following are some of the search fields that can be customized to meet the user's specific needs:

- Keyword allows the user to search a set of indexed terms associated with NCBI's accessible databases (e.g., GenBank, EMBL, PDB, DDBJ, SWISS-PROT, PIR, or PRF).
- Accession allows the user to search accession numbers assigned to proteins, nucleotide sequences, structures, or genomic records.
- Author Name has information about the authors of published papers. These are typically MEDLINE articles.
- Affiliation is used to search for the author's institutional affiliation and address.
- Journal Title is used to search for the name of the journal where the record was published. The user may utilize the List Terms mode to browse the list of abbreviated journal names (e.g., the Journal of Biological Chemistry is abbreviated as "J Biol Chem").
- E. C. Number is a designation number assigned to enzymes by the Enzyme Commission.
- Feature Key can be used to search keywords denoting a particular DNA feature.
- Gene Symbol can be used to search standard names for given genes.
- MEDLINE UID is used to search citations using a MEDLINE identifier.

- MeSH Terms are used to search Medical Subject Headings. These are a set of keywords used to index MEDLINE.
- MeSH Major Topic includes all the terms in MeSH tagged by the indexers as being of major importance.
- Publication Date is used to search for the date the article or sequence was published or submitted.
- Modification Date is the date the record was placed into Entrez.
- Page Number contains the page number of the published article.
- Property tells the user what type of sequence the citation contains.
- PubMed ID is PubMed's identifier for a given citation.
- Organism is used to search for the common and scientific names of the organisms associated with the protein or nucleotide sequence entries.
- Protein Name is used to search for the name of the protein a sequence is associated with.
- SeqId is a string identifier for a given sequence.
- Substance is used to search for the names of chemicals associated with the records from Chemical Abstract Service (CAS) registry.
- Title Words is used to search for words that are only found in the title line of a record.
- Text Words is used to search for "free text" associated with a given record. For protein and nucleotide sequence records, this includes the definition, comment, name, and description of the given sequence. For MEDLINE entries, this includes the title and the abstract of the given record.
- Volume is used to search for the number of the journal volume that contains the article of interest.

If the specified search field does not find the records of interest, it is helpful to repeat the search using "All Fields" or "Text Words." The intersection symbol is translated as AND in Entrez, and will only seek records that contain all of the given terms separated by AND or AND symbols. Entrez recognizes the union symbol as OR, which allows the user to find documents that contain any of the given terms. Finally, the difference, or BUTNOT, option enables the user to find all the documents that contain the uppermost terms, but not the lower terms.

After a successful search, the user is given retrieval options in the list of documents meeting the given criteria. The list of search results appears in chronological order from the most recent records to the oldest relevant records on file. The user can either retrieve all the documents or select the most relevant reports from the list of records found. The following are several different viewing formats for the relevant retrieved files: PubMed articles can be viewed as Citation, Abstract, MEDLINE, or ASN.1 type formats. Citation formats display the title, abstract, MeSH terms, and the substance information of an article. The Abstract formats display only the title and the abstract of the article. ASN.1 is a special format used by PubMed articles, while MEDLINE displays the article in MEDLARS format. GenBank/GenPept, Report, ASN.1, Graphic view, and FASTA are some of the viewing format options for protein and nucleotide records. GenBank/GenPept format is the standard GenBank or GenPept database file. Report allows the user to view the sequence record as a GenBank report format. Graphic View enables the user to display a graphical view of the sequence entry including alignment information. The FASTA format is most useful for further analysis of the given entry.

Many of the alignment tools (e.g., BLAST) require the user to input the order of interest in a FASTA format. The viewing options for structural information are the Structure Summary and the ASN.1 formats. The Structure Summary format is used to gain access to a summary of the structural data available for a given molecule. For instance, in crystallized protein structures, this view allows the user to gain access to information regarding the resolution of the given structure, author information, date of submission, complexing ligands, and other basic information. This format also allows the user to view the 3-D representation of the molecule. The graphical view is also an option for genomic records.

All the formats described can be saved as documents in a user's file. The three primary save options are: Text, HTML, and MIME. The MIME format is of particular use if the user has access to GenBank's MIME viewer. Otherwise, the output file must be saved in a text or HTML format to be of use. The HTML format is useful if the results will be viewed through a Web browser. The Text format lacks the HTML tags and breaks but can be viewed with standard word processing software such as Microsoft Word.

2.2.1.5.4 BankIt

BankIt is GenBank's World Wide Web sequence submission server. It allows the user to submit new sequences to GenBank via a user-friendly Web browser. The sequence and all relevant information are pasted into a submission box and sent to GenBank. GenBank's staff then contacts the submitting party and assigns an accession number to the sequence.

2.2.1.5.5 OMIM (Online Mendelian Inheritance in Man)

This database of human genes and gene disorders is maintained by Dr. Victor A. McKusick, his colleagues at Johns Hopkins University, and other contributors.[13] The OMIM Morbid Map is also supported on this site and maps genetic locations based on and organized by genetic disorders. Entrez, GDB, the Davis Human/Mouse Homology Map (archive only; now accessible as HomoloGene and TaxPlot), the Online Mendelian Inheritance in Animals (OMIA), the Human Gene Mutation Database (HGMD), the Alliance of Genetic Support Groups, the Cedars-Sinai Medical Center Genetics Image Archive, the Jackson Laboratory, RetNet (retinal genetic disorders),

HUM-MOLGEN, and the locus-specific mutation databases are some of the resources available at OMIM. This site is typically used by physicians and medical investigators concerned with genetic disorders. Having a solid understanding of scientific concepts and procedures is necessary for optimal interpretation of the images and text found at OMIM.

2.2.1.5.6 Taxonomy

NCBI's taxonomy home page contains organismic databases with the scientific and common names of organisms for which some sequence information is known. The server allows the user to access species' genetic information and how it ties in with related and not-so-related species. Trees are typically a representation of such relationships. These relationships can be based on similar proteins or nucleotide sequences. This page also has links to other NCBI servers (e.g., Structure and PubMed).

2.2.1.5.7 Structure

The Structure home page at NCBI supports the Molecular Modeling Database (MMDB) and a variety of software tools relevant to structural analysis.[14] The MMDB information is obtained from the Protein Data Bank (PDB). This includes the x-ray crystallography or Nuclear Magnetic Resonance (NMR) determined structures of important biological macromolecules. Cn3-D[15] is NCBI's structural visualization software for MMDB and is available at the Entrez/Cn3-D FTP site. Structure also offers research tools such as PKB and Threading. This software is available through the FTP site and requires Splus. The site's Entrez/PubMed link facilitates the search for applicable and related information to the molecules of interest.

References

1. Barker, W.C. et al., The PIR—International Protein Sequence Database. *Nucleic Acids Res.*, 1998, 26(1), 27–32.
2. Bairoch, A. and Apweiler, R., The SWISS-PROT protein sequence data bank and its supplement TrEMBL in 1998. *Nucleic Acids Res.*, 1998, 26(1), 38–42.
3. Sussman, J.L. et al., Protein Data Bank (PDB): database of three-dimensional structural information of biological macromolecules. *Acta. Crystallogr. D. Biol. Crystallogr.*, 1998, 54(1 [Pt 6]), 1078–1084.
4. Benson, D., Lipman, D.J., and Ostell, J., GenBank. *Nucleic Acids Res.*, 1993, 21(13), 2963–2965.
5. Boguski, M.S., Lowe, T.M., and Tolstoshev, C.M., dbEST—database for "expressed sequence tags" [letter]. *Nat. Genet.*, 1993, 4(4), 332–333.
6. Moyzis, R.K. et al., The distribution of interspersed repetitive DNA sequences in the human genome. *Genomics*, 1989, 4(3), 273–289.
7. VanBogelen, R.A. et al., The gene-protein database of *Escherichia coli*: edition 5. *Electrophoresis*, 1992, 13(12), 1014–1054.
8. Martin, A.C., Accessing the Kabat antibody sequence database by computer. *Proteins*, 1996, 25(1), 130–133.

9. Payne, W.E. and Garrels, J.I., Yeast protein database (YPD): a database for the complete proteome of *Saccharomyces cerevisiae*. *Nucleic Acids Res.*, 1997, 25(1), 57–62.
10. Cavin Perier, R., Junier, T., and Bucher, P., The Eukaryotic Promoter Database EPD. *Nucleic Acids Res.*, 1998, 26(1), 353–357.
11. Altschul, S.F., et al., Basic local alignment search tool. *J. Mol. Biol.*, 1990, 215(3), 403–410.
12. McEntyre, J., Linking up with Entrez. *Trends Genet.*, 1998, 14(1), 39–40.
13. Rashbass, J., Online Mendelian Inheritance in Man. *Trends Genet.*, 1995, 11(7), 291–292.
14. Ohkawa, H., Ostell, J., and Bryant, S., MMDB: an ASN.1 specification for macromolecular structure. *Ismb*, 1995, 3, 259–267.
15. Hogue, C.W., Cn3-D: a new generation of three-dimensional molecular structure viewer. *Trends Biochem. Sci.*, 1997, 22(8), 314–316.

2.2.2 European Bioinformatics Institute (EBI)

EBI (Figure 2.12) is an outstation of the European Molecular Biology Laboratory (EMBL) located at Hinxton, England. Fourteen European countries and Israel support EMBL and its outstations. EBI's main purpose is to conduct research and provide information about bioinformatics to the world's scientific community. As of September 1994, EBI has assumed all the activities that were previously handled at EMBL's Data Library in Heidelberg, Germany. The EBI[1] is comparable to the NCBI in the United States and is the main bioinformatics server for the European community. Its tasks and goals are similar to those of NCBI and include:

- Bioinformatics tracking technology
- Research and development of bioinformatics software
- Training and supporting its subscribers
- Relevant bioinformatics services

2.2.2.1 Who is Employed by EBI?

Like NCBI, EBI's staff consists of computer scientists, molecular biologists, mathematicians, biochemists, research physicians, and structural biologists. Their collaborative efforts allow them to study the molecular basis of disease using mathematical and computational tools.

2.2.2.2 What Kind of Research is Conducted at EBI?

The staff at EBI is involved in many facets of the bioinformatics world. Their research tasks include:

- Developing more robust comparison algorithms
- Creating more elaborate, but user-friendly, networked information systems
- Designing more efficient databases

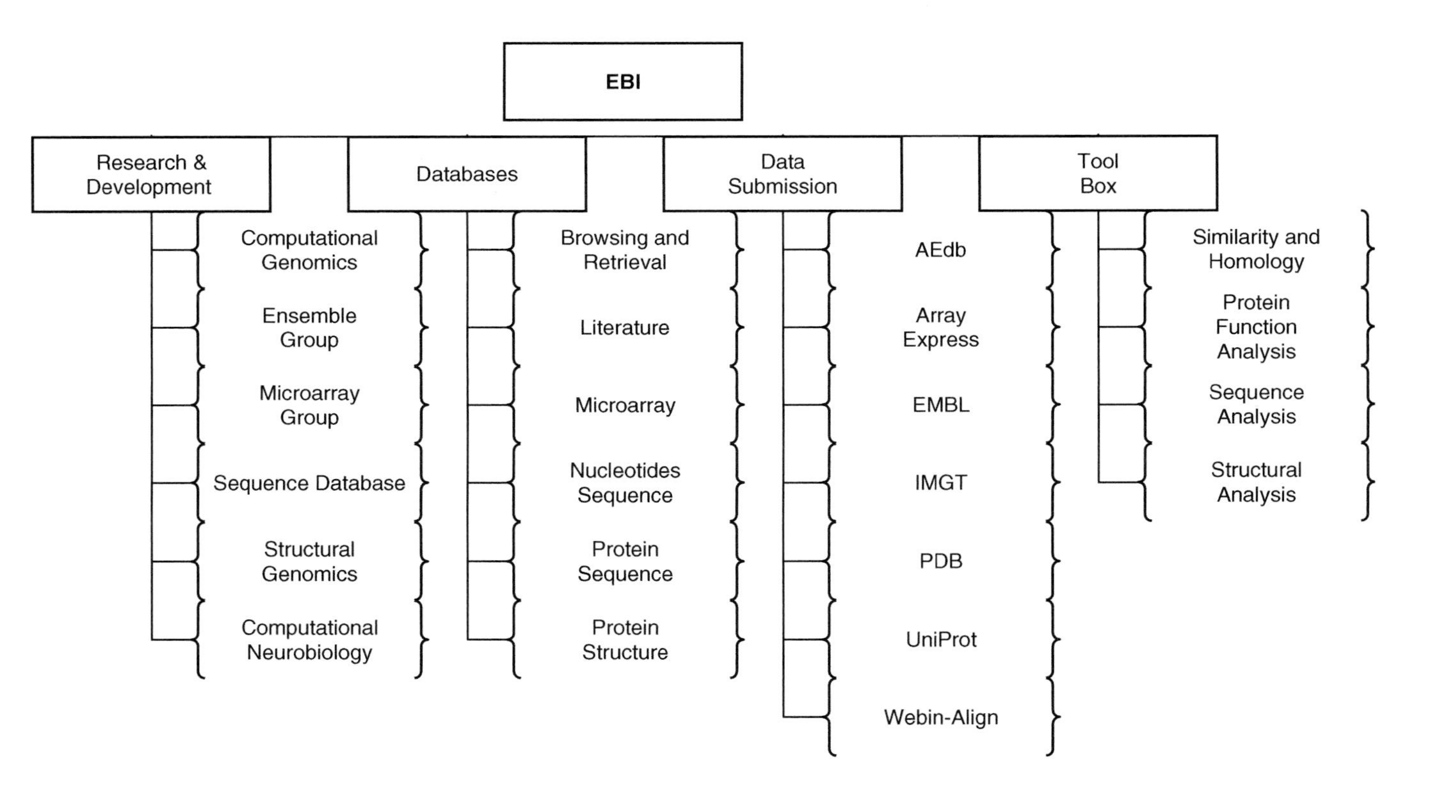

FIGURE 2.12
Overview of the European Bioinformatics Institute.

2.2.2.3 What Are Some Of The Services Offered By EBI?

A. Databases
B. Data submission
C. Query databases and similarity searches (e.g., FASTA[2] and BLITZ[3])
D. Online applications
E. FTP archives
F. Research and development

2.2.2.3.1 Databases at EBI

Following are the main databases supported by EBI's Web server:

1. EMBL Nucleotide Sequence Database[4]
2. UniProt Knowledgebase (UniProt)/SWISS-PROT[5]
3. Ensembl Genome Browser
4. MSD/ePDB Macromolecular Structure Database[6]
5. IMGT database[7]
6. ArrayExpress for microarray data
7. Databases on EBI FTP server
8. Software Bio Catalog

The first five are discussed below. Other EBI services and tools can be accessed through their website at www.ebi.ac.uk/services.

EMBL Nucleotide Sequence Database

This is a comprehensive database of nucleotide sequences (e.g., DNA and RNA). The nucleotide sequences at EMBL[4] are from a variety of sources. Some are from scientific literature and patent applications, but a large portion of the database includes sequences submitted directly by the sequencing source (researchers or sequencing groups). The database is a collaboration between the American GenBank nucleotide database at NCBI and the DNA database of Japan (DDBJ). The EMBL database communicates with the other two databases through its daily exchange program and constantly updates its contents. This allows EMBL to offer the worldwide scientific community an updated nucleotide database of all known public domain nucleotide sequences. In addition, EMBL's collaboration with various genomic sequencing groups allows it to introduce large-scale nucleotide sequences.

What types of information are in an EMBL nucleotide sequence file?

- The sequence
- A brief description
- Source of the sequence (the organism to which the sequence belongs)

- Bibliographic and citation information
- Locations of coding regions in the sequence (e.g., signal sequence, alpha chain, beta chain)
- Biologically significant sites in the sequence (an EST entry has very little biological information compared to a sequence entry that has been extensively studied by the researcher who reported the entry). Note: EST entries are "single pass" sequences. These are typically derived from random clones and there is little functional and biological information known. It is important to know that sequences submitted by sequencing groups are extensively annotated, but the information is based on their similarity to other known sequences, not on a detailed experimental analysis of the sequence.

What are some of the genome projects that collaborate with EBI?

- Human (*Homo sapiens*)
- Nematode (*C. elegans*)
- Fruit fly (Drosophila)
- Mouse (*M. musculus*)
- Yeast (*S. pombe*)
- *Mycobacterium leprae*
- *Mycobacterium tuberculosis*
- *Methanococcus jannaschii*

UniProt Knowledgebase (UniProt) and SWISS-PROT Protein Sequence Database[5]

UniProt stands for "universal protein resource" and is a comprehensive protein information database. It serves as the central repository of protein sequence at EBI. UniProt is comprised of three parts dedicated to distinct uses. The UniProt Knowledgebase (UniProt) is the central access point for extensive curated protein information, including function, classification, and cross-reference. The UniProt Nonredundant Reference (UniRef) databases combine closely related sequences into a single record to speed searches. The UniProt Archive (UniParc) is a comprehensive repository, reflecting the history of all protein sequences. The sequences and information in UniProt is accessible via text search, BLAST similarity search, and FTP.

The University of Geneva and EBI's EMBL Data Library together maintain the SWISS-PROT protein database. The translated DNA sequences at EMBL are directly submitted to SWISS-PROT, which is an adaptation of the PIR (Protein Identification Resource) database.[8] The SWISS-PROT database is nonredundant and has cross references to some of the other relevant libraries. For example, its cross-references to the EMBL database allow the user to gain access to the nucleotide sequence. It also has reference material from the PDB[6] and the PROSITE[9] libraries. PDB references are found only in sequence files whose 3-D structures are known and present in the PDB. The PROSITE reference is found in sequence files whose sequence contains a characterized motif present in the PROSITE motif database.

Are all the sequence files in SWISS-PROT a result of peptide sequencing projects?

Even though the SWISS-PROT database is a protein sequence library, many of its sequence files are results of translated DNA and RNA sequence files from EMBL. The coding sequence (CDS) files at EMBL are translated into amino acid sequence files and maintained in TREMBL. TREMBL is the host of all translated EMBL nucleotide sequences and incorporates its data into the SWISS-PROT database. SP-TREMBL contains the translated sequence files, which will eventually be incorporated into the SWISS-PROT database.

How are the SWISS-PROT accession numbers assigned?

Upon submission, the EBI assigns a SWISS-PROT accession number to protein sequences whose sequence has been directly determined by a peptide-sequencing project. For translated sequence files, the accession number is assigned upon its incorporation into the TrEMBL[5] library. The directly sequenced proteins can then be distinguished from those that were derived from a nucleotide sequence. A SWISS-PROT file with a cross reference to TrEMBL is a certain sign of a translated nucleotide sequence. Currently, the SWISS-PROT database[5] holds about 158,000 sequence entries, while TrEMBL is the host to approximately 1,400,000 (UniProt Release 2.5).

Ensembl Genome Browser

Ensembl is a joint project between the EMBL-EBI and the Wellcome Trust Sanger Institute. It functions as a navigatable genome map assembled automatically from annotation of large eukaryotic genomes. It is a comprehensive database of completed genomes containing stable annotation with confirmed gene predictions that have been integrated from external data sources.

Where does information for Ensembl come from?

Ensembl annotates known genes and predicts new ones, with functional annotation of gene families from InterPro (protein families, domains, and functional sites), OMIM (Online Mendelian Inheritance in Man), and SAGE (Serial Analysis of Gene Expression). The raw data for genome mapping originally comes from STS, BLAST scores, experimental settings, and cross-references in the radiation hybrid maps collected in the RHdb database[10] are used to construct radiation hybrid maps. These are chromosome maps constructed from radiation hybrid score vectors, and are alternatives to genetic maps. They can be used to mark nonpolymorphic markers, as well as to order unresolved clusters of polymorphic STSs (see Table 2.2). The presence of precise STS maps is invaluable for studying multifactorial genetic disorders in humans.

TABLE 2.2

Radiation Hybrid Map Statistics

Genetic Element/Marker	Human	Mouse
STSs	58579	316
Radiation Hybrid Entries	74337	328
ESTs	44623	1
cDNA Sequenced (whole)	338	0
Genetic Markers	4200	0
Alternative STSs (from genetic loci)	2186	
Markers in CpG Islands	1	0
STSs (unknown polymorphic or expressed elements)	3331	0
Entries Describing Experimental Conditions	107	107
Maps	23	0
Cross-References to Other Databases	219,411	651

Note: The above is reported data as of June 6 1998.

It is a database of submitted maps that were calculated prior to their submission. The following are some of the programs used to calculate hybrid maps:

- RHMAP[11]
- RHMAPPER: (Whitehead Institute/MIT Center for Genome Research)
- RADMAP/MULTIMAP[12]

Other sequence information is obtained from dbEST[13] and dbSTS database at NCBI. The database entries are composed mainly of ESTs, single-pass cDNA sequence entries, and STSs, short genomic landmark sequences. All of these entries are primarily maintained by NCBI.

What types of information can be retrieved from dbEST?

- Positionally cloned human disease genes
- cDNA sequence data and mapping data
- Identification of coding regions and tagging human genes

What types of information can be retrieved from dbSTS?

- More comprehensive and detailed annotations and contact data
- Experimental conditions
- Genetic map locations
- Putative homology to other entries through BLAST[14]

The EBI SRS interface can be used to search the dbEST and dbSTS libraries.

Macromolecular Structure Database (MSD/ePDB)

The Macromolecular Structure Database incorporates protein structure data from the Protein Data Bank (Figure 2.13). PDB[6] is a library of all known

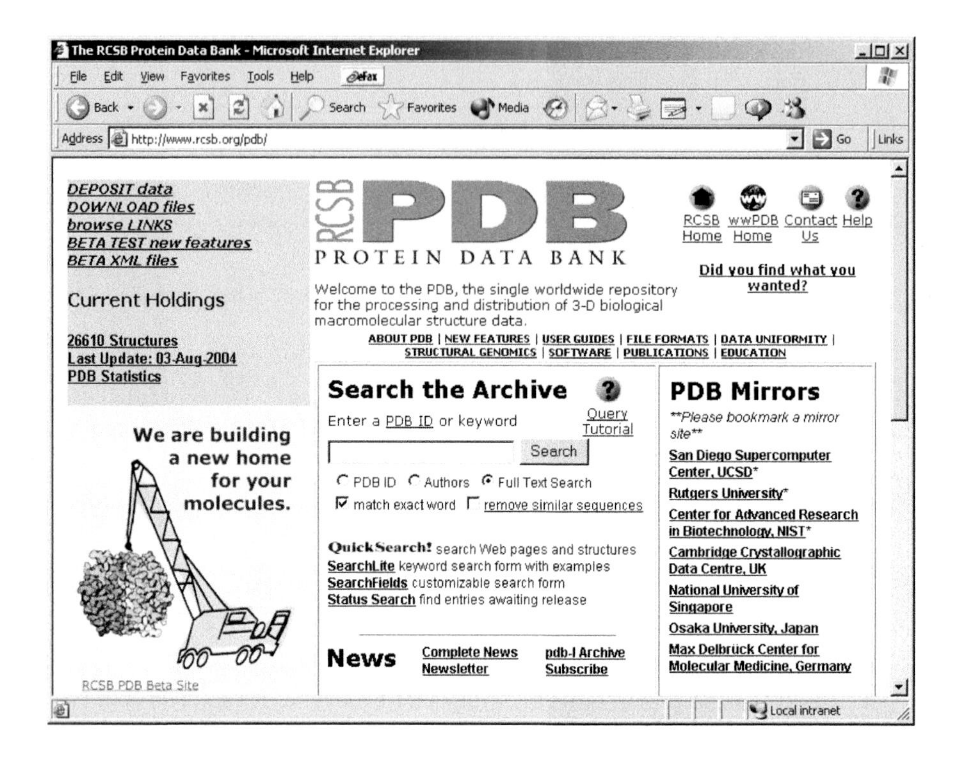

FIGURE 2.13
Protein Data Bank.

public domain 3-D structures. PDB[6] is physically located at the San Diego Super Computer Center (www.sdsc.edu). This service is supported by the U.S. National Science Foundation, the U.S. Public Health Service, National Institute of Health, National Center for Research Resources, National Institute of General Medical Sciences, National Library of Medicine, and the U.S. Department of Energy. Until July 1999, the Protein Data Bank was maintained by the Brookhaven National Laboratory in Long Island, New York. A global protein structure database, the Worldwide Protein Data Bank, or wwPDB (http://www.wwpdb.org), was established in 2003 to maintain a single Protein Data Bank Archive of macromolecular structural data that is freely and publicly available to the global community.

What types of structures are maintained at PDB?

- Proteins
- Proteins + nucleotide sequence (e.g., DNA)
- Proteins complexed with metals
- Proteins complexed with inhibitors

How were the 3-D structures determined?

- Nuclear magnetic resonance (NMR)
- X-ray crystallography

What is the difference between the two techniques?

- In NMR, the structure of the molecule is determined in solution. This technique yields a great deal of dynamic information. We gain insight into how the molecule behaves in an aqueous (solution) environment. The structural features of the molecule give us insight into its functional characteristics.
- X-ray crystallography gives us a 3-D static picture of the molecule. The structure of the molecule is determined in its crystallized form; therefore, the structures solved through this technique lack dynamic data. In other words, we do not know how the molecule behaves in solution, its natural state.

What types of information are in a PDB file?

- Atomic coordinates determined either by NMR or x-ray crystallography
- Bibliographic citations
- Primary structure information (e.g., the amino acid sequence)
- Secondary structure information (e.g., alpha helix, beta strand)
- Crystallographic structure factors and NMR experimental data

If solving structures through NMR gives more information and insight into the molecule, why aren't all the structures solved by this method?

Solving structures through NMR is limited by the size of the molecule. Many of the proteins are outside this size range. Therefore, there is a need for alternative techniques in solving the 3-D structures of the larger molecules (e.g., x-ray crystallography).

IMGT Database: The International ImMunoGeneTics Database

This is a nucleotide database of immunologically important genes,[7] many of which belong to the immunoglobulin superfamily. Most of the molecules in the immunoglobulin superfamily are involved in immune recognition and response. T-cell receptors (TcRs), B-cell receptors (Ig), and major histocompatibility complex (MHC) molecules are some of the classic examples of the immunoglobulin superfamily (Figure 2.14).

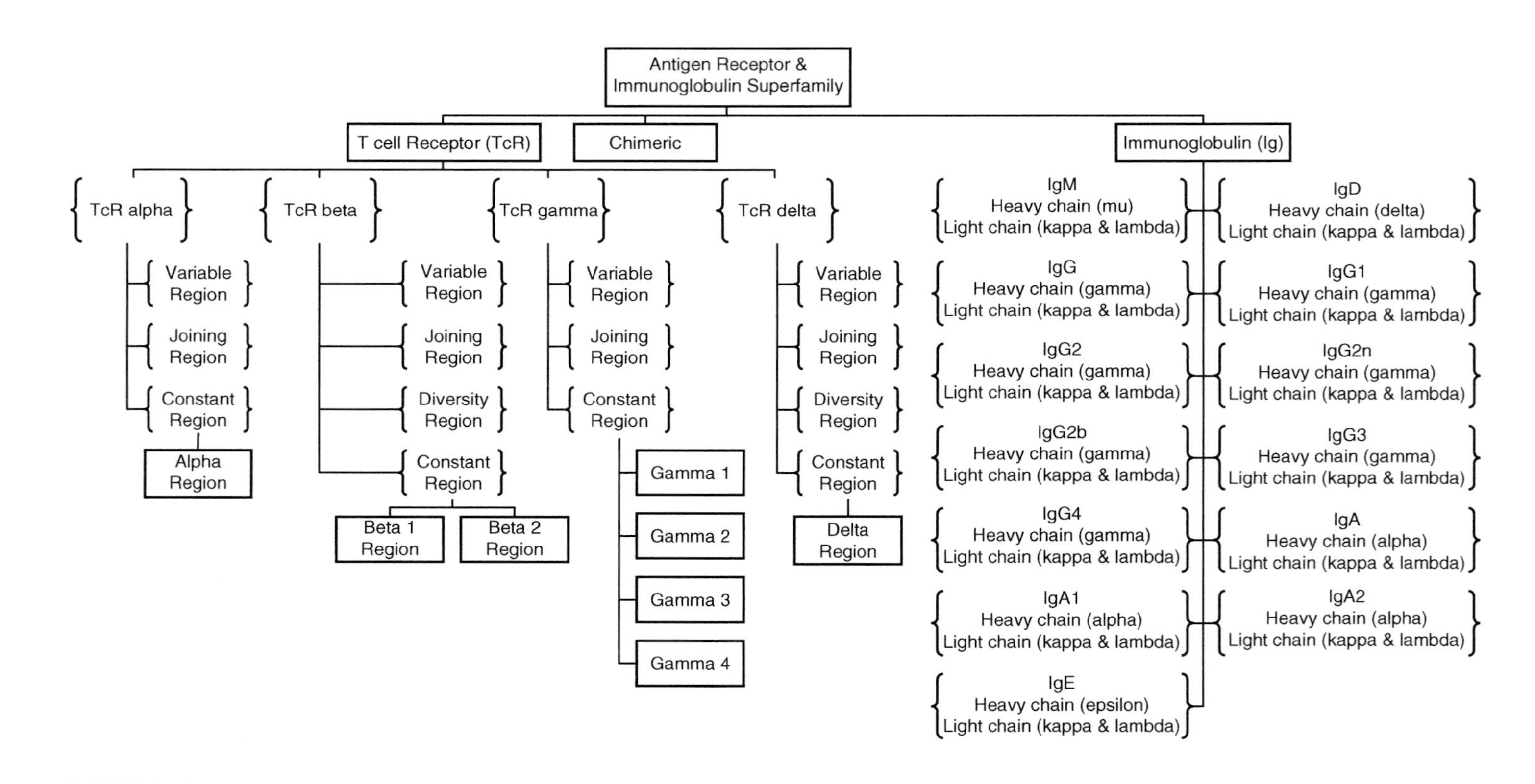

FIGURE 2.14
IMGT'S classification scheme.

What types of information can be retrieved from IMGT?

- Nucleotide sequences
- Protein sequences
- Sequence alignments
- Allele, polymorphism, and STS information
- Genetic maps
- Relevance to known disorders

Are all the immunologically significant molecules at IMGT stored in one database?

IMGT data are stored in two separate databases according to their molecular identity.

1. LIGM-DB: This database maintains Immunoglobulin (Ig) and T-cell receptor (TcR) molecules. LIGM-DB stands for the Laboratoire d'ImmunoGenetique Moleculaire.[15]
2. MHC/HLA-DB: This database is primarily concerned with the major histocompatibility complex molecules.[16] In humans, these molecules are referred to as human leukocyte antigens (HLA).

Who is this service designed for and what are some of its potential applications?

- Medical researchers (e.g., HIV/AIDS research, cancer research, and autoimmune disease research)
- Therapeutics and immunohistochemistry studies (e.g., antibody production and design as assay markers and potential fighting agents in certain therapies, grafts, and immunotherapy)
- Evolutionary biologists and bioinformaticians: the science of evolution and its connection to genomic diversity. The molecule's relatedness to other genes in different species could serve as a powerful tool in finding the genes of interest that are relevant in a pathogenic event.

Who are the primary contributors to the IMGT database?

LIGM: Laboratoire d'ImmunoGenetique Moleculaire, CNRS, Université Montpellier II, Montpellier, France

- CNUSC: Centre National Universitaire Sud de Calcul, Montpellier, France
- ICRF: Imperial Cancer Research Fund, London, England
- EBI: European Bioinformatics Institute, Hinxton, England
- IFG: Institut fur Genetik, Köln, Germany

- BPRC: Biomedical Primate Research Centre, Rijswijk
- EUROGENETEC: Seraing

What are some of the tools and services offered by IMGT's server?

- Sequence alignment tools (e.g., DNAPLOT)
- Modeling tools
- Tools for mapping data
- Tools useful in classification of the query sequence
- Links to other biologically relevant databases
- Direct data submission through the Web interface

IMGT also has its own unique numbering scheme. The increased variability observed in immunologically significant molecules necessitates a more lenient or different approach in the classification and analysis of these molecules. IMGT's numbering scheme accounts for the framework (FR) of the molecule, its complementarity determining regions (CDRs), its structural data when applicable, and the characterization of its hypervariable loops.

Why is this numbering scheme useful?

- It facilitates sequence comparison between the less-conserved regions of the molecules (e.g., Ig or TcR variable chains).
- The position of the conserved residues remains constant (e.g., Leu 89).
- Framework residues and residues of CDRs of same length maintain their position and can easily be identified in the absence of alignment tools.
- It allows a unique characterization of the variable regions of FRs and CDRs based on their length.
- It allows a comparative approach of the germ lines (V-GENE, D-SEGMENT, J-SEGMENT, and C-GENE) in identifying mutations, polymorphisms, and somatic hyper mutations. The DNAPLOT alignment tool uses this numbering scheme in its comparative steps with the different germ line, functional, and ORF sequence sets.

The following keyword categories can be used to search the IMGT database for a particular sequence:

- Receptor type (e.g., chimeric, T-cell receptor, and immunoglobulin)
- Receptor class (e.g., TcR alpha, TcR beta, IgM, and IgG)

- Chain type (e.g., TcR alpha chain, TcR beta chain, Ig heavy chain, Ig light chain, and Ig kappa chain)
- Region type (e.g., TcR constant region, Ig constant region, TcR alpha constant region, and Ig delta constant region)
- Descriptive keywords (e.g., Fab, Fc, lambda 5, and transgene)

Other services and tools can be accessed through EBI's website at www.ebi.ac.uk/ebi_home.html.

References

1. Emmert, D.B. et al., The European Bioinformatics Institute (EBI) databases. *Nucleic Acids Res.*, 1994, 22(17), 3445–3449.
2. Pearson, W.R., Using the FASTA program to search protein and DNA sequence databases. *Methods Mol. Biol.*, 1994, 25, p. 365–389.
3. Brenner, S.E., BLAST, Blitz, BLOCKS and BEAUTY: sequence comparison on the net. *Trends Genet*, 1995, 11(8), 330–331.
4. Stoesser, G., et al., The EMBL nucleotide sequence database. *Nucleic Acids Res.*, 1998, 26(1), 8–15.
5. Bairoch, A. and Apweiler, R., The SWISS-PROT protein sequence data bank and its supplement TrEMBL in 1999. *Nucleic Acids Res.*, 1999, 27(1), 49–54.
6. Sussman, J.L. et al., Protein Data Bank (PDB): database of three-dimensional structural information of biological macromolecules. *Acta. Crystallogr. D. Biol. Crystallogr.*, 1998, 54(1 (Pt 6)), 1078–1084.
7. Lefranc, M.P. et al., IMGT, the International ImMunoGeneTics database. *Nucleic Acids Res.*, 1998, 26(1), 297–303.
8. Barker, W.C. et al., Protein sequence database of the protein identification resource (PIR). *Protein Seq. Data. Anal.*, 1987, 1(1), 43–98.
9. Bairoch, A., Bucher, P., and Hofmann, K., The PROSITE database, its status in 1997. *Nucleic Acids Res.*, 1997, 25(1), 217–221.
10. Rodriguez-Tome, P. and Lijnzaad, P., The radiation hybrid database. *Nucleic Acids Res.*, 1997, 25(1), 81–84.
11. Boehnke, M., Lange, K., and Cox, D.R., Statistical methods for multipoint radiation hybrid mapping. *Am. J. Hum. Genet.*, 1991, 49(6), 1174–1188.
12. Matise, T.C., Perlin, M., and Chakravarti, A., Automated construction of genetic linkage maps using an expert system (MultiMap): a human genome linkage map [published erratum appears in *Nat. Genet.* 1994 Jun, 7(2), 215]. *Nat. Genet.*, 1994, 6(4), 384–390.
13. Rodriguez-Tome, P., Searching the dbEST database. *Methods Mol. Biol.*, 1997, 69, 269–283.
14. Altschul, S.F. et al., Basic local alignment search tool. *J. Mol. Biol.*, 1990, 215(3), 403–410.
15. Lefranc, M.P. et al., LIGM-DB/IMGT: an integrated database of Ig and TcR, part of the immunogenetics database. *Ann. N.Y. Acad. Sci.*, 1995, 764, 47–49.
16. Newell, W.R., Trowsdale, J., and Beck, S., MHCDB — database of the human MHC. *Immunogenetics*, 1994, 40(2), 109–115.

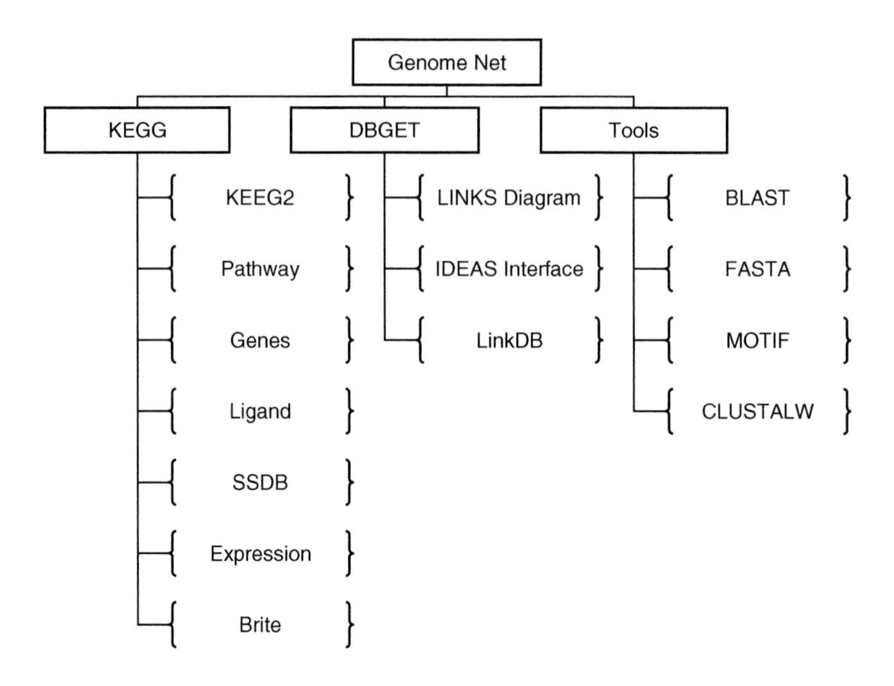

FIGURE 2.15
GenomeNet Organization.

2.2.3 Kyoto Encyclopedia of Genes and Genomes (KEGG)

Kyoto Encyclopedia of Genes and Genomes or KEGG (Figure 2.15) is part of GenomeNet,[1] Japan's network of database and computational services for genome research and related areas in molecular and cellular biology. GenomeNet was established in September 1991 under the Human Genome Program (HGP) of the Ministry of Education, Science, Sports, and Culture (MESSC). GenomeNet services are operated jointly by the Supercomputer Laboratory (SCL), Institute for Chemical Research (ICR), Kyoto University and the Human Genome Center (HGC), in the Institute of Medical Science, of the University of Tokyo. GenomeNet can be accessed at GenomeNet Services at www.genome.ad.jp and provides the molecular and pathway database KEGG,[2] including binary relations for network computation and logical reasoning database BRITE, the Integrated Database Retrieval System DBGET[3] and sequence and protein motif search and retrieval functions BLAST, FASTA, and MOTIF.

The KEGG database is the central architecture of GenomeNet and provides a unique metabolic database use and interactivity that also connects to sequence and structure databases mirroring the content of GenBank and EBI. KEGG is structured around metabolic pathways in various organisms (Table 2.3).

KEGG also provides tools for software development, manuals, education, and links to other database via XML, API, and DBGET (Figure 2.16). The

TABLE 2.3

Kyoto Encyclopedia of Genes and Genomes Services

Database	Content and Search Function
KEGG2	Table of contents
PATHWAY	Current knowledge on molecular interaction networks, including metabolic pathways, regulatory pathways, and molecular complexes
GENES	The universe of genes and proteins in complete genomes containing the information about ortholog groups and conserved gene clusters
LIGAND	The universe of chemical reactions involving metabolites, enzymes, reactions, complex carbohydrates, as well as xenobiotic compounds
SSDB	Sequence Similarity Database for exploring the universe of protein-coding genes in the complete genomes
EXPRESSION	For mapping gene expression profiles to pathways and genomes
BRITE	A database of binary relations for network computation and logical reasoning involving genes, proteins, and other biological molecules

Source: From http://www.genome.ad.jp/. With permission.

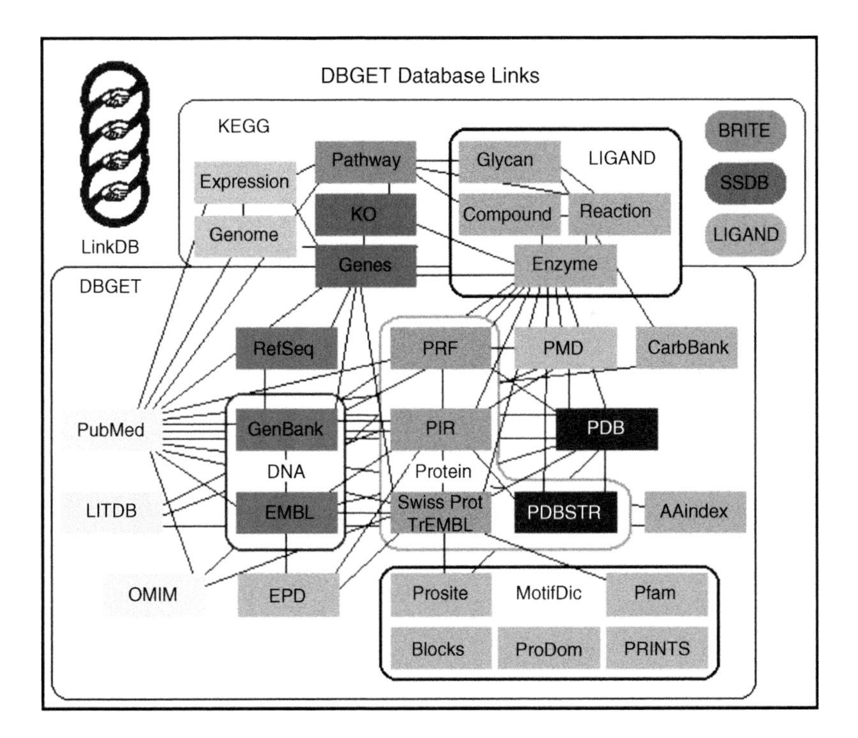

FIGURE 2.16
DBGET database links diagram. DBGET's links diagram allows easy access to the appropriate databases that contain information organized and selected for specific categories. The lines connecting the databases refer to the links between databases found within the annotated information of each entry. (From Kyoto Encyclopedia of Genes and Genomes. With permission.)

Japanese database system provides more than just links to DNA and protein sequence databases in the U.S., Europe, and Japan. It was created to ensure the biological quality of the information content in these databases.

GenomeNet allows sequence analysis and provides a collection of sequence interpretation tools as summarized in Figure 2.17. GenomeNet's tools include BLAST[4] and FASTA,[5] which are sequence similarity search programs, and MOTIF, a sequence motif search program developed at Kyoto University. Motifs are sequence patterns rather than linear strings that correlate directly to structural features of encoded proteins. CLUSTAL W[6] is a multiple sequence alignment program similar to BLAST and FASTA, which are restricted to pairwise comparison only.

Other tools search for functional elements in sequences as they relate to protein structure. TFSEARCH identifies transcription factor binding sites. Transcription factors are proteins that control gene activity by directly binding to DNA close to gene sequences. Gene recognition and assembly Internet link (GRAIL) is used to identify novel genes in newly sequenced DNA for which no biological data is otherwise known.[7]

Because genes have certain structures that allow their control (see transcription factors) and the expression of proteins, specific patterns or motifs of short DNA sequences (10 to 50 base pairs) are indicative of the presence of functional

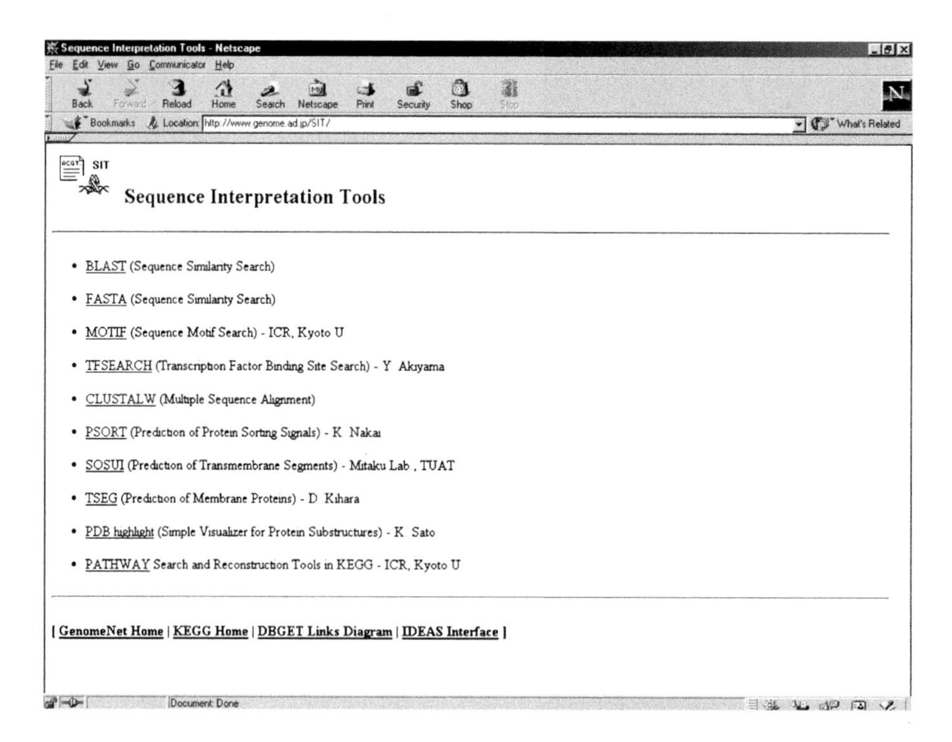

FIGURE 2.17
Sequence interpretation tools at GenomeNet. The Institute for Chemical Research, Kyoto University (www.genome.jp/kegg/).

genes. In addition, eukaryotic genes are often split into functional and nonfunctional (or coding and noncoding) regions called exons and introns. Only exon DNA sequences are translated into protein sequences and thus are relevant to interpreting or predicting associated protein structures and functions. To predict these functions, programs like PSORT (prediction of protein sorting signals)[8] and SOSUI (prediction of transmembrane segments)[9] can be used.

Although the majority of database information pertains to sequences, the number of protein structures is continuously increasing. This is important from an evolutionary point of view, because the structure of proteins is better conserved (it is the phenotype) than the corresponding amino acid (and DNA) sequence. The PDB highlight (simple visualizer for protein substructures) emulates the PDB information and is used to compare structures, if available. Finally, a tool referring to the metabolic function of organisms can be found in KEGG. This is a tool at Kyoto University for pathway search and reconstruction and will be explained here in more detail.

LIGAND[10] is a chemical database that allows the search for a combination of enzymes and metabolic compounds. It is maintained at the Institute for Chemical Research at Kyoto University and contains 9317 entries: 4327 for enzymes and 6067 for enzyme reactions, and 11,058 for metabolic compounds (chemical compound database). BRITE is a biomolecular relations information transmission and expression database at the Institute for Chemical Research of Kyoto University that contains 278 entries.

LIGAND lets you search for metabolic compounds, enzymes, glycan (complex carbohydrate structures) or reaction of interest, the "Search and compute with KEGG" link on the KEGG home page will bring up a page featuring search tools for pathway maps, genome maps, coloring tools, prediction tools, and sequence similarity tools. These links are useful only when knowing exactly what to look for because this search mode requires the exact entry number, e.g., enzyme nomenclature E.C. 2.7.1.1 for hexokinase, or chemical compound number, e.g., C00417 for *cis*-Aconitate. Their KEGG entries are shown in Figure 2.18. For a more general keyword search, or if only the name or even partial name of an enzyme, compound, or pathway is known, the LIGAND search mode or the Pathway Maps and Molecular Catalogs link is better suited. The latter provides many search categories, among them "pathway" and "enzyme."

How is a standard metabolic pathway found?

From the KEGG table of contents click on the "Metabolic pathways" under pathway category to see a list of all the pathways (Figure 2.19). To find the pathway link for "lysine biosynthesis," scroll down to the group of pathways called "amino acid metabolism" and click on the link. You should now see the standard pathway MAP00300 for lysine biosynthesis (Figure 2.20).

How are species-specific pathway maps found?

Once in the standard pathway map, select the species name (e.g., *Mus musculus*) in the "Go To" window and click on "Go." You should now see the same pathway and it should show the species name in the window (e.g.,

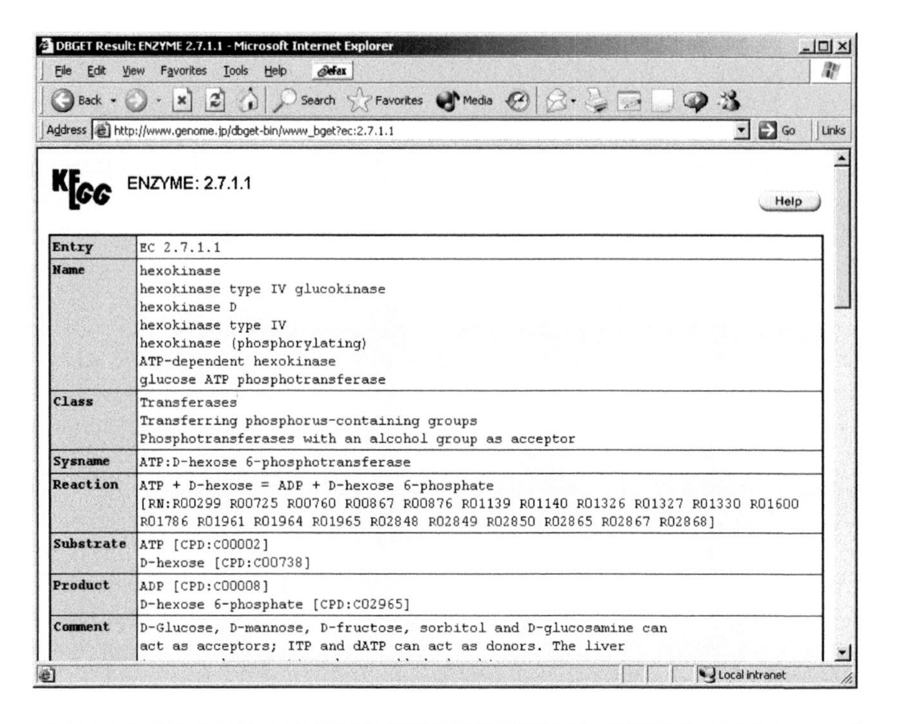

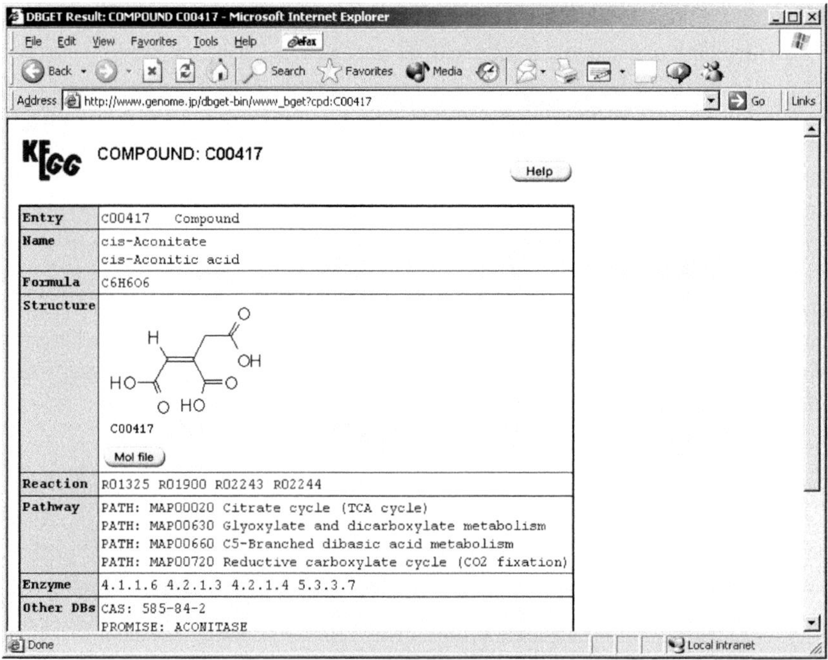

FIGURE 2.18
KEGG entry for enzyme hexokinase and metabolite *cis*-aconitate. How to use the KEGG Metabolic Database.

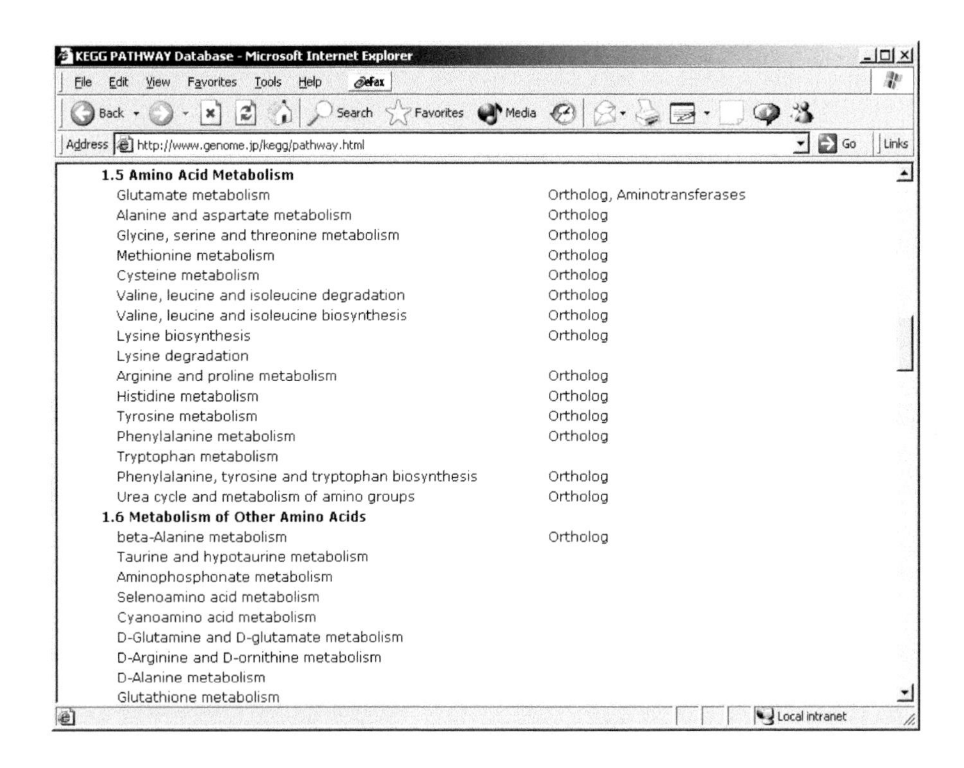

FIGURE 2.19
KEGG metabolic pathways, amino acid metabolism: the links provided allow access to the corresponding site map for the metabolisms of the amino acids listed. (From Kyoto Encyclopedia of Genes and Genomes, The Institute for Chemical Research, Kyoto University, www.genome.ad.jp:80/kegg/. With permission.)

Mus musculus). All known mouse enzymes for which a database entry exists are labeled in green. The mouse example shows only two such enzymes, one of which is lysyl-tRNA synthetase (EC:6.1.1.6). The corresponding map for *Homo sapiens* shows five marked enzymes. Clearly, despite the complete genome information for both mammals, no enzymes can be found responsible for lysine biosynthesis. L-lysine is an essential amino acid for humans and mice. L-lysine has to be part of our diet, because we lack the necessary enzymes for its biosynthesis (Figure 2.21).

Now, compare this with the corresponding map for *E. coli* and you will find that this Gram-negative bacteria can synthesize L-lysine from L-aspartate through a series of enzymatic reactions marked in green (Figure 2.22).

How is a chemical structure or metabolite information found from a pathway map?

Note that the immediate precursor of L-lysine also serves as a substrate for peptidoglycan synthesis (Enzymes 6.3.2.10 and 6.3.2.13) to form an activated precursor molecule of this bacterial cell wall component. The structure of this molecule can be found by clicking on the small circle next to the compound name. You will see an information page for KEGG entry C04882 (Figure 2.23).

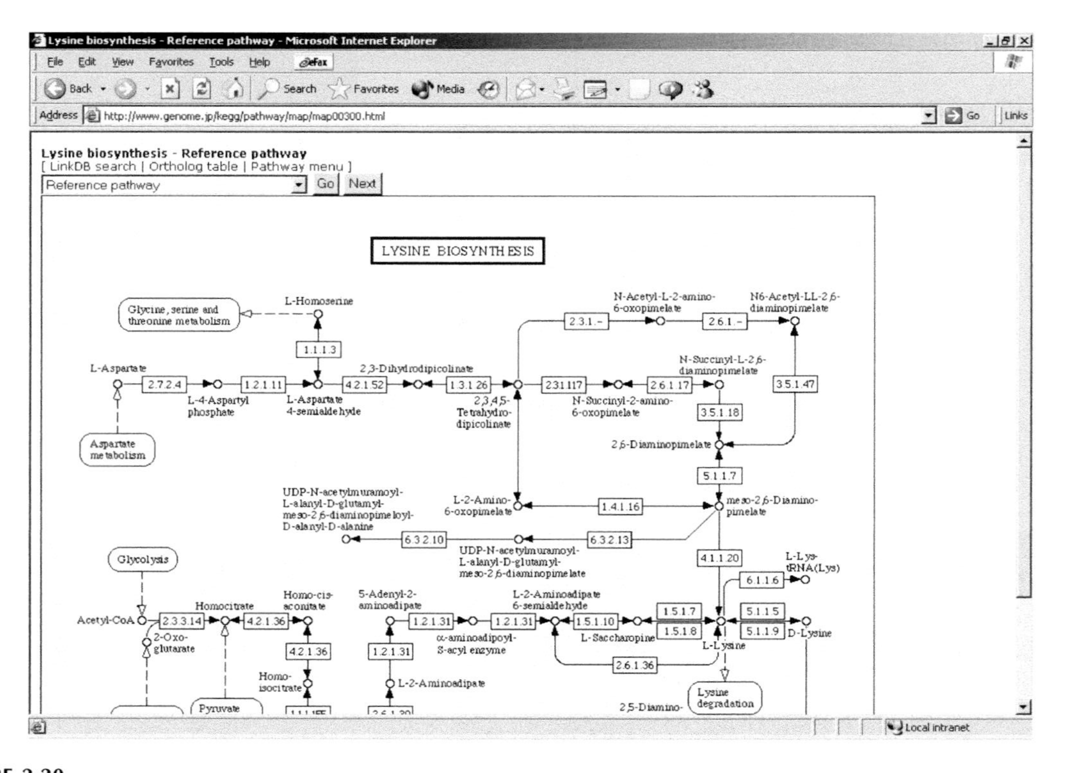

FIGURE 2.20
Lysine biosynthesis, standard metabolic pathway. The standard metabolic pathway map for lysine biosynthesis shows all enzymes and metabolic intermediates on a generic map collected from 33 organisms with current genome projects (completed or incomplete). This allows for comparison of pathways among different species. (From Kyoto Encyclopedia of Genes and Genomes, The Institute for Chemical Research, Kyoto University, www.genome.ad.jp:80/kegg/. With permission.)

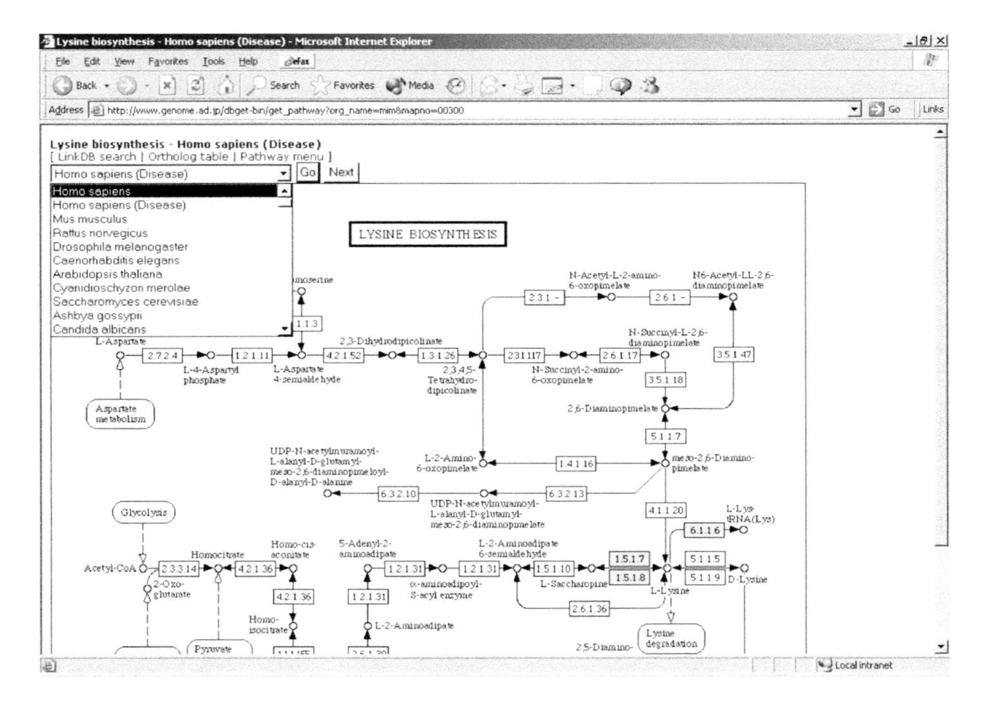

FIGURE 2.21
Lysine biosynthesis, *Homo sapiens* (disease). Same as pathway shown in Figure 2.19, except that all known enzymes in lysine metabolism linked to *H. sapiens* diseases are in color (shaded enzyme boxes). No shading means that the pathway does not exist or has not been described. Note that lysine is an essential amino acid and no enzymes are marked, indicating that humans do not have the capacity to synthesize this amino acid, but do have follow up reactions that use the amino acid as substrate. Menu selection of organisms in upper left corner shown. (From Kyoto Encyclopedia of Genes and Genomes, The Institute for Chemical Research, Kyoto University, www.genome.ad.jp:80/kegg/. With permission.)

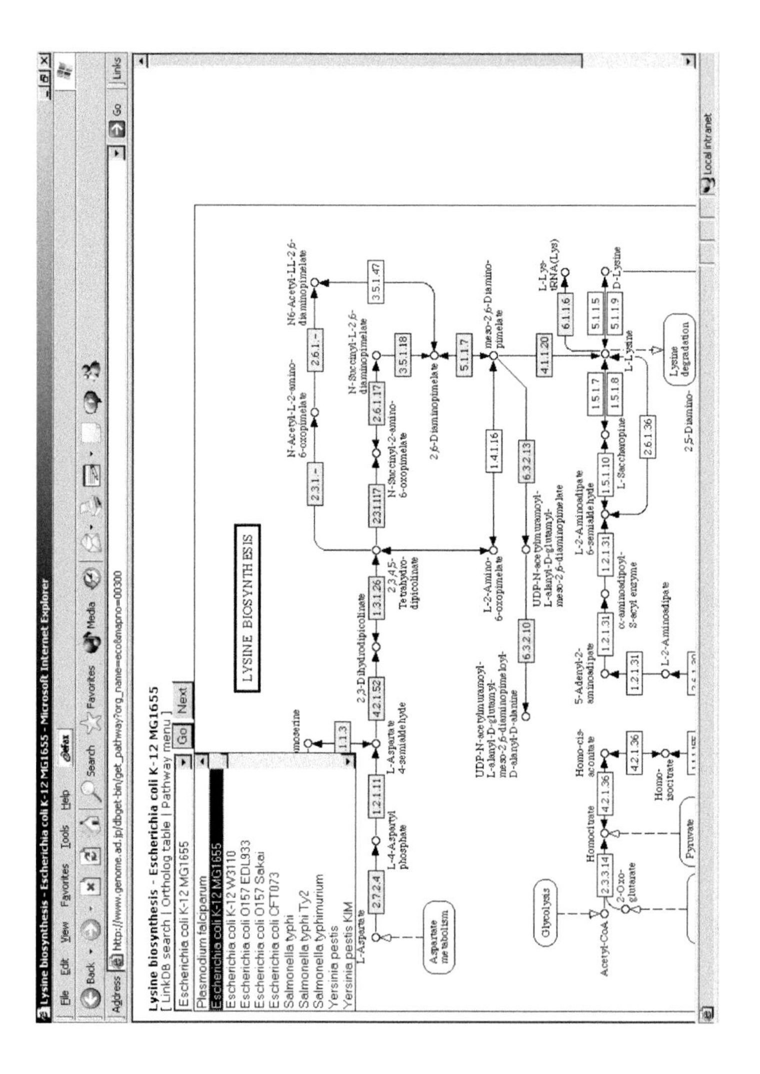

FIGURE 2.22
Lysine biosynthesis, *E. coli* K-12 strain MG1665. Lysine biosynthesis pathway for the eubacteria *E. coli*. Microorganisms can synthesize all twenty amino acids used for protein synthesis. All enzymes with database entries are in color. (From Kyoto Encyclopedia of Genes and Genomes, The Institute for Chemical Research, Kyoto University, www.genome.ad.jp:80/kegg/. With permission.)

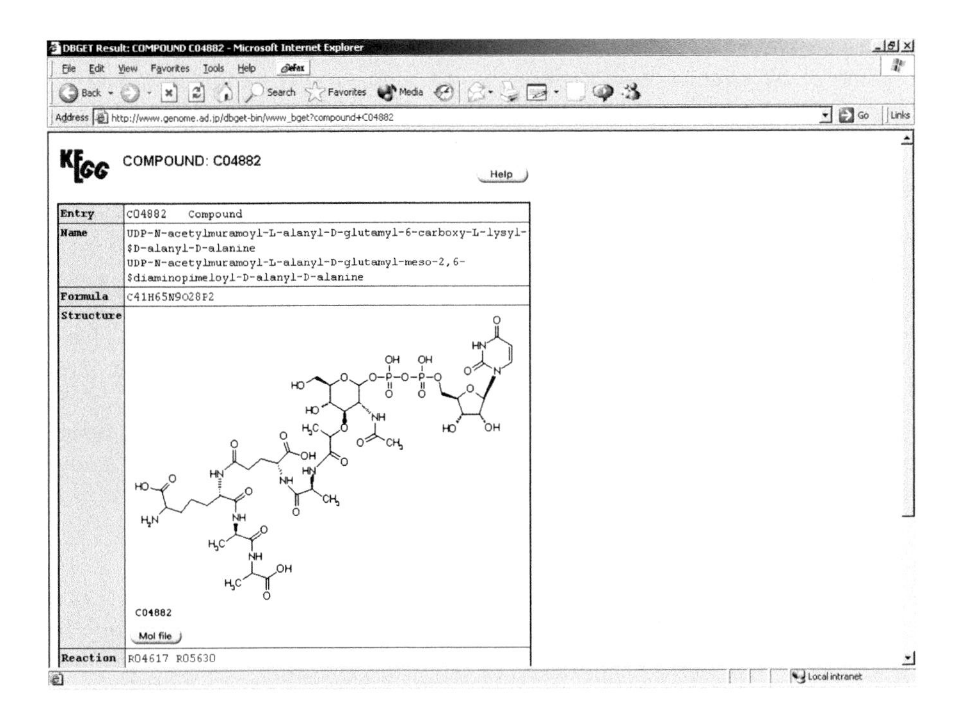

FIGURE 2.23
DBGET result, compound C04882: This entry shows the chemical structure and information of UDP-N-acetylmuramoyl-L-alanyl-D-glutamyl-meso-2,6-diaminopimelotl-D-alanyl-D-alanine, a biosynthetic precursor of the bacterial cell wall component peptidoglycan. (From Kyoto Encyclopedia of Genes and Genomes, The Institute for Chemical Research, Kyoto University, www.genome.ad.jp:80/kegg/. With permission.)

The preceding problem gives an example of how to find information about a metabolite—namely, its chemical structure, formula, KEGG entry number, and the pathway map numbers for which it is an intermediate. Clicking on a substrate name (or its circle) on a pathway map page is the easiest way to find relevant chemical information about a substrate and its shared pathways. The same can be done for an enzyme by clicking on the E.C. number box or for an intersecting pathway indicated in a box with smooth edges. For example, there is a link from the lysine biosynthesis map to the "lysine degradation" pathway map. Clicking on the box marked "lysine degradation" will bring up the corresponding catabolic processes. Note that the species selection will not change (we selected *E. coli* pathways). The new pathway map number is MAP00310.

How is a chemical structure or metabolite information found by keyword search?

To find a pathway metabolite or enzyme, the table of contents offers a direct link to the DBGET Ligand database at KEGG. This search mode can be found on the "table of contents" page under the "enzyme" category, DBGET search. Click on the link "Ligand" to access a generic search mode that allows a keyword entry. Note that having an exact enzyme number or compound number is not necessary in the DBGET database. To find information about lysine or L-lysine, type in "lysine" and hit the return (enter) key. You will receive a return list with 161 hits — the search will have returned all KEGG entries that contained the word "lysine" anywhere in the enzyme or compound name. The list contains 49 enzyme links (ec: x.x.x.xx) and 50 compound links (cpd: Cxxxxx), one being L-lysine (cpd:C00047) and all others being derivatives. There are also 62 reactions listed for lysine (rn:Rxxxxx) (Figure 2.24).

Clicking on the cpd number will bring you to the chemical structure information sheet. This sheet lists compound entry numbers for L-lysine (note that D-lysine has a different entry), common name(s), formula, structure, all pathway maps that contain L-lysine as a metabolite (five maps for L-lysine including synthesis and degradation, biotin metabolism, alkaloid biosynthesis II, and amino acyl-tRNA biosynthesis), and a list of all known enzymes that use L-lysine as a substrate (Figure 2.25).

2.2.3.1 Classification of Biological Molecules

One additional helpful feature is the molecular catalog entry, more specifically, the "compound classification." This leads to a catalog of metabolites classified according to their functional classes, e.g., carbohydrates, fatty acids, phospholipids, neurotransmitters, etc. If you want to find the structures of a class of molecules such as amino acids or various hexoses, this link will give you the best and most comprehensive results, and can be used as a reference for structure information. If, for example, you are interested in the general structure of steroid hormones, a link in the category "Lipids" will connect to a page containing the names and chemical structures of seven cholesterol-derived steroid hormones (Figure 2.26).

Clicking on the name link "aldosterone" connects to a structure information page providing a link to the pathway map for C21 steroid hormone metabolism (MAP00140). Following the pathway map link results in the standard pathway for steroid hormone metabolism with the aldosterone position marked as a red circle since we started our search from aldosterone. Selecting the *Homo sapiens* version of the map shows a variety of pathways. Organisms that lack any pathway elements of steroid hormone synthesis are not listed in the selection. Thus, no link for *E. coli* is provided, or any other eubacteria, as these prokaryotic microorganisms lack monooxygenases, the enzymes required for sterol synthesis.

It is important to understand the limitations of databases such as KEGG. Sometimes an enzyme is not marked where you would expect it (like in the aldosterone pathway above). This pathway map shows all known reactions summarized in a standard pathway map. Species-specific enzymes are marked green. Missing enzymes that appear to interrupt a pathway occur when no entry for this enzyme (gene, amino acid sequence, protein structure) exists in any database, including KEGG. The enzyme with the entry EC:1.14.15.5 is corticosterone 18-monooxygenase and converts corticosterone into aldosterone. Following the E.C. link for this enzyme to the entry in GenBank (mirrored

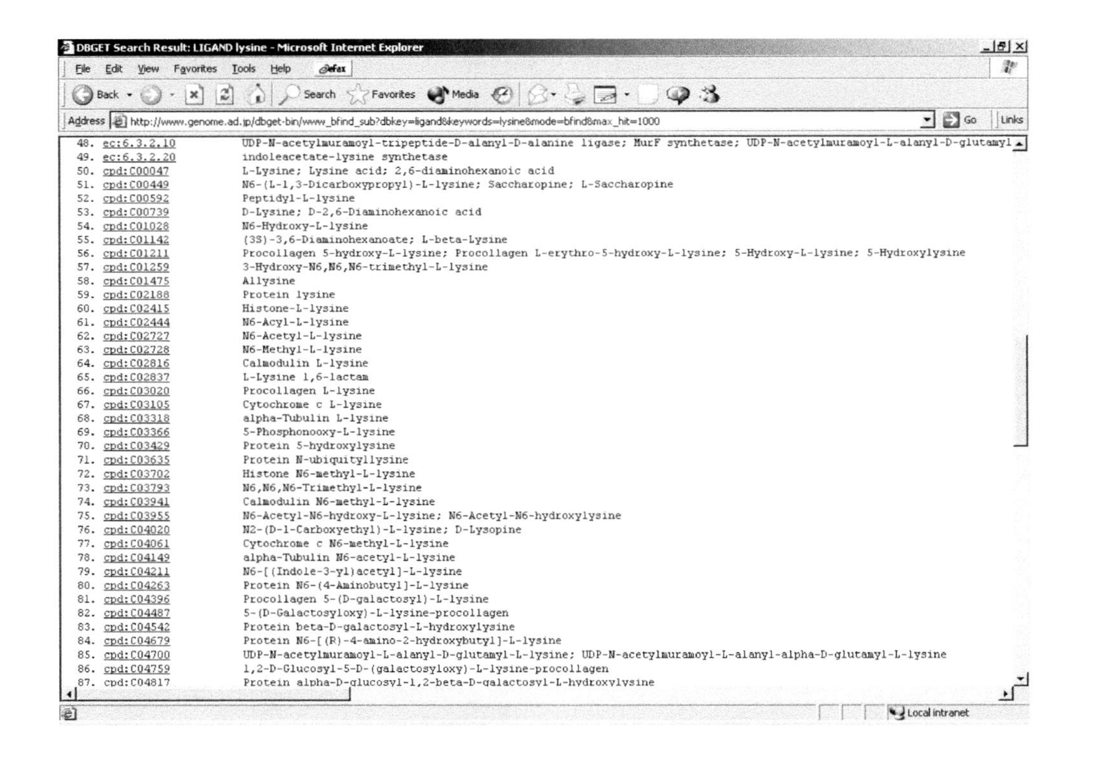

FIGURE 2.24
DBGET search result, LIGAND key word Lysine. Note the long list of entries. The list contains lysines and biosynthetic derivatives thereof (exceptions are proteins in general, since all proteins contain lysine residues), as well as all enzymes involved in the metabolic pathways of these compounds. (From Kyoto Encyclopedia of Genes and Genomes, The Institute for Chemical Research, Kyoto University, www.genome.ad.jp:80/kegg/. With permission.)

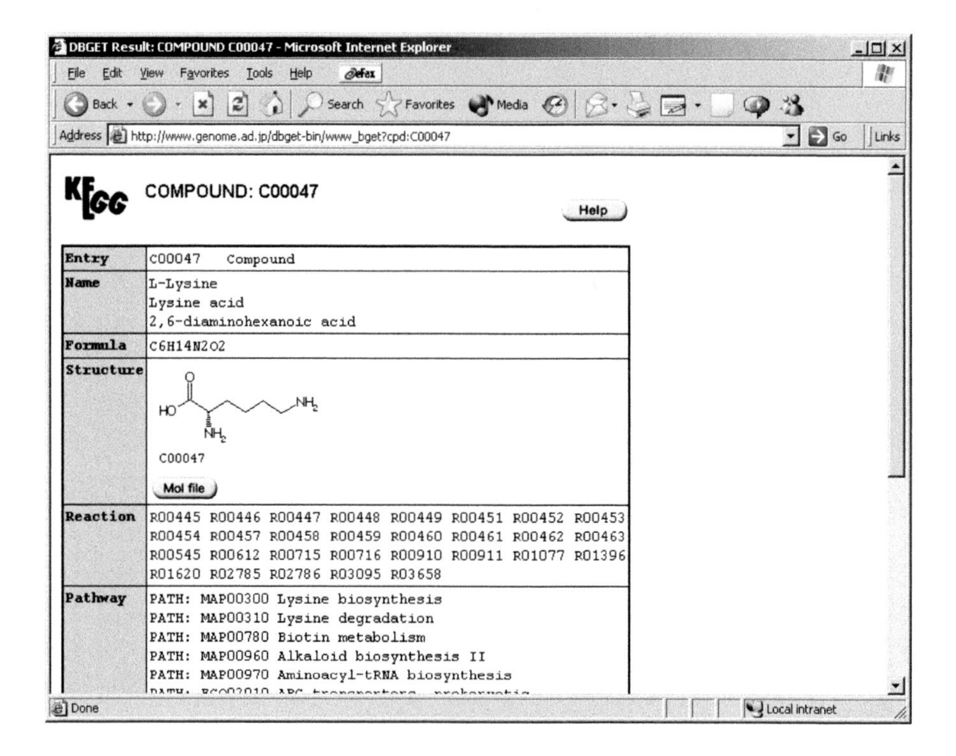

FIGURE 2.25
DBGET result: Compound C00047 (Lysine). (From Kyoto Encyclopedia of Genes and Genomes, The Institute for Chemical Research, Kyoto University, www.genome.ad.jp:80/kegg/. With permission.)

from NCBI) shows one nucleic acid sequence report for rat (exon 9 of rat CYP11B2 gene for aldosterone synthase). A human homolog is likely to exist for this monooxygenase, but no sequence has yet been reported.

2.2.3.2 Cellular Processes at KEGG

KEGG pathway database has recently been expanded to go beyond metabolic pathways. It includes a growing number of Genetic Information Processing pathways, Environmental Information Processing, Cellular Processes, and Human Diseases (Figure 2.27).

1. Metabolism
   Carbohydrate
   Energy
   Lipid
   Nucleotide
   Amino acid
   Other amino acid

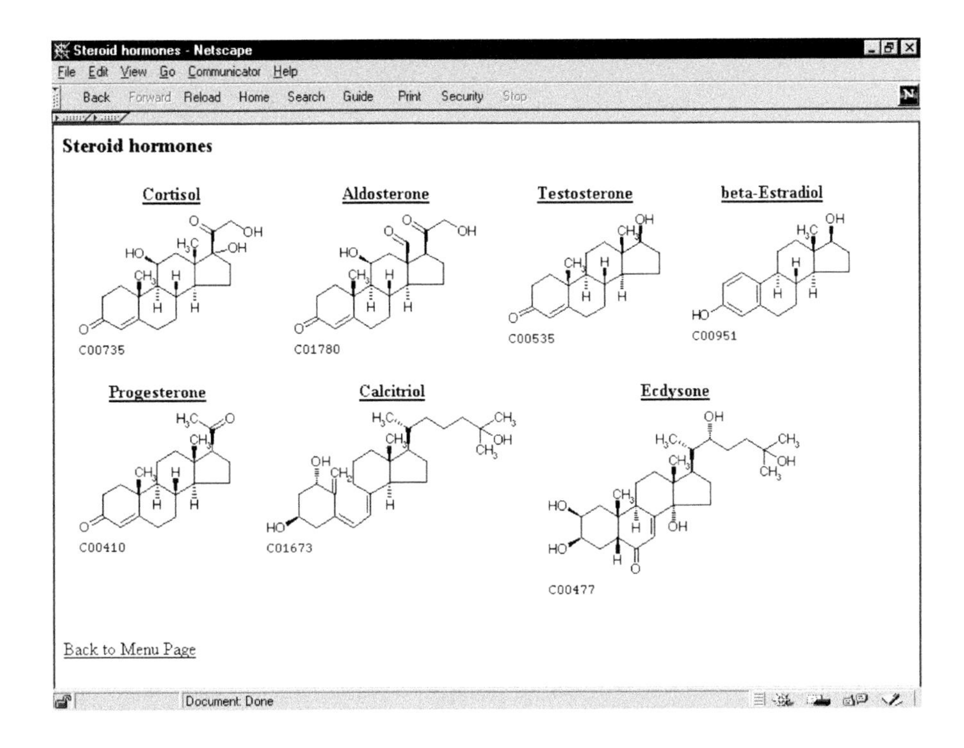

FIGURE 2.26
Steroid hormones (compound classification). (From Kyoto Encyclopedia of Genes and Genomes, The Institute for Chemical Research, Kyoto University, www.genome.ad.jp:80/kegg/. With permission.)

Glycan
PK/NRP
Cofactor/vitamin
Secondary metabolite
Xenobiotics

2. Genetic Information Processing
   Transcription
   Translation
   Sorting and degradation
   Replication and repair
3. Environmental Information Processing
   Membrane transport
   Signal transduction
   Ligand–receptor interaction
   Immune system

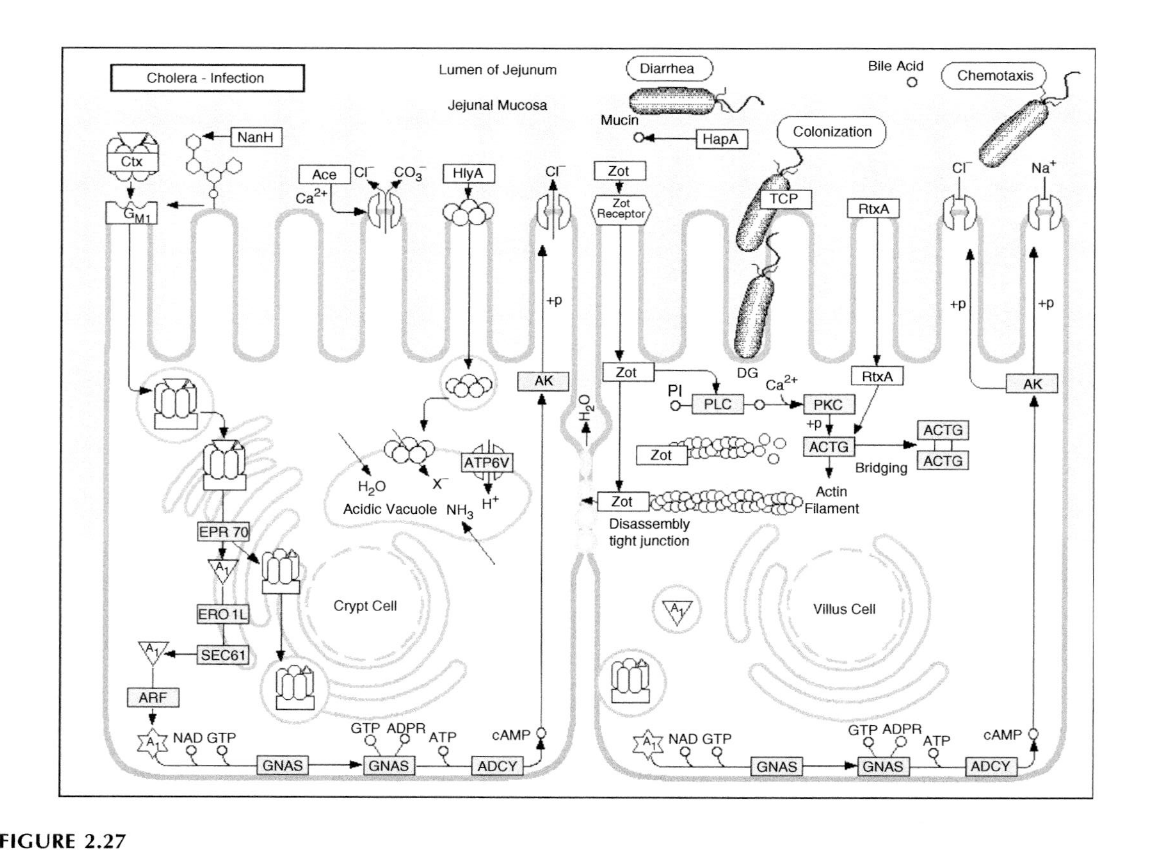

FIGURE 2.27
Cholera mechanism at KEGG: human diseases; infectious diseases: cholera. The page also links to *Vibrio cholerae* agent and provides maps for infection, colonization, diarrhea, environment, and life cycle of the bacterium. (From KEGG, with permission.)

4. Cellular Processes
   Cell motility
   Cell growth and death
   Cell communication
   Development
   Behavior
5. Human Diseases
   Neurodegenerative disorders
   Infectious diseases

Information processing includes transcription and translation, nucleic acid repair mechanism, and protein sorting and turn over. Environmental information processing includes membrane transport, signal transduction pathways, ligand–receptor interaction, and immune system information. Cellular processes include cell mobility, growth, communication, development, and behavior. Human diseases include so far neurodegenerative disease models and cholera as an example of infectious disease mechanism.

References

1. Kanehisa, M., Linking databases and organisms: GenomeNet resources in Japan. *Trends Biochem. Sci.*, 1997, 22(11), 442–444.
2. Ogata, H. et al., KEGG: Kyoto Encyclopedia of Genes and Genomes. *Nucleic Acids Res.*, 1999, 27(1), 29–34.
3. Fujibuchi, W. et al., DBGET/LinkDB: an integrated database retrieval system. *Pac. Symp. Biocomput.*, 1998, 683–694.
4. Altschul, S.F. et al., Basic local alignment search tool. *J. Mol. Biol.*, 1990, 215(3) 403–410.
5. Pearson, W.R., Using the FASTA program to search protein and DNA sequence databases. *Methods Mol. Biol.*, 1994, 24, 307–331.
6. Thompson, J.D., Higgins, D.G., and Gibson, T.J., CLUSTAL W: improving the sensitivity of progressive multiple sequence alignment through sequence weighting, position-specific gap penalties and weight matrix choice. *Nucleic Acids Res.*, 1994, 22(22), 4673–4680.
7. Roberts, L., GRAIL seeks out genes buried in DNA sequence [news]. *Science*, 1991, 254(5033), 805.
8. Nakai, K. and Horton, P., PSORT: a program for detecting sorting signals in proteins and predicting their subcellular localization. *Trends Biochem. Sci.*, 1999, 24(1), 34–36.
9. Hirokawa, T., Boon-Chieng, S., and Mitaku, S., SOSUI: classification and secondary structure prediction system for membrane proteins. *Bioinformatics*, 1998, 14(4), 378–379.
10. Goto, S., Nishioka, T., and Kanehisa, M., LIGAND database for enzymes, compounds and reactions. *Nucleic Acids Res.*, 1999, 27(1), 377–379.

2.3. Database Mining Tools

2.3.1 Sequence Similarity Search Tools: BLAST and FASTA

This section deals mainly with BLAST[1] and FASTA,[2] two of the most popular, user-friendly sequence similarity search tools on the Web. The BLAST server is supported through NCBI in the U.S., while FASTA is maintained by the European Bioinformatics Institute (EBI) in the U.K. The focus of this section is predominantly on BLAST. The BLAST mirror site at EBI provides its subscribers with the option of BLAST or FASTA, along with a few other useful search programs. NCBI subscribers, however, are limited to the BLAST server, which is quite effective and adaptable to various search tasks. BLAST programs are further discussed in this chapter. Additional information on alternative sequence similarity search tools can be obtained from the EBI and NCBI websites (Figure 2.28).

A basic understanding of sequence alignments is necessary to comprehend BLAST or any other sequence similarity search tool. The next several pages discuss sequence alignments on which most sequence similarity search tools are based.

What is a sequence alignment?

Sequence alignments are, for the most part, used to find potential homologs, which are then used to predict potential functions for the query sequence, aid in phylogenetic analysis, or to help in modeling its 3-D structure.

How are the sequence alignment tools classified?

They are classified as either global or local alignment tools.

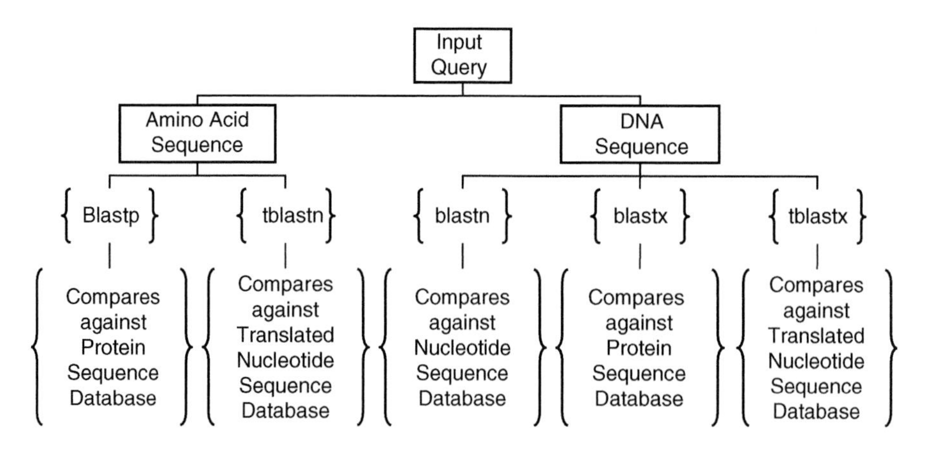

FIGURE 2.28
Overview of BLAST programs.

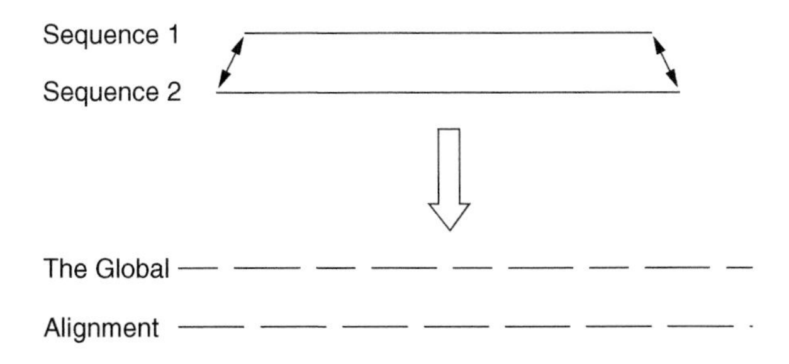

FIGURE 2.29
Schematics of a global alignment for sequences 1 and 2.

What is a global alignment?

A global alignment[3] is the best overall alignment over the entire length of the specified sequences. The introduction of gaps within the two sequences allows for their alignment over their full length.

What are the advantages of using a global alignment tool?

The main advantage of a global alignment is its optimization of sequences that share a high degree of sequence similarity. This is useful in the alignment step of a structural modeling prediction scheme based on sequence homologs with known 3-D structures (Figure 2.29).

What is a local alignment?

A local alignment[4] finds the optimal alignment between subregions or local regions of the specified sequences.

What are the advantages of using a local alignment tool?

A local alignment is most suitable for sequences that display localized similarity regions. A local alignment search tool is used to find sequence motifs, domains, and other types of repeats within the sequence, and is also useful for finding similar sequences for the query in a given database. In summary, a local alignment tool is best used to identify shorter regional segments with a very high degree of similarity. Often these regional segments can be used to find full-length sequence similarities (Figure 2.30).

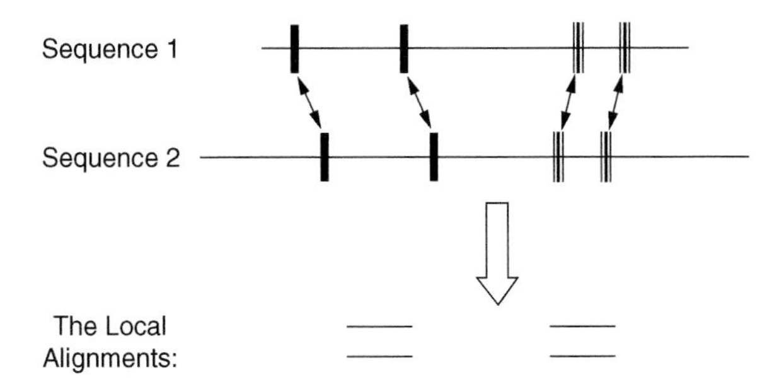

FIGURE 2.30
Schematics of local alignments for sequences 1 and 2.

2.3.1.1 Shared Characteristics in Both Sequence Alignment Tools

All sequence comparison algorithms rely on some sort of scoring scheme. Before moving on it is important to have a fundamental understanding of a scoring matrix.

First we need to learn on how an alignment is scored. The score (S) of an alignment is the sums of all aligned positions and their respective gap scores. Most alignments use a scoring matrix to assign an overall score to each of the alignments. The alignment score is a summation of smaller scores assigned to each of its paired amino acids or nucleotide pairs. The criteria that differentiate the scoring matrices depends on the type of score it is based on. The substitution matrix is how a score is assigned to each pair of alignments. Most of the matrices are based on one of the following scoring schemes:

A scoring scheme based on "identity."
In this scoring scheme, the paired identical residues or nucleotides are assigned a positive score, while the nonidentical pairs are given a score of zero. Generally, the positive score assigned to the identical pairs is one. The overall identity score is simply converted to a percent identity.

Strengths: it is nonheuristic and simple. It works well for sequences that have a high degree of sequence similarity.

Weaknesses: this scoring scheme is generally inferior to those that incorporate outside knowledge. This is predominantly due to the inequality of the nonidentical pairs. For instance, an alanine/valine pair is biologically more acceptable than an alanine/aspartate pair. In this example, the inequality is a result of the relative hydrophobicity of the residues involved. Therefore, the identity-scoring scheme is less effective in detecting sequences or sequence regions that have a low degree of sequence similarity.[5,6] Hence, a scoring scheme that incorporates some sort of weighting step for its nonidentical pairs is biologically more significant and effective than one that utilizes simple identity scoring. Finally, the percent identity reported from the alignment is not always an accurate indicator of the degree of homology that is present. This is predominantly due to the length dependency of the percent identity score.

A scoring scheme based on "chemical similarity."
This is basically an effort to overcome some of the limitations associated with the identity scoring scheme. With this method, the residue pairs are weighted according to some of their chemical and structural characteristics. McLachlan's[7] and Feng's[5] scoring schemes both incorporate amino acid properties such as polarity, charge, size, and structural features into the scoring scheme.

Strengths: it agrees, to a certain extent, with the true selection pressures involved in protein structures at the amino acid level. It is a fact that certain mutations are more devastating to the functions of proteins than others. Generally, these mutations involve a drastic change in the characteristics of the amino acid involved. A polar to nonpolar residue change or vice versa is generally more effective in altering the structure and function of the protein than a mutation involving residues with similar properties.

Weaknesses: the observed mutations in nature are not always explainable through simple scoring schemes that involve our basic understanding of such natural phenomena. Certain evolutionary mutations in nature are yet to be explained.

A scoring scheme based on the "genetic code."
This method considers the minimum number of base changes at the genomic level required to convert one amino acid into another.[8]

Strength: it is based on the principle of molecular biology.

Weakness: the element of chance produces an obstacle to the reliability of the method. A lower number of base changes does not always correspond to a greater degree of similarity between the changed residues.

A scoring scheme based on "observed mutations."
This approach is based on the frequency of mutations observed in aligned sequences.

Strength: it is based on what really happens in nature and minimizes certain otherwise faulty expectations.

Weaknesses: the scoring matrix is based on mutation frequencies found in a set of aligned sequences. The initial alignment requires human intervention, which could potentially alter the true frequencies observed. Aligning the sequences by eye could introduce matching errors that would eventually lead to less-natural mutation frequencies. Scoring schemes based on observed mutations are generally better representative of natural events than those that try to explain relationships through a chemical similarity, identity, or genetic code scoring matrix.

2.3.1.2 How are Sequence Alignments Useful?

Phylogeny: increased sequence homology between sequences is usually an indication of a closer evolutionary relationship.

Structure prediction: a sequence alignment with proteins whose structures are known is used to predict the three-dimensional structure of a sequence whose structure is yet to be experimentally identified. This is based on the assumption of a direct relationship between sequence homology and structural similarity in related proteins.

Sequence motif identification: local sequence alignments could identify potential sequence and functional motifs in proteins and nucleotide sequences.

Function prediction: a high degree of sequence similarity between the proteins is usually an indication of shared functionality among the homologous sequences analyzed.

What is the fundamental concept behind most protein sequence algorithms?

They are based on the 210 possible pairs of amino acids that are represented by a 20×20 scoring matrix. 210 is a summation of the 20 identical and 190 mismatched amino acid pairs. The total possible pairs of characters in a given alphabet is represented through the $(n - 1)i$ formula. Hence, proteins with 20 amino acid characters have $(20 - 1)i$ which corresponds to 210 possible pairs of amino acids. As discussed earlier, the identical amino acid pairs (e.g., leucine and leucine) are assigned the highest scores within the scoring matrix, followed by those that share some degree of similarity (e.g., leucine and isoleucine) and, finally, by those residues that lack similarity traits (e.g., leucine and arginine).

2.3.1.3 Basic Local Alignment Search Tool (BLAST)

BLAST[1] is capable of searching all the available major sequence databases (e.g., SWISS-PROT,[9] PDB,[10] etc.). The default database in a typical BLAST run is the NR (nonredundant) database. The nr database is maintained by NCBI and its lack of redundant sequences for the same species expedites the analysis of the BLAST output file. Even though the nonredundant NR database happens to be the default database for a simple BLAST run, the user still has the option of selecting the database of his choice. For instance, if the user is interested in finding protein homologs whose structures are known for modeling a protein sequence whose structure is yet to be determined, then the PDB (Protein Data Bank) database would be the most logical choice.

Who created BLAST?

BLAST's statistical theory was developed by Samuel Karlin and Steven Altschul.[1]

What type of scoring matrix is used by BLAST?

All BLAST programs use a substitution scoring matrix. The substitution matrix is used in both the scanning phase and the extension phase of the alignment process. The matrix is used to score matches. Substitution matrices are known to dramatically enhance the sensitivity of the alignment process. This is vital to BLAST's attempt to find patches of regional or local sequence similarities.

The default scoring matrix for BLAST is BLOSUM 62.

Why BLOSUM 62 matrix?

This is chosen as default because it has an intermediate divergence pattern.

Note: Low numbered BLOSUM matrices (e.g., BLOSUM 45) are similar in function to high numbered PAM matrices (e.g., PAM 250) while high numbered BLOSUM matrices (e.g., BLOSUM 80) are similar in function to low numbered PAM matrices (e.g., PAM 1).

For distantly related sequences it is best to use low numbered BLOSUM and high numbered PAM matrices such as BLOSUM 45 or PAM 250. For closely related sequences the opposite holds true.

What are substitution matrices?

A substitution matrix is a scoring method used in the alignment of one residue or nucleotide against another. The first substitution matrices used in the comparison of protein sequences in evolutionary terms were developed by the late Margaret Dayhoff and her coworkers. These matrices were derived from global alignments of closely related sequences. These matrices were also used to extrapolate other matrices for less similar or evolutionarily more distant sequences.

These matrices are typically referred to as the Dayhoff, MDM, or PAM[11] set of matrices. The numbers that accompany these matrices (e.g., PAM 40, PAM 100, etc.) correspond to the relative evolutionary distance between the respective sequences. Smaller numbers represent evolutionarily less distant sequences, while larger numbers signify a greater evolutionary distance. The major objection to the PAM series of matrices is their incorrect assumption that the set of selection pressures on closely related sequences is the same as those that are less related. In contrast to the PAM series of matrices which are based on global alignments of closely related sequences, the BLOSUM matrices developed by Steve Henikoff and his coworkers, are derivatives of local alignments of distantly related sequences. Unlike the PAM series of matrices, the BLOSUM matrices do not extrapolate the closer-related sequences based on previously calculated matrices of a less-related set (Figures 2.31 and 2.32).

The matrices in this method are all directly calculated. In contrast to the PAM series, the numbers that accompany BLOSUM matrices (e.g., BLOSUM 62, BLAST's default matrix) refer to the minimum percent identity used to construct the matrix. Therefore, the smaller numbers correspond to blocks that are evolutionarily more distant.

Which matrix should I use?

The PAM series are generally more suitable for global similarity searches, while the BLOSUM series are found to perform better in finding regional or local similarity regions. Both series of matrices have their share of strengths and weaknesses. The current approach is to incorporate the two methods so that their strengths complement each other. Such a chimeric matrix could enhance the performance level of future similarity searches (Figure 2.33).

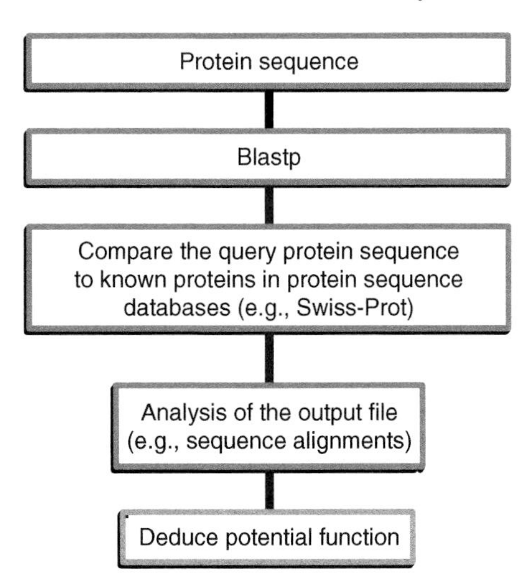

FIGURE 2.31
BLASTp diagram.

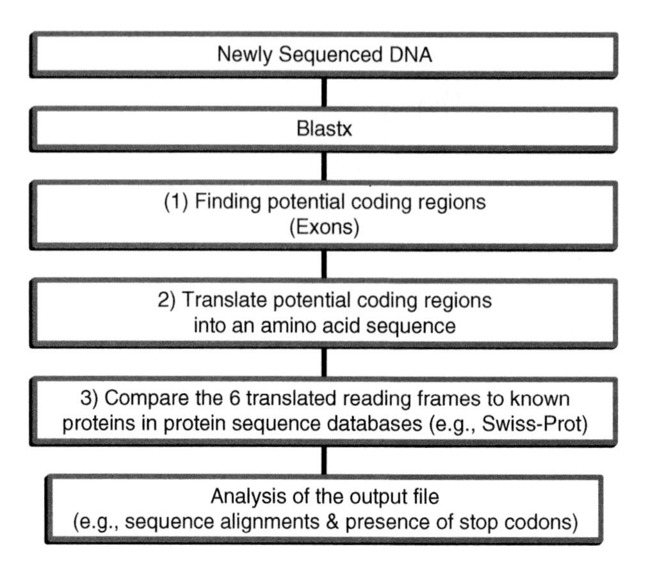

FIGURE 2.32
BLASTx diagram.

BLAST programs are designed to enhance speed while maximizing the sensitivity of the distance sequence relationships. This allows the program to find the closest sequence homologs in a time-efficient manner. The BLAST programs use a heuristic algorithm that identifies local alignments. In contrast to algorithms that seek global alignments, BLAST's local alignment search finds isolated regions of sequence similarities. The BLAST server supports a variety of analysis programs that are either accessible through a web interface or can be installed on local networks to expedite the analysis step. The Standard BLAST is the original BLAST program that searches for simple sequence similarities in NCBI's database network.

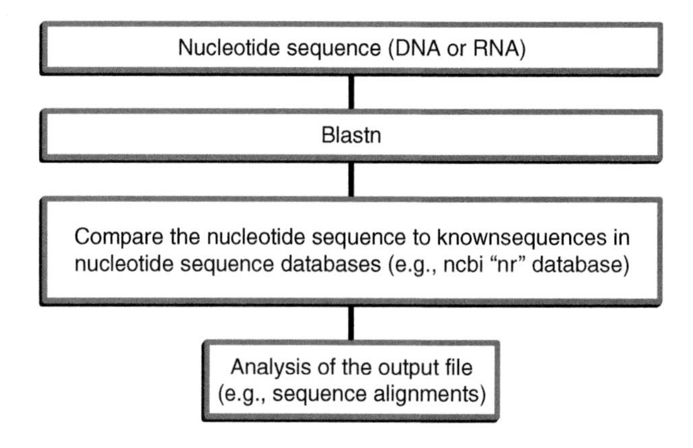

FIGURE 2.33
BLASTn diagram.

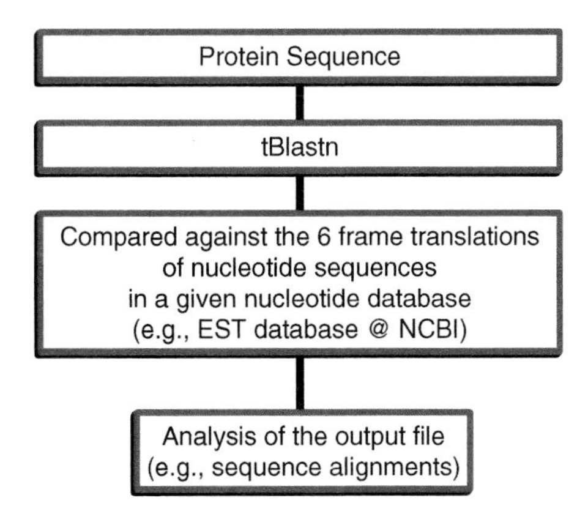

FIGURE 2.34
tBLASTn diagram.

What are some of the limitations associated with a basic BLAST search?

The basic BLAST program does not allow gaps in its alignments. This will, in theory, reduce the sensitivity of the search. However, the output file reports multiple regional alignments that can be used to anticipate the gaps between the query sequence and the database entry (Figure 2.34).

What are the different BLAST programs and how are they useful?

- BLASTp: This program allows the user to search a protein query sequence against a protein database. This can be used to find all possible sequence homologs for a given protein query sequence.
- BLASTx: This program allows the user to search a translated nucleotide sequence against a protein database. The query nucleotide sequence is initially translated in all of its six possible reading frames. This program is particularly useful in finding nucleotide sequencing errors by comparing the translated nucleotide query sequence to its potential protein homologs in a protein sequence database. The information in a BLASTx output file can also help to identify unclear nucleotides in a given reported nucleotide sequence.
- BLASTn: This program allows the user to search a nucleotide query sequence against a nucleotide database. A newly sequenced nucleotide query can be compared with itself or its homologs for identification and potential contamination of the query sequence.
- tBLASTn: This program allows the user to search translated nucleotide sequences in a given nucleotide database against a protein query sequence. The nucleotide sequences in a given nucleotide database are initially translated into each of its six possible reading frames and then are compared with the protein query sequence. This program is particularly useful in finding protein sequencing errors by comparing the protein query sequence to its potential translated

nucleotide homologs in a given nucleotide database. The information in a tBLASTn output file can also help clarify unclear amino acid residues in the given query sequence. tBLASTn is similar to BLASTx in terms of its six-reading-frame translation comparison approach, but instead of a nucleotide query sequence (BLASTx) it uses a protein sequence query.

- tBLASTx: This program allows the user to search the six frame translations of a nucleotide query sequence against the six frame translations of nucleotide sequence entries in a given nucleotide database. The tBLASTx program has similarities to both BLASTx and tBLASTn programs and can be used to complement a BLASTx search.

The new BLAST programs are called BLAST 2.0. The Gapped BLAST and the PSI-BLAST[12] are just two programs supported by the new BLAST 2.0 server. The new BLAST 2.0 server has been redesigned to optimize speed and sensitivity, while adding new capabilities that allow it to support the Gapped BLAST and PSI-BLAST programs (Figure 2.35).

What is Gapped BLAST?

The Gapped BLAST algorithm allows the introduction of gaps in the alignments returned in BLAST's output file. Gaps are deletions and insertions introduced into the sequence. This prevents the similar sequence regions from being broken into segments. The heuristic input of the algorithm allows

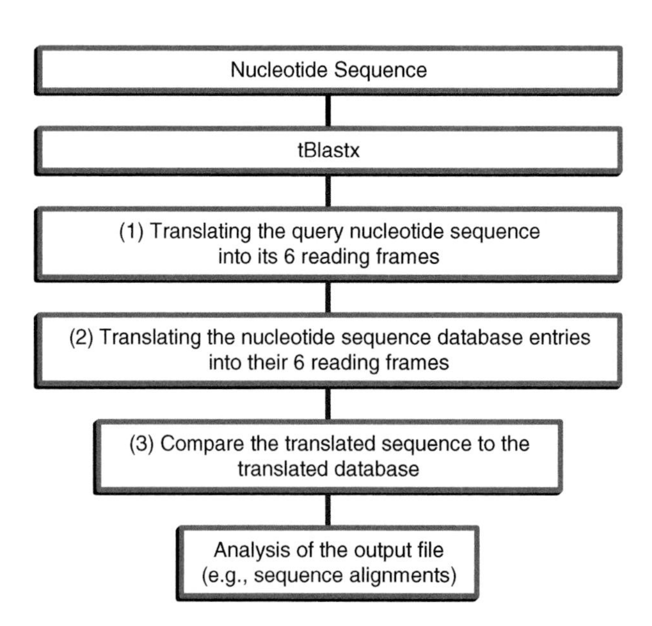

FIGURE 2.35
tBLASTx diagram.

the score to reflect biological relationships associated with the alignment. This generally reflects active sites and binding sites of the sequence that tends to be better conserved. Hence, the introduction of gaps prevents these regions from being split up into less meaningful sequence fragments.

What is PSI-BLAST?

PSI-BLAST stands for *position-specific iterated* BLAST. The PSI-BLAST initially performs a Gapped BLAST and uses the output alignment as its input file. PSI-BLAST then constructs a position-specific score matrix that replaces the original query sequence and is used to find profiles in the next several iterative database search runs. Profile is an increased sensitivity pursuit for homologous sequences.

Some of the BLAST tools can be installed on local machines. This is the BLAST network client software. The network client software on a local computer communicates with the distant BLAST server at NCBI. BLAST-2 and PowerBLAST are some of the basic network services offered through BLAST. What is BLAST-2?

BLAST2 is the standard BLAST service that provides output files in HTML format. Its filtering capabilities enable a user to find sequences with low-complexity regions.

What is PowerBLAST used for?

This is a network BLAST client that is designed to conduct large-scale analyses of genomic information. This and other network client software can be retrieved via FTP from the NCBI home page under the network directory at BLAST.

The BLAST server can be accessed in several ways, but the easiest way is through the Web (http://www.ncbi.nlm.nih.gov). A very user-friendly Web interface is used to perform the BLAST run. Following are the general steps a user needs to follow for a successful BLAST run:

1. The query sequence of interest must be in the correct format (e.g., FASTA format). If the query sequence was retrieved from NCBI's Entrez, the easiest route is to copy and paste the FASTA format of the sequence directly from Entrez into the BLAST interface.
2. The proper formatted sequence can then be pasted into the "input sequence" box on the BLAST web interface.
3. Depending on the type of sequence analyzed, the appropriate BLAST program is selected (e.g., BLASTp for protein sequences, BLASTn for DNA or RNA sequences, etc.).
4. Finally, the appropriate database must be selected. The default database on BLAST is the NCBI's nr database. The nr database will search for all the available nonredundant sequences present. For example, if the user is only interested in finding sequence homologs whose structures are known, then it would be wise to search a database that is specific to molecules with known structures. Therefore, instead of using nr, the user would select PDB[10] as the preferred database. The

sequence is now ready to be submitted to the BLAST server. The results of the search can be obtained either by e-mail or seen interactively on the BLAST web interface. The e-mail route is preferable when analyzing multiple sequence files. This allows the user to analyze the sequences of interest in a time-efficient manner, while being able to analyze the result sections later (BLAST figures are saved as GIF files).

As discussed earlier, BLAST can also be accessed through the network BLAST client. In order to do this, the user must initially install the appropriate BLAST client software through FTP (ftp://ncbi.nlm.nih.gov). A BLAST search can also be done through NCBI's e-mail server (blast@ncbi.nlm.nih.gov). This is mainly for people without convenient Internet access. The query sequence must still be in the proper format in order for BLAST to conduct the proper operations. Another way to run BLAST is to install a fully executable version on the local machine that searches against the user's local databases. This version of BLAST is found under the "executables" directory of BLAST and can be obtained via FTP (ftp://ncbi.nlm.nih.gov). The executable versions of BLAST are available for IRIX 6.2, Solaris 2.5, DEC OSF1, and Win 32 operating systems. BLAST's output files are designed to complement other servers at NCBI. The sequences retrieved from the results section of a BLAST run are either directly or indirectly linked to NCBI's Entrez and PubMed servers.

What is the significance of the expected (E) value in a BLAST output file and its relationship to the Probability (P) value?

E represents the number of expected matches found in a given database by chance. Therefore, the significant hits are those that are assigned lower *E* values. For simplicity, the BLAST programs just report the *E* value rather than *P* values.

The mathematical relationship of *P* and *E* values with respect to the High Segment Pairs in the alignment score can be expressed as:

$$P = 1 - e^{-E} \quad \text{and} \quad E = Kmn\ e^{-\lambda S}$$

K and λ represent statistical parameters of the high segment pairs. The remainder of this paragraph will give us a better sense of why these two values, especially the *E* value.

An *E* value of zero means that the particular hit has a zero probability of being a match by random chance. In contrast, an *E* value of one means that for the given database there is a probability of finding a similar scoring match by chance. The *E* value is basically the random noise for a given match. This value decreases exponentially with increasing score (*S*) values. The default *E* value can be increased to find statistically less-significant hits or shorter peptide or nucleotide sequences whose matching scores and *E* values are, in many cases, statistically less significant.

Note: For *E* values that are <0.01, *E* and *P* values are almost identical and sequence pairs with low *E* values are the best overall representation of the sequence similarity searches. Hence the value of reporting *P* and *E* is of less significance.

2.3.2 An Overview of Database Sequence Searching

Objectives:

1. Finding sequence homologs to deduce the identity of the query sequence.
2. Identifying potential sequence homologs with known 3-D structures for predicting the three-dimensional structure of the target sequence and deducing functional features.

Potential problems: being able to distinguish between a true sequence homolog (common ancestor) and one that has been found by chance in the given database can be a problem. These indistinguishable hits must undergo further examination to understand their true relationship to the query sequence.

How should a sequence database search be started?

1. A query sequence is needed. This is the target sequence that needs to be analyzed. The query sequence could be either a newly determined sequence whose identity is yet to be determined or one whose identity is known. A database search can help in determining the identity of the newly determined query sequence or finding possible sequence homologs for a known query sequence entry.
2. Select the appropriate server. The server must be reliable, regularly updated, and powerful. These characteristics are generally associated with government or government-funded bioinformatics servers such as NCBI. The NCBI is a collection of several public domain databases and search tools that is readily available through the Internet and is compatible with most Web browsers.
3. Select the appropriate program or set of programs in a given server. If NCBI is the server of choice and a program is needed to conduct a simple sequence similarity search, then one of the BLAST[1] programs might be appropriate.
4. Which BLAST program should be used for a simple sequence similarity search? If the query is a protein sequence, then BLASTp is the appropriate tool. If the query sequence is DNA or RNA, then the BLAST program must be utilized. These are just two of several programs available at the BLAST server. Other BLASTn programs (e.g., tBLASTn, tBLASTx) can be utilized for finding sequence homologs for the query sequence, and also to perform more advanced tasks. For instance, the BLASTx program can be used to find potential coding regions in a gene, while other BLAST programs

can be utilized to check for potential sequencing errors in a newly determined sequence. (BLAST was explained in detail in the Sequence Similarity Search Tools section in Section 2.2.)

5. Select the appropriate database. There are two ways to select the appropriate database:
   - Search all relevant databases: this is a nonredundant database of all possible submitted entries. Selecting this option will enable the user to search through all the available sequence entries.
   - Search through a specific database: in this case, the user is merely interested in a specific type of database. For instance, if the user is solely interested in finding sequence homologs with known 3-D structures, then the PDB (Protein DataBank) database[10] would be the most logical choice since the 3-D structures of all its sequence entries are known.
6. Select the appropriate filter. For the convenience of its subscribers, BLAST has incorporated a set of filter options in each of its programs. The filter option excludes sequences with low complexity regions. Due to the repetitive nature of these sequences, the probability of false positive hits or random hits within the search increases and ultimately obscures the result section. It is recommended that a filter be used in a given search to reduce the number of false positives.

Can the filter option exclude a true positive from the result section?

Yes. A true positive hit with low complexity regions may be excluded from the output file. Because of this, the filter option may reduce the sensitivity of the search.

How can the sensitivity of the search be maximized while minimizing false positives?

This can be achieved by conducting two separate searches for the same query sequence. In one search, a filter is utilized to minimize false positive hits, while the other search is done without a filter to maximize sensitivity. The output files from both search results are then compared and contrasted to find possible true positive hits that were excluded from the initial filtered run.

7. Reading, comprehending, and analyzing the output file. In order to derive a possible hypothesis from the result section of the searched query, the user needs to be familiar with the terminology used in the output file. The key subjects of the output file are its assigned score for each of the found entries, and the databases and accession numbers associated with each of those entries. The score assigned to each of the sequences found is typically an indication of its homology to the query sequence. In a BLAST output file the score is also related to the expected, or *E*-value assigned to each entry. The *E*-value in a BLAST output file is the probability of the sequence being a random or chance hit. The closer it is to zero, the smaller the chance of it being a random hit from a given database.

Now that the general steps in a simple sequence database search have been covered, it is important to ask the following fundamental questions:

What are the advantages of using a network database as opposed to a local database?

- Network databases are regularly updated. NCBI's collaboration with the European EBI and the Japanese DDBJ keeps their databases updated on a daily basis. These daily updates provide a dependable and nonredundant resource for their subscribers.
- Local database maintenance is not trivial and, in most cases, is beyond the scope of its average user. Utilizing and maintaining a personal database can be time consuming and expensive. These obstacles increase the value of the public domain network databases at the NCBI, the EBI, and the DDBJ. (These resources were explained in detail in Section 2.2.)
- Network databases provide their subscribers with the appropriate search tools. The NCBI provides its subscribers with the BLAST server, as do the EBI and the DDBJ. The search tools provided by the public domain servers are regularly updated and, therefore, are more useful to their subscribers.

What are the disadvantages of using a network database instead of a local database?

- A local database is readily available in the event of a network failure.
- The subscriber is limited to the search tools provided in a network database. The scanning methods used by network servers are not always the most effective tools. Local scanning methods could be employed through a local server and could be more appropriate for that particular search.

The BLAST programs described in this chapter can be accessed through the BLAST server at www.ncbi.nlm.nih.gov (NCBI's home page).

References

1. Altschul, S.F. et al., Basic local alignment search tool. *J. Mol. Biol.*, 1990, 215(3), 403–410.
2. Pearson, W.R., Using the FASTA program to search protein and DNA sequence databases. *Meth. Mol. Biol.*, 1994, 24, 307–331.
3. Huang, X., On global sequence alignment. *Comput. Appl. Biosci.*, 1994, 10(3), 227–235.
4. Altschul, S.F. and Gish, W., Local alignment statistics. *Methods Enzymol.*, 1996, 266, 460–480.
5. Feng, D.F., Johnson, M.S., and Doolittle, R.F., Aligning amino acid sequences: comparison of commonly used methods. *J. Mol. Evol.*, 1984, 21(2), 112–125.

6. Schwartz, R.M. and Dayhoff, M.O., Origins of prokaryotes, eukaryotes, mitochondria, and chloroplasts. *Science*, 1978, 199(4327), 395–403.
7. McLachlan, A.D., Repeating sequences and gene duplication in proteins. *J. Mol. Biol.*, 1972, 64(2), 417–437.
8. Fitch, W.M., An improved method of testing for evolutionary homology. *J. Mol. Biol.*, 1966 16(1), 9–16.
9. Bairoch, A. and Boeckmann, B., The SWISS-PROT protein sequence data bank. *Nucleic Acids Res.*, 1992, 20 Suppl, 2019–2022.
10. Sussman, J.L. et al., Protein Data Bank (PDB): database of three-dimensional structural information of biological macromolecules. *Acta. Crystallogr. D. Biol. Crystallogr.*, 1998 54(1 [Pt 6]), 1078–1084.
11. Wilbur, W.J., On the PAM matrix model of protein evolution. *Mol. Biol. Evol.*, 1985, 2(5), 434–447.
12. Altschul, S.F. et al., Gapped BLAST and PSI-BLAST: a new generation of protein database search programs. *Nucleic Acids Res.*, 1997 25(17), 3389–3402.

2.3.3 Pattern Recognition Tools (Prosite)

Prosite[1] is one of the most widely used databases containing biological motifs and signatures. Prosite is a collection of functional sites and sequence patterns found in many proteins.

What kind of information is stored in the Prosite database and how is it useful to its subscribers?

- Many of the characterized binding sites and motifs are gathered and maintained by Prosite. The entries in Prosite are, for the most part, cross-referenced and linked to other appropriate sites. For instance, the calcium-binding EF-hand signature is well documented and its entries are further characterized by their SWISS-PROT file names. These entries are generally linked to SWISS-PROT[2] or other relevant databases.
- The Prosite file includes the sequence entries that share the matched sequence motif or signature of interest. The file will also inform the subscriber of the reported false positives, false negatives, and matched sequences whose identity is questionable.
- What is a false positive sequence? This is a sequence that contains by mere chance the signature or motif of interest. A false positive generally lacks the functional characteristics associated with the sequence motif of interest.
- What is a false negative sequence? This is a sequence that shares functionality with the true hits but lacks the specified signature sequence.
- What are the questionable sequences? These are sequences that share the motif characteristics but whose functional significance has yet to be proven through experiments. Experimentation would place these questionable sequences in either the false positive or the true positive category.

This type of information provides the investigator with a powerful tool that can enhance the efficiency of the research.

- Prosite hinders redundancy. The characterized signatures are well documented in order to minimize redundant motifs.

Prosite has search tools for matching patterns. The PROMOT[3] search tool can be used to match a sequence against the Prosite database. It can also be used to match the sequence of interest against a set of predefined patterns. Prosearch[4] is another tool that can search the SWISS-PROT and TrEMBLE databases with a given sequence pattern or signature. Through Prosearch, novel sequence signatures and patterns can be found efficiently in all SWISS-PROT and TrEMBL sequence entries.

How is the data presented in Prosite files?

The documentation of each pattern is presented in a ".doc" file, while the actual pattern is presented in a separate file labeled ".dat." The dat file also contains information on pattern scanning programs and other sequence pattern compilations.

2.3.3.1 The Significance of Embedded Symbols within Each Signature and How to Read and Construct Signatures

The calcium-binding EF-hand sequence motif is used as an example in order to better understand the symbols used in each of the signature sequences in the Prosite database.

The following represents Prosite's calcium-binding EF-hand signature:
D-X-[DNS]-{DENSTG}-[DNQGHRK]-{GP}-[LIVMC]-
[DENQSTAGC]-X(2)-[DE]-[LIVMFYW]

1. The hyphen is used to separate each position within the sequence motif.
2. [ ]: residues within each bracket denote the allowed residues at that particular position in the sequence motif. For instance, in [DNS] the allowed residues at that particular position are aspartate, asparagine, and serine.
3. { }: the characters within curly brackets represent residues that are not allowed for that particular position in the sequence motif. In other words, all other residues are allowed for that particular position of the sequence motif.
4. X: the letter X denotes any of the twenty amino acids.
5. (n): this denotes a repeat of a particular residue or X. For instance, X(2) can also be represented as –X-X-.
6. (*n*,*m*): this denotes a repeat of a length between *n* and *m*. For instance, A(2,5) means that a continuous stretch of two, three, four, or five alanines are all equally likely at that particular position within the sequence motif.

References

1. Bairoch, A., PROSITE: a dictionary of sites and patterns in proteins. *Nucleic Acids Res.*, 1991 19 Suppl., 2241–2245.
2. Bairoch, A. and Boeckmann, B., The SWISS-PROT protein sequence data bank. *Nucleic Acids Res.*, 1991, 19 Suppl., 2247–2249.
3. Sternberg, M.J., PROMOT: a FORTRAN program to scan protein sequences against a library of known motifs. *Comput. Appl. Biosci.*, 1991, 7(2), 257–260.
4. Kolakowski, L.F., Jr., Leunissen, J.A., and Smith, J.E., ProSearch: fast searching of protein sequences with regular expression patterns related to protein structure and function. *Biotechniques*, 1992, 13(6), 919–921.

2.3.4 Multiple Alignment and Phylogenetic Tree Analysis

Studying a gene means understanding its variations within populations and across taxonomic groups. Sequence alignment, the pair-wise comparison of sequences, is the first step in assessing the property of a newly sequenced gene, finding homologs in other organisms, or identifying a new sequence as novel. NCBI allows a comparison between two sequences using a "BLAST 2 sequences" tool (Figure 2.36). This is a specialized version of the general

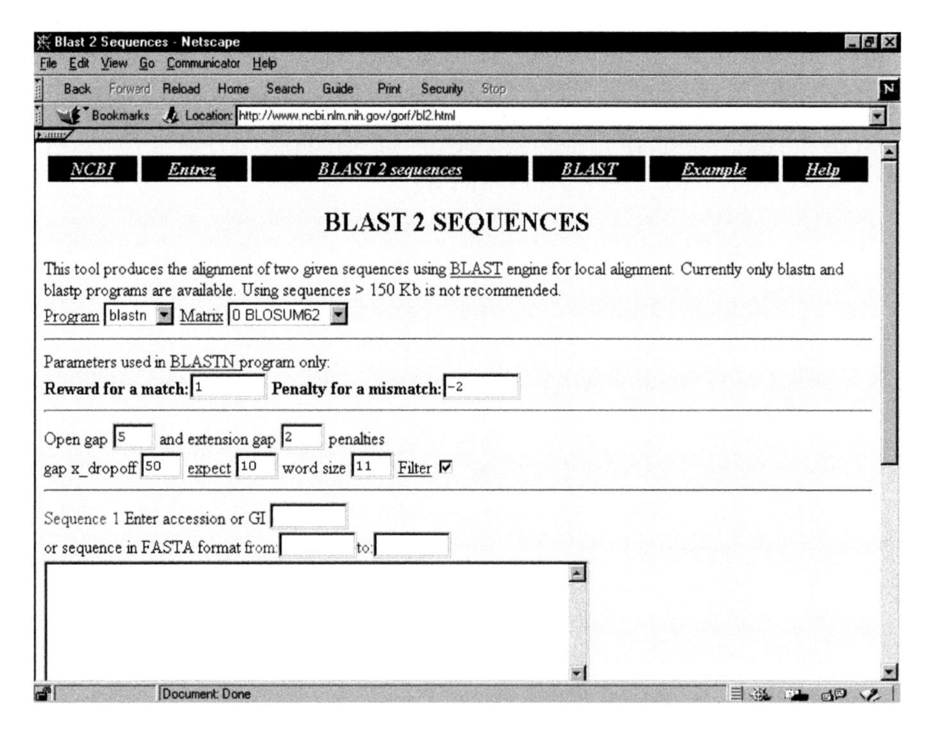

FIGURE 2.36
Blast 2 sequences; entry form. (www.ncbi.nlm.nih.gov/blast/bl2seq/bl2.html).

BLAST algorithm (see Section 2.2) to search for similar sequences and to retrieve them from databases. The BLAST 2 algorithm allows nucleotide (BLASTn) or amino acid sequences (BLASTp) to be compared. Several different matrix algorithms can be selected. Basically, sequences can be compared for their entire length, or for best local alignments. Distantly related genes or functional similarities in evolutionary unrelated proteins can be identified with the second approach. Sequences can be entered by accession number (GI) or in FASTA format.

To understand a novel sequence for its potential functionality in an organism, multiple sequence alignment provides biological information through evolutionary related genes and proteins. While BLAST-2 searches for local similarities and motifs through pairwise comparison for structural and functional analysis, ClustalW provides a hierarchical output of best alignments of multiple sequences highlighting both best matches and differences among sequences. The alignment hierarchy can be used as input information for cladograms and phylogenetic trees, i.e., an analysis of evolutionary divergence of the sequences.

For multiple sequence alignment, the program ClustalW is available at many bioinformatics websites. The EBI can be accessed for this purpose. The ClustalW interactive site can be found on the home page at www.ebi.ac.uk under the "Services" link and from there under "On-line applications." This site summarizes all available links provided by the EBI, as well as non-EBI servers.

Several sequences can be submitted and different output settings can be selected. The results include information about the identities from pair-wise alignments and the order of most-identical to least-identical sequence pairs. An output description for creating a graphical representation (phylogenetic tree) is also included. Programs that create trees can be downloaded as stand-alone versions and ClustalW output files can be saved on the local hard drive for later reference, analysis, and representations.

Many proteins are organized in families of related sequences. They often have very similar function, are found as subunits in larger, functional protein complexes, or are expressed at different times during development or in a cell specific manner. The Superfamily of ligand gated receptor is a well-studied group of proteins that has been functionally and structurally characterized. Families contain groups of proteins where members are closely related and respond to the same cellular stimuli (ligands), whereas superfamilies contain families with sometimes physiologically distinct behavior (different ligands). Here we compare a few members of two superfamilies of ligand gated receptor channels, the nicotinic acetylcholine receptor subunit alpha and the human serotonin receptor 5HT-3A and 5HT-3B.

Databases have been reorganized over the last few years providing cluster analysis with precalculated results. At NCBI the cluster of orthologous groups, Homologene, and BLAST link (BLink), which displays the results of BLAST searches that have been done for every protein sequence in the Entrez Proteins data domain. Blink allows display of multiple protein

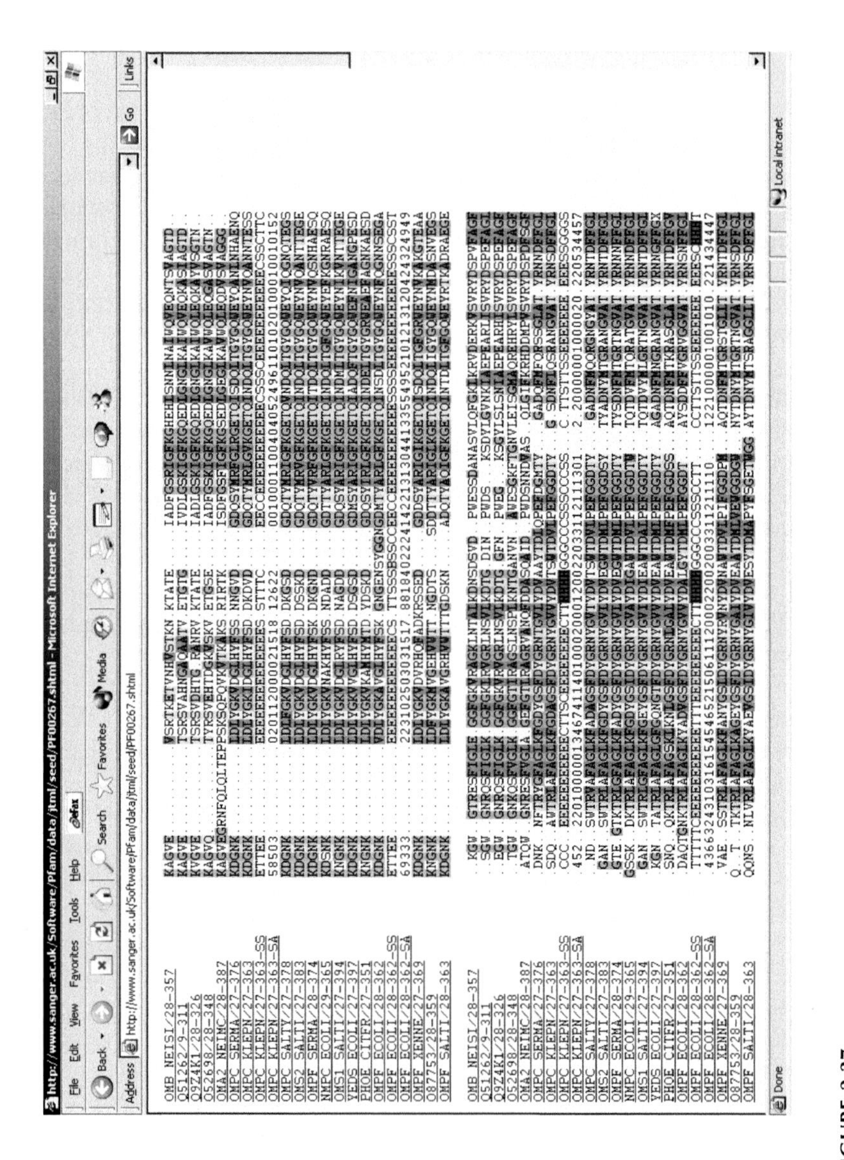

FIGURE 2.37
PFAM alignment for Gram-negative porins (Porin-1).

sequence alignments for all hits, best hits only, taxonomic group relationship, cladogram (common tree), and structural features through the Molecular Modeling Database (MMDB) and Conserved Domain Database (CDD). More information can be found in Section 3.2 on comparative genomics.

A large collection of multiple sequence alignments and hidden Markov models covering many common protein domains and families can be found at PFAM, the protein family databases of alignments and HMM (hidden Markov models). This database is provided by the Sanger Institute (www.sanger.ac.uk/Software/Pfam). PFAM contains curated motif information on 7459 protein families. Seventy four percent of all known protein sequences in databases have at least one match to the PFAM database. This means that from all putative proteins identified from genome sequencing projects, one fourth are completely unknown and do not contain a single structural motif described in PFAM. With PFAM, the interested biologist can view multiple alignments, protein domain architectures, known protein structures, examine species distribution and follow links to other databases (Figure 2.37).[1]

A structural alignment of protein motifs often reveals evolutionary relationships among proteins that cannot be found with sequence comparison alone. The reason for this lies in the redundant set of divergent amino acid sequences that can fold into similar or identical protein structures. As for evolutionary conservation, protein structures are more conserved that corresponding amino acids, and amino acid sequences are more conserved that nucleotide sequences because of the underlying redundancy at both levels, the protein folding level and the degeneracy of the genetic code.

References

1. Bateman, A. et al. The PFAM protein families database. *Nucl. Acids. Res.*, 2004, 32(90001), D138–141.

3

Genome Analysis

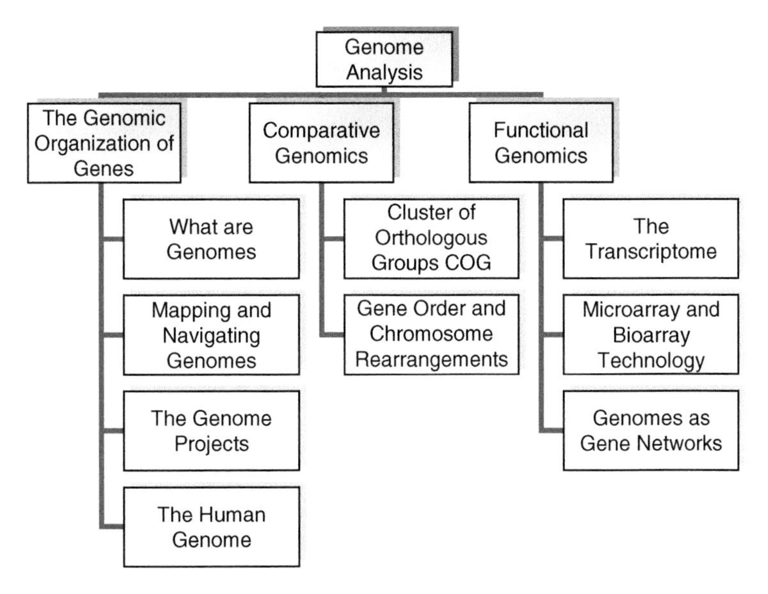

FIGURE 3.1
Chapter overview.

3.1 The Genomic Organization of Genes

3.1.1 What are Genomes?

A genome is the entirety of the genetic material of an organism that is passed along from generation to generation. A genome is composed of thousands to tens of thousands of genes, which are the hereditary units of traits or phenotypes. Genomes are organized in physical units called chromosomes. Chromosomes contain genes at specific locations (called locus). Often, individuals of a population carry minor sequence differences called allelic variants or alleles. Alleles contribute to genetic variability in a population of like individuals and differing traits like hair color, blood type, and susceptibility for diseases. Genes specify nucleotide sequences of chromosomal DNA that *code* for proteins and structural RNAs. Genes are not the only information contained in chromosomes. In fact, various amount of *noncoding* sequence information is found in genomes ranging from a few percent in viruses and bacteria to a whopping 98% in humans. Some of these noncoding sequences appear to be degenerated into long stretches of repeat sequences, which have been labeled junk DNA in the 1970s. Even genes contain noncoding sequences that include regulatory sequences and introns (intervening sequences; 50 to 20,000 nucleotides) fragmenting the protein coding sequences of genes into discontinuous stretches known as exons. While exons often correlate with the coding of functional protein domains, the function of intervening sequences is largely unknown, although they may control alternative splicing and allow exon shuffling, thus contributing to the complexity and evolution of the proteome. Modern genome projects will contribute to a better understanding of their biological roles.

Whole genomes are of interest not only because they contain all the hereditary genetic component of an organism, but also because they reflect the complexity of an organism, something analysis of a list of individual genes fails to do. Surprising or not, complexity is not just a reflection of the number of genes, but the ability of combining the activity of genes in cell-type specific manner, a phenomenon called epigenetics. With genome projects of many organisms being completed one by one, an understanding of the differences of genomes from the three domains of life (bacteria, archaea, and eukarya) is in reach. In addition, an understanding of the relationship of genome organization to the form and function of an organism is beginning to take shape.

The complex fragmentation of genomes into coding and noncoding sequences is typical for higher eukaryotes, particularly plants and mammals. Prokaryotes, however, have a very different genome structure. While their names refer to the absence or presence of a nuclear compartment within the cell, the differences do not stop there. The relative density of coding vs. noncoding regions differs, as does the organization of genes on the chromosomes. While bacteria have a compact genome often with a single circular chromosome (and some small satellites or plasmids) with little noncoding DNA (e.g., *E. coli* has 934 genes per 1,000,000 base pairs), eukaryotic chromosomes are often extremely large, divided into several linear chromosomes, and have a low gene density (humans have 11 genes per 1,000,000 base pairs). In bacteria and archaea, genes code for one protein or structural RNA, although they have self-splicing introns in ribosomal and transfer RNA similar to those found in mitochondria and chloroplasts. In animals and plants, a protein-coding gene has an average of three gene products or splice variants, the result of splicosome activity in eukaryotic nuclei stitching together exons in different combinations. The increase in number and use of splice variants in higher organisms (all eukarya) may be an indicator that this more complex gene structure supports the evolution of cellular complexity. Single-celled yeast, for instance, does not make extensive use of splice variants.

The importance of genome structures is also highlighted by the observation that genomes from related organisms are conserved, including the relative position of genes, amount of noncoding sequences, and appearance of chromosomes. This indicates that the genome structure is a form of genetic information important for the function of an organism. Thus it is not only important what kind of genes there are, but how they are organized in a physical or spatial dimension.

For example, genes coding for enzymes of metabolic pathways in bacteria that are needed only under certain conditions but always at the same time are often physically clustered together, forming a gene expression unit called *operon*. Regulating a pathway through an operon instead of through each gene individually contributes to metabolic efficiency. An entire pathway for the synthesis of an amino acid is upregulated in a coordinated fashion, avoiding the individual regulation of every enzyme needed for a pathway. In eukaryotes, this physical closeness of proteins forming functional complexes is often not found, as has been shown for the alpha and beta hemoglobin gene clusters. The beta globin gene is found on chromosome 11, locus p15.5, while the alpha globin cluster is found on chromosome 16, locus p16.3. Despite the physical separation of α and β genes in chromosomes, functional hemoglobins are complexes of two α and two β proteins. Two protein copies of both genes have to be made as they cannot function independently.

Still, genes are not randomly distributed on the genome, and their location is associated with the size and morphological organization of a chromosome. Several databases with access to complete genome information for various organisms provide information about nucleotide sequences, gene organization in physical maps, and sometimes functional information such as coregulation.

Figure 3.2 shows the large-scale architecture (ideogram) of human chromosomes with the typical G-banding patterns produced from staining GC rich DNA segments. Chromosomes are partitioned into two arms, the p and q arm, and are either metacentric with the constriction or centromere

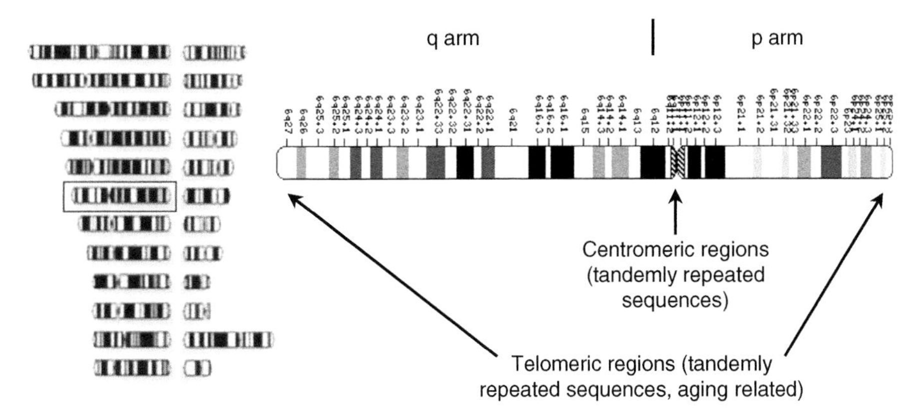

FIGURE 3.2
Chromosome structure (ideogram) with G-banding pattern.

separating the arms close to the physical center of the chromosome. Acrocentric chromosomes have the centromer close to one end of the chromosome producing a very short p arm or satellite structure. The following distinct morphological (structural) features of chromosomes have been established:

- Telomeric regions contain tandemly repeated sequences, aging related
- Centromeric regions contain tandemly repeated sequences
- Nucleolar organizer [genes for ribosomal RNA on p arms (microsatellites) of acrocentric chromosomes]

Genome databases play an increasingly important role in understanding novel genes whose functions have yet to be determined, because a gene's chromosomal location is indicative of its function. Yet chromosome locations, like DNA sequences, are subject to changes (e.g., mutations) from generation to generation. In eukaryotes, rearrangements of chromosomal fragments (homologous recombination and reciprocal crossing over) are an important source of genetic variability among individuals. Rearrangements can influence and alter gene expression in an orderly and programmed fashion. While genetic polymorphism based on chromosomal rearrangements makes individuals genetically unique, the overall genomic contents (the whole of all genes and noncoding portions inherited) *must* remain constant.

Not surprisingly, many of these rearrangement processes cause diseases, an additional motivation to understand the relationship between gene expression and chromosomal morphology. This is reflected in the growing size of genome databases related to medical issues, as well as information sites about inheritable diseases and their relationship to genetics (NIH health information about genes and diseases by chromosome).

Figure 3.3 shows three genes on chromosome 17 associated with hereditary diseases. Two genes are causing cancer when mutated, while the third is involved in a neurodegenerative disease that can result in central and peripheral neurological defects such as loss of motor skills (muscle control) and deafness.

3.1.2 Mapping and Navigating Genomes

A genome map consists of contiguous chromosome sequences annotated by markers that serve as *map elements* and indicate the location of signature elements along the length of a chromosome. Several types of markers or map elements ranging from simple banding patterns (ideograms) obtained from staining metaphase chromosomes to sequence information are used to navigate genomes in a hierarchical fashion from broad overview to single nucleotide sequence. Map elements are derived using physical identifiers from cytogenetic analysis, hybridization, and sequencing and linkage identifiers from determining recombination frequency between hereditary units.

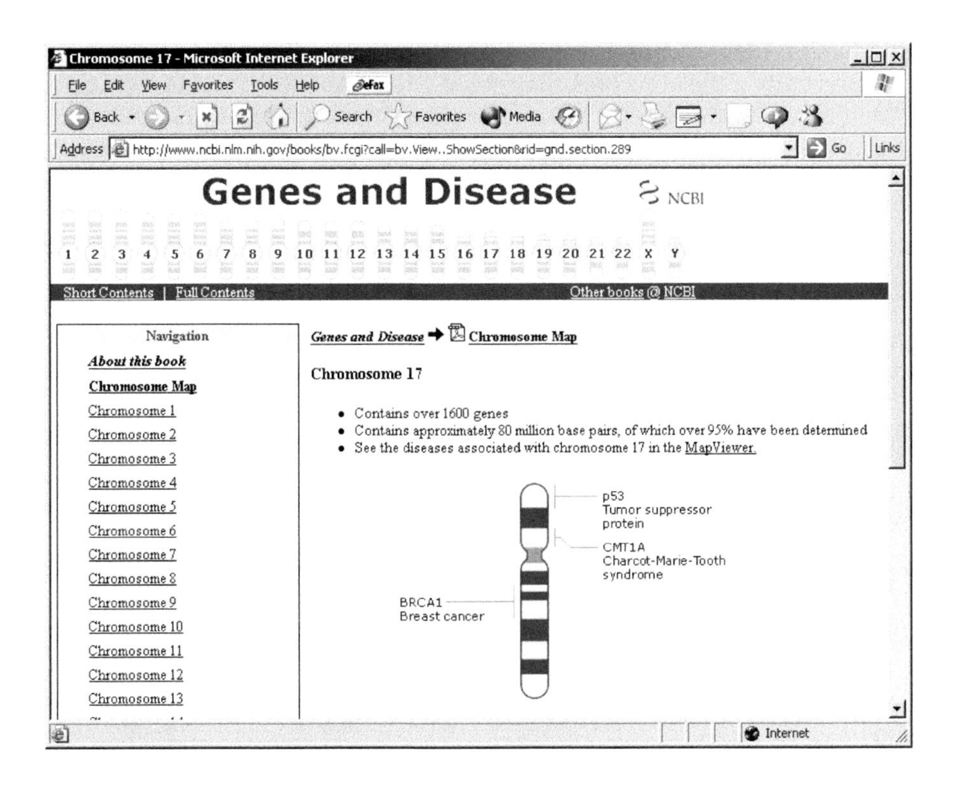

FIGURE 3.3
Hereditary disease marker on the metacentric human chromosome 17. (From www.ncbi.nlm.nih.gov/books/bv.fcgi?rid=gnd.chapter.272. With permission.)

The types of maps found at GenBank and other repositories include sequence, cytogenetic, genetic linkage, radiation hybrid, and yeast artificial chromosome (YAC) contig maps. For instance, the Whitehead-YAC STS content map (sequence tagged sites) includes 10,850 STS markers placed onto 16,494 YACs with an average intermarker distance of 276 kilobases.

Navigating is essentially a top-down approach of localizing segments of chromosomes (length measured in Mb) and genes (length measured in kb) on a chromosomal map with increasing resolution. *Physical maps* indicate the actual physical location on the ideogram of a chromosome arm. *Linkage maps* give information about recombination activity that can be assessed by showing how likely it is that two genetic markers are inherited together or can segregate independently (Mendelian inheritance). Independent segregation is an indication that genes are located on different chromosomes. However, chromosome segments can be swapped (recombined) between two sister chromatids during meiosis. The closer two markers are physically located on a chromosome, the less likely they are to be swapped and segregated during sexual reproduction. Linkage maps thus give an idea about genomic organization of genes derived from a purely functional analysis (population genetics).

Physical maps and linkage maps must show the same order of different markers along the arms of chromosomes. However, the calculated distances can be quite different. The reason for this is the existence of recombination hotspots on chromosomes that results in the observation that certain nearby alleles are more often swapped among sister chromatides than other alleles that are physically separated by an equal distance, but found on different chromosomes or different regions of a chromosome. Physical or cytogenetic maps are measured in base pairs (bp; e.g., Mb, megabases), while linkage maps are measured in centiMorgans (cM). For whole genome sequences, the physical map is the central navigatable element. An overlapping map from cytogenetic and linkage analysis allows identification of recombination hotspots.

Genome maps are navigated starting at the level of the ideogram (Figure 3.4) of chromosomes (low resolution) down to the nucleotide sequence (high resolution). Ideograms serve as the entry points into MapViewers at various Genome Browsers.

The genome information accessible through the Internet is as varied as the people who are interested in particular projects. Many individual sequencing centers provide websites that highlight interests pertaining to a particular disease, animal model, or other organisms (Table 3.1). Some centers focus

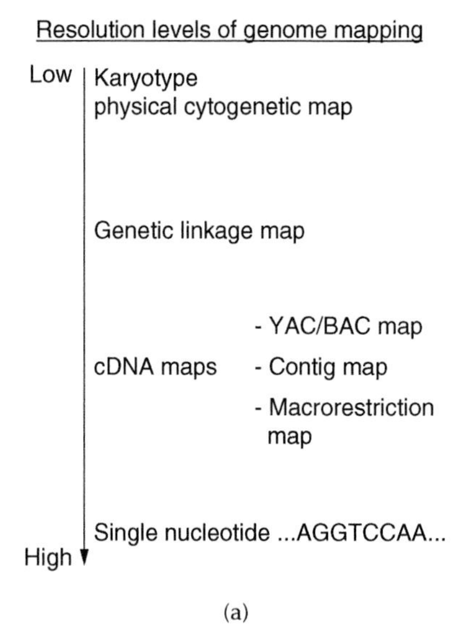

(a)

FIGURE 3.4
Resolution levels of genome maps. The genome projects progress through several levels of increasing resolution of the genetic information contained on chromosomes. The division of the genome in physically individual pieces (chromosomes) is the karyotype. Genetic markers are linked if they are on the same chromosomes (genetic linkage map). The relative position of genes can be determined by correlating physical and genetic linkage maps. Smaller fragments allow the fine resolution of markers located very close to each other on the genome in cDNA maps. Smaller and smaller fragments are amenable to complete sequence analysis at the single nucleotide level.

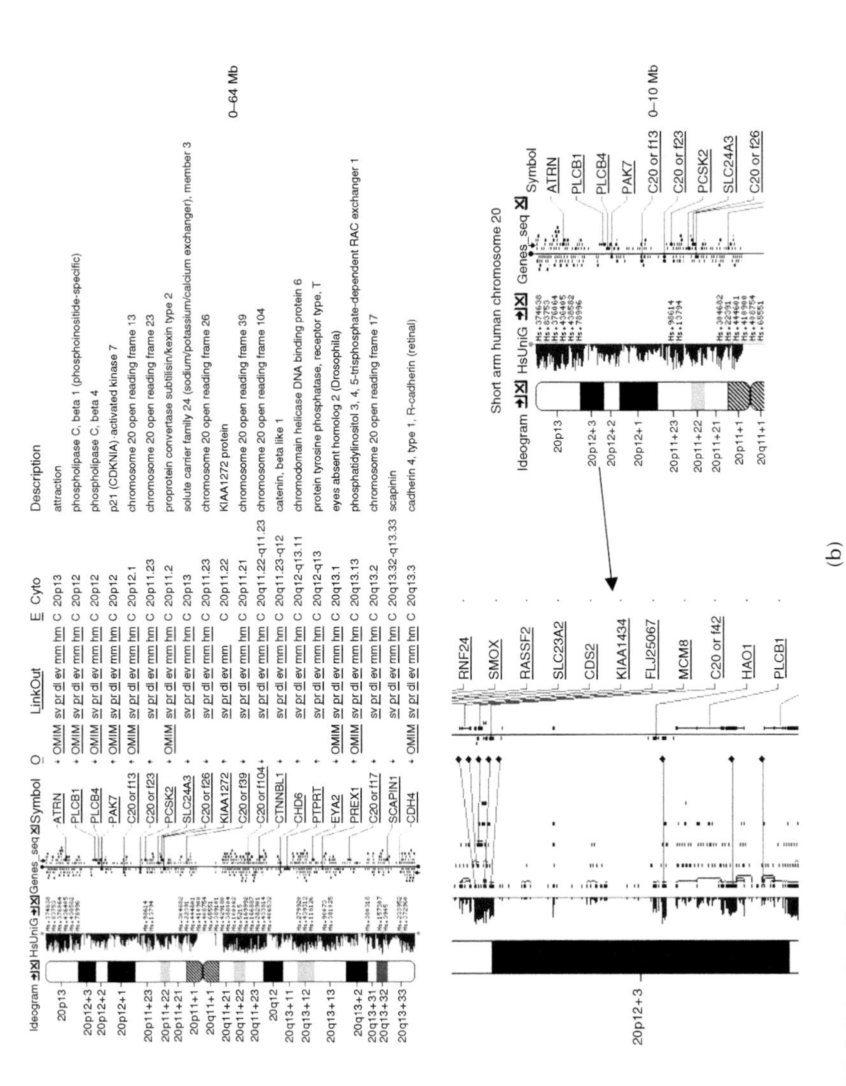

(b)

FIGURE 3.4 *(continued)*

TABLE 3.1

Major Genome Sequencing and Database Centers

Center	Institute
Genome Sequencing	ACGT (Univ. Oklahoma)
	Baylor College of Medicine
	Biotechnology Center, Univ. Illinois (BCUI)
	CBC University of Minnesota
	DOE Joint Genome Institute
	Fruit Fly (Berkeley Drosophila Genome Project)
	Genome Sequencing Center, (GSC) Vancouver, Canada
	Genoscope
	Göttingen Genomics Laboratory (GGL)
	Japan Marine Science and Technology Center (JMSTC)
	Kazuza DNA Research Institute, Japan
	Lawrence Berkeley Laboratory
	Los Alamos National Laboratory
	Lawrence Livermore National Laboratory
	National Human Genome Research Institute (NHGRI)
	NIH Intramural Sequencing Center (NISC)
	National Institute of Technology and Evaluation (NITE), Japan
	Rice Genome Research Program, Japan
	Riken GSC, Japan
	The Sanger Center, UK
	Stanford Genome Technology Center
	TIGR
	Univ. of Cincinnati (UCinn)
	Univ. Minnesota (UMinn)
	Univ. Texas Houston Health Science Center (UTex/HHSC)
	Univ. Utah Genome Center (Utah)
	Univ. Washington Genome Center (UWGC)
	Univ. Wisconsin-Madison Genome Sequencing (UWisc)
	Uppsala Genome Sequencing Lab (GSL)
	Washington Univ. School of Medicine Genome Sequencing Center
	The Whitehead Institute (MIT)
Genome Database	CEPH-Genethon, France
	Cooperative Human Linkage Center (CHLC)
	DNA Data Bank of Japan (DDBJ)
	European Bioinformatics Institute (EMBL)
	FlyBase
	Généthon, France
	The Jackson Laboratory
	Kyoto Enc. of Genes and Genomes (KEGG), Japan
	MaizeDB
	Mosquito Genomics Database
	PlasmoDB
	Nematode Database (Nembase)
	Stanford Genomic Resources (SGR)
	YPD, PombePD, WormPD

Source: From www.ncbi.nlm.nih.gov/genomes/static/links.html. With permission.

TABLE 3.2

Genome Browsers

Genome Browser	WWW	Organisms[a]
Generic Genome Browser (CSHL)	www.wormbase.org/db/seq/gbrowse	Worm, fly, human, yeast
NCBI Map Viewer	www.ncbi.nlm.nih.gov/mapview/	Mammals, fish, invertebrates, plants, fungi, protozoa
Ensembl Genome Browser (EBI and Sanger Institute)	www.ensembl.org	Mammals, birds, fish, invertebrates, yeast
UCSC Genome Browser	http://genome.ucsc.edu/cgi-bin/hgGateway?org=human	Mammals, birds, invertebrates, yeast, SARS
Apollo Genome Browser	www.bdgp.org/annot/apollo	Stand-alone genome annotation viewer and editor

[a] Does not include prokaryotes.

on individual chromosomes of specific organisms rather than the whole genome. Some are interested in mapping entire genomes or developing resources, while others focus on automation, data handling, and analysis. Still others are involved in developing new tools to analyze sequences, compare genomes, study the structure and expression of genes, identify polymorphism, and study chromatin structure in relation to function. All these studies put together will help generate new insights into the biological function of genomes.

However, the three main public databases—the National Center for Biotechnology Information, the European Bioinformatics Institute, and the National Institutes of Genetics in Japan—provide a complete picture of available data (Table 3.2).

3.1.2.1 Genetic Linkage Maps

Genetic linkage maps give the specific location of a gene by measuring "distances" to other genes based on observed recombination frequencies. These maps are used when wanting to obtain an overall view of a large portion of DNA on a chromosome and to select for genes that influence certain traits.[1] Genetic linkage maps depict the relative chromosomal locations of DNA markers (genes and other identifiable DNA sequences) by their patterns of inheritance. That is to say, they answer the question are they inherited always from the same parent or not? A relatively easy way of answering this question is in relation to rare hereditary diseases that allow phenotypic screening (the disease) of a genetic marker (mutated gene). Linkage mapping has been used to find the exact chromosomal location of several important disease genes, including cystic fibrosis, sickle cell disease, Tay-Sachs disease, fragile X syndrome, and myotonic dystrophy.

The distance between markers measured in cM on the map indicates the frequency of how often they are inherited together (inverse relationship). Two markers are 1 cM apart if their recombination rate is 1%. A genetic distance of 1 cM is roughly equal to a physical distance of 1 million bp (1 Mb). Consensus maps of chromosomes have averaged 7 to 10 cM between genetic markers. Radiation hybrid mapping has been a particularly important tool to generate more detailed linkage maps *in vitro* than can be obtained for relying on natural recombination rates. Here, normally associated markers can be separated by inducing artificial chromosome breaks using a high-radiation energy source (x-rays). Assessing the frequency of marker sites remaining together after radiation-induced DNA fragmentation can establish the order and distance between the markers at high resolution (fractions of cM). Because only a single copy of a chromosome is required for an *in vitro* analysis, even nonpolymorphic markers are useful in radiation hybrid mapping.[2]

Links to radiation hybrid maps are listed in Table 3.3 and can be accessed via NCBI's Entrez Genome website. GenBank linkage maps are based on sequence markers coming from three sources. The first is dCODE, the high-resolution genetic map from Iceland's population. This is a straightforward cytogenetic map analysis from a tightly knit human population including many previous generations. The second set of markers comes from the Généthon linkage map constructed on the basis of nearly 1 million genotypes from eight Centre d'Étude du Polymorphisme Humain (CEPH) families using the same family resources to collect linkage data. They incorporated >8000 short tandem-repeat polymorphisms (STRPs), primarily from Généthon, the Cooperative Human Linkage Center, the Utah Marker Development Group, and third, the Marshfield Medical Research Foundation, a human linkage map incorporating >8000 polymorphic markers. (More on CEPH http://locus.umdnj.edu/nigms/ceph/ceph.html.)

TABLE 3.3

Radiation Hybrid Maps at Entrez Genome

Map	Content
GeneMap99-G3	7061 STS markers mapped onto the G3 RH panel by the International Radiation Hybrid Consortium
GeneMap99-GB4	45,758 STS markers mapped onto the GB4 RH panel by the International Radiation Hybrid Consortium
NCBI RH	NCBI Integrated Radiation Hybrid Map contains 23,723 markers from both the G3 and GB4 RH panels of GeneMap 99.
Stanford G3	Includes 11,458 STS markers (both gene-based and nongene-based) mapped onto the G3 RH panel
Stanford TNG	The TNG map includes over 37,000 markers
Whitehead-RH	Includes 6193 STS markers mapped onto the GB4 RH panel

The RH maps described above are static and will not be updated with additional markers.

Source: From Gyapay, G. et al. *Hum. Mol. Genet.*, 1996, 5(3), 339–346.

3.1.2.2 Physical Maps

A physical map, otherwise known as a cytogenetic map, offers a linear alignment of genetic markers along the physical length of a chromosome. It may be identified graphically on ideograms by segment number, e.g., G-banding pattern (see Figure 3.2), associated marker, contig sequence, or clone library including bacterial and yeast artificial chromosome (BAC and YAC) libraries. The cytogenetic maps listed in Table 3.4 can be distinguished based on the methodology of obtaining the sequence information, e.g., derived from genomic DNA fragments (STS and SNPs from restriction mapping) or expressed sequence tags (ESTs) from RNA samples. It should be noted that no map stands alone and together form a complete set of genomic organization.

The different maps are structured to navigate the chromosomes based on various biological information. The basic scaffold of whole genome information comes from clone libraries where individual short fragments have been sequenced and the sequences aligned in the correct order based on overlapping ends. Such contigs are then ordered based on the location of known markers located on ideograms (from which band did the fragment, clone originate?) or determined by linkage analysis (distance in cM). Thus, cytogenetic maps are constructed by ordering individual clone sequences into a contiguous map. Entry points into finding sequences of interest on genome maps come from sequence data including both finished and draft high-throughput genomic sequences (HTGSs). The quality and accuracy of these sequence maps really depends on the assembly process of stringing the large number of individual short sequences into the complete and contiguous genome sequence of an organism. There are several types of sequence maps for each human chromosome, showing various features that have been annotated on the human genome sequence assembly (e.g., contig, component, and gene sequence) as described in Table 3.5.

Physical maps describe the molecular organization of genes or markers within a genome or chromosome. Depending on the technique available or

TABLE 3.4

Cytogenetic Map Elements at Entrez Genome

Map Format	Content
Ideogram	Ideogram of the G-banding pattern at the 850 band resolution
FISHClone	BAC clones that were mapped to cytogenetic bands using fluorescent *in situ* hybridization (FISH) (Note: These clones have also been aligned to the genomic sequence data)
Genes_Cytogenetic	Cytogenetic locations of genes as reported in LocusLink
Mitelman Breakpoint Map	Genome-wide map of chromosomal breakpoints, based on the Mitelman Database of Chromosome Aberrations in Cancer
Morbid	Cytogenetic map locations of disease genes described in OMIM

Note: The genes on all cytogenetic maps are ordered based on cytogenetic bands. At present, order within a band is not being calculated. Completeness of a genome sequence is defined as the percentage of the euchromatin sequence in the physical map (assembly).

TABLE 3.5

Sequence Map Elements at Entrez Genome

Map Elements	Description
Ab initio	mRNA alignments were used to segment the genomic sequence by putative gene boundaries, and Gnomon was executed on these segments to predict genes. Please note that this process predicts exons and not all possible mRNAs, so there is only one model per putative gene.
Assembly	The assembly map allows users to visualize all of the sequence data available for a given region of the genome, and separates the data by assembly. The assembly map also acts as a filter through which all of the other sequence maps are viewed, allowing you to see the annotations that have been placed on the sequence data from each assembly.
BES_Clone	Alignment of BAC end sequences to the assembled genomic sequence. During the alignment process, at least 50% of the BAC end had to align to the genome with >96% identity. (If a BAC end sequence hit two places on the genome with the same high bit score, both of those hits are shown.)
Clone	Localization of FISH mapped clones. These clones were placed onto the genomic sequence by using their clone insert sequences either in finished or draft form. (Note: These clones have also been placed onto the cytogenetic map.)
Component	Components of the human genome assembly. Shows the placement of individual GenBank sequence entries that were used to generate the genomic contigs. (Note: The Component map shows the individual GenBank records used in assembling the contigs [next].)
Contig	Shows the chromosomal placement of contigs that have been assembled at NCBI using finished and draft high-throughput genomic (HTG) sequence data.
CpG Island	Shows regions of high G + C content on the assembled genome sequence (CpG islands; 50% or higher G + C content).
dbSNP_Haplotype	The dbSNP haplotype map displays intervals of contig sequence when there is information in dbSNP to define chromosomal haplotypes (the set of alleles revealed in a particular experiment as a dbSNP Haplotype Set). The dbSNP Haplotype is an annotation of regions of the genome in high linkage disequilibrium. The current data is spotty, mostly focused on chromosome 21. When the International HapMap Project is completed all chromosomes will have linkage measures to show region of strong (or weak) disequilibrium.
Haplotype	The haplotype type map shows alternative reference sequences, or haplotypes, for regions of the genome where humans are polymorphic for the specific sets of genes they carry. Any single chromosome, or chromosome region, might contain data from only a subset of available haplotypes. There is currently no chromosome that contains data from all the haplotypes.
FES_Clone	Alignment of fosmid end sequences to the assembled genomic sequence (df: clone insert into F factor of *Escherichia coli*).

TABLE 3.5
(continued)

Map Elements	Description
GenBank DNA	Shows the placement of human genomic DNA sequences from GenBank that were *not* used in the assembly of contigs. The placement is based on the alignment of the sequences to the components of the contigs. It includes human genomic sequences longer than 500 bp that have at least 97% identity to the components for at least 98 base pairs. If a sequence extends beyond a contig, that portion of sequence is not shown.
Gene_Sequence	Genes that have been annotated on the genomic contigs. This includes known and putative genes placed as a result of alignments of mRNAs to the contigs. If multiple models exist for a single gene, corresponding to splicing variants, the gene_sequence map presents a flattened view of all the exons that can be spliced together in various ways. For example, if one splice variant uses exons 1, 3, 4, and another splice variant uses exons 2, 3, 4, the gene_sequence map shows exons 1, 2, 3, 4. (In comparison, the transcript (RNA) map shows what combinations of exons are valid based on mRNA sequences from RefSeq and GenBank.)
Phenotype	Shows the placement of loci associated with phenotypes on the assembled human genome sequence. Phenotypes include those described in Online Mendelian Inheritance in Man (OMIM), and quantitative trait loci (QTLs). Note: While the OMIM resource itself shows the location of phenotypes (when known) in cytogenetic coordinates, the phenotype map shows the location in sequence coordinates. Thus it is now easier, when querying by a disease name, to know if it has been placed on a sequence map at all. If the phenotype is placed by linkage or association to mapped markers (QTLs), the phenotype is placed by the position of that marker or markers.
SAGE_tag	Map of SAGE tag sequences aligned to the genomic contigs, providing connections to SAGEmap, NCBI's resource for *serial analysis of gene expression* data. (Note: the SAGE technique quantifies a "tag" which represents the transcription product of a gene.)
STS	Placement of sequence tagged sites (STSs) from a variety of sources onto the genomic data using Electronic-PCR (e-PCR). The markers are from RHdb, GDB, GeneMap'99 (gene-based markers), Stanford G3 RH map (both gene and nongene markers), TNG mapWhitehead RH map and YAC maps (both gene and nongene markers), Généthon genetic map, Marshfield genetic map, and several chromosome-specific maps, such as the NHGRI map for chromosome 7 and the Washington University map of chromosome X.
TCAG-RNA (assembly-specific)	Map of annotated transcripts provided by The Center for Applied Genomics (TCAG) at the Hospital for Sick Children on their assembly of chromosome 7.

(continued)

TABLE 3.5
(continued)

Map Elements	Description
Transcript (RNA)	Diagrams of the RNAs that are predicted on the genomic contigs. The transcript map and gene_sequence map are built in the same way, using the same types of evidence, described above. However, the gene_sequence map shows a view of all the exons in a gene, while the transcript map shows the combinations of exons (i.e., *splice variants*) that are valid, based on mRNA sequences.
UniGene Maps	The UniGene maps show mRNA and EST sequences from a given organism aligned to the assembled human genomic sequence. Only ESTs supplied with orientation are used. Each alignment is the single best placement for that sequence in the current build of the human genome. The UniGene maps include mRNAs and EST clusters aligned to the assembled genomes of *Homo sapiens* (human), *Mus musculus* (mouse), *Rattus norvegicus* (rat), *Sus scrofa* (pig), and *Bos taurus* (cow).
EST Maps	The EST maps show mRNA and EST sequences from a given organism aligned to the assembled human genomic sequence. The display for EST maps differs from those labeled as Xx_UniGene in what is displayed here are the alignments and putative introns of ESTs and longer mRNAs best placed at that position. In contrast, the "UniGene" map is a summary of probable splicing events, with connections to UniGene for the clusters that contain those sequences.
Variation	Alignment of genetic variation data from dbSNP onto the genomic sequence.

Source: From www.ncbi.nlm.nih.gov/mapview/static/humansearch.html. With permission.

used, the resolution of the map can vary widely. Early methods relied on microscopic techniques analyzing banding patterns on condensed forms of chromosomes. Bands often correlate to active regions of genomes.

Development of marking chromosomes using DNA probes with fluorescent markers (fluorescence *in situ* hybridization [FISH]) have been particularly useful to establish location and distribution of individual genes and larger genome regions. In a method known as chromosome painting, sequences specific for one chromosome are converted to fluorescence probes (Color Figure 3.5). Probes of various wavelengths (color) are available allowing for specific mixture of dyes to target different sequences. Subsequent hybridization of individual probes (FISH) with colors specific for certain chromosome regions results in specific labeling of a chromosome. Simultaneous hybridization with all probe sets results in a chromosome spread in which each of the chromosomes in a human haploid chromosome set appears a different color when viewed with a fluorescence microscope. This technique is used to detect chromosome translocations, breakages and other anomalies simply by following color patterns and changes among chromosomes. Use of the same chromosome paints for chromosomes of different

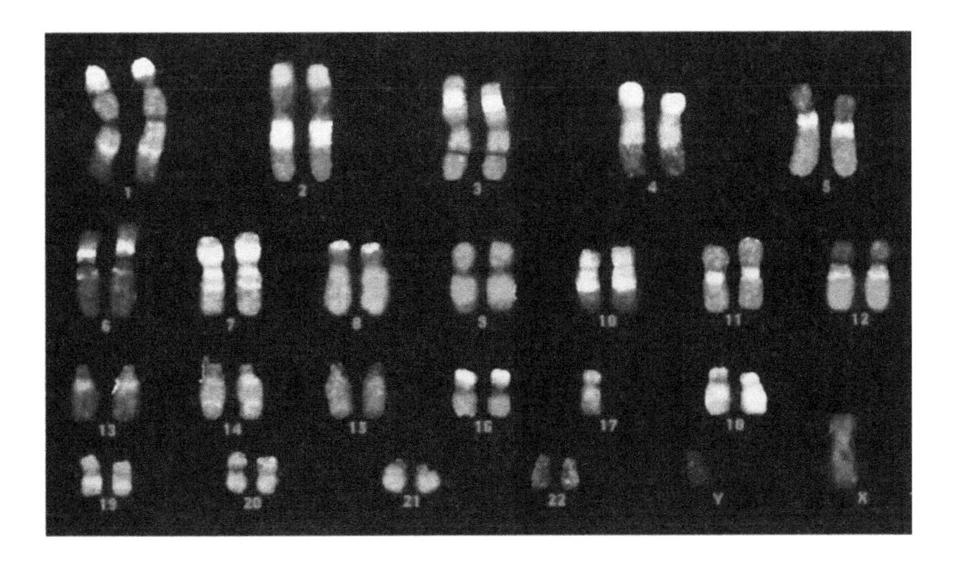

FIGURE 3.5 (See color insert following page 44.)
Fluorescent *in situ* hybridization (FISH) identification of human chromosomes—chromosome painting. (From Image © Applied Imaging, Hylton Park. With permission.)

species reveals the extent of chromosome rearrangements since divergence of the species. Such studies reveal extensive synteny between fairly divergent species. Synteny means that the gene order in chromosomes is conserved over wide evolutionary distances.[3]

High-resolution physical maps (Figure 3.4) make use of the increasingly detailed sequence information available, combining microscopic data with genetic linkage maps and DNA sequences around those markers (genes). The ultimate physical map, then, will be the entire, contiguous DNA sequence of all chromosomes in a genome.

The Sanger Institute, in collaboration with European Bioinformatics Institute (EBI), maintains the Ensembl genome database (www.ensembl.org), allowing access to animal genomes. Ensembl is an automated database that tracks and incorporates all sequences published in major repositories along the known genome structure. It then identifies genetic elements from known sequences and through prediction such as genes, simple repeats, larger homologous elements found in other genomes, and SNP markers. Together, these markers constitute an interactive map that can be accessed to read any high-resolution nucleotide sequence of interest. The Sanger Institute provides the computation of the genome assembly of published sequences, while EBI hosts the genome database and provides public access. The automated genome assembly is based on known maps and single nucleotide polymorphism markers. The interface thus follows the hierarchical organization of chromosomes, where the interested user can zoom into an incrementally detailed map until the nucleotide sequence level of a gene of interest is reached.

The rapid pace of sequencing often results in partial or unfinished sequences. The high-throughput genome division of GenBank (www.ncbi.nlm.nih.gov/HTGS) tries to accommodate this fact and coordinates submission of such fragments not only to GenBank, but also to the Japanese and European depositories. This is a coordinated effort among the three international nucleotide sequence databases: DDBJ, EMBL, and GenBank.

Noncoding DNA is sequenced fragment by fragment through PCR technology. The creation of the dbSTS depository includes specific sequence tags that can be used to uniquely identify chromosome locations (they serve as markers for genes when co-segregating). By using electronic PCR, STS with known chromosomal positions can be searched and compared with new sequences and the genomic position of the latter can be determined. In this way, e-PCR can be used for the creation of various types of genomic maps.

3.1.2.3 From Sequence Maps to Gene Function Maps

Sequencing genomes is driven by the desire to understand how the blueprint of life determines an organism's physical and functional identity. The identification of structural genes precedes the Human Genome Project, while the complete genome information will add insight into the function of genes within the context of an organism's life cycle. Identifying novel genes is often accomplished by analyzing messenger RNA sequences for the construction of so-called expression maps. Most commonly, these maps are based on expressed sequence tags (ESTs) and are included on genome maps as ESTs, UniGene or transcript RNA markers (see Table 3.5). Transcript markers from eukaryotic genes always differ from genomic sequences (e.g., based on STS markers, clones, etc.) because of the exon/intron organization of genes. Transcripts only reflect the exon sequences, which can be expressed in various combinations known as splice variants.

Because genes are composed of both coding and noncoding flanking regions containing regulatory sequences, both ESTs and STSs have been instrumental in creating high-resolution genome maps. The identification of ESTs is a shortcut for identifying human genes since they are derived from active genes. ESTs can be obtained without any knowledge of their function. Since genes are not expressed all the time and often in a cell type-specific manner (as well as being dependent on different states of the development of an organism), the entire life cycle and all physiologically relevant tissues have to be probed for the presence of mRNA and its subsequent sequencing. This approach misses a major portion of a eukaryotic genome, but reveals interesting physiological and medical conditions (see functional genomics).

References

1. Furey, T.S. and Haussler, D., Integration of the cytogenetic map with the draft human genome sequence. *Hum. Mol. Genet.*, 2003, 12(9), 1037–1044.

2. Gyapay, G. et al., A radiation hybrid map of the human genome. *Hum. Mol. Genet.*, 1996, 5(3), 339–346.
3. Melcher, U., Oklahoma State University, http://opbs.okstate.edu/~melcher/MG/MGW1/MG1226.html.

3.1.2 The Genome Projects

Over a thousand genomes have now been sequenced in their entirety, of which 95% are viral, organellar, and bacterial chromosomes and plasmids. Most are very small genomes ranging from several thousand to hundreds of thousands of base pairs. The number of completed larger eukaryotic genomes with base pair numbers ranging in the billions are rapidly closing in. NCBI's Entrez Genome lists several organisms of interest including those of the honey bee, cat, chicken, chimpanzee, cow, dog, frog, mouse, pig, rat, and sheep. In addition several agriculturally important crop plants are also sequenced including tomato, soybean, barley, wheat, corn, oat, cotton, potato, and rice. The list is growing rapidly.

As of summer of 2004, the NCBI Entrez Genome database lists 157 bacterial and 18 archaea genomes, of which 146 are considered complete. In addition, the genomes of 23 eukaryotic organisms from the protozoan, fungi, plant, and animal kingdom are also completed. The first cellular genome (*H. influenzae*) was finished in 1995. Many smaller genomes have been sequenced, however, long before that time and number in the thousands today from viruses (1536) and bacteriophages (224) to plasmids (594) and organellar genomes (559). Some of the viral and organellar genomes can be very small, such as the 2981 base pair long segment S of ϕ-13 (NC 004170), a dsRNA containing cystovirus infecting Pseudomonas bacteria.

As new genome projects are started and old ones completed, the monitoring of sequencing progress is a fascinating activity. Many web sites contain information and links to various sequencing (genome) projects and databases; they all focus on specific programs and are relatively complete. It is still, however, up to the interested scientist to validate the amount and timeliness of data contained in a database.

Monitoring the sequencing progress of a human DNA clone can be found at the Sanger web site (www.sanger.ac.uk/Info/Statistics[1]). The Sanger site also provides links to FTP sites through a FASTA sequence format so that a status summary for a clone or a sequence (i.e., is the protein or gene of interest cloned? Is it homologous to other species?) can be obtained. It can be overwhelming to see how fast the numbers of sequences are added to the many databases. Figure 3.6 shows an overview of Sanger's "Progress Statistics" that can be accessed. The site provides information on unfinished and finished numbers of nucleotides sequenced at the Sanger Center of the British Medical Research Council. Unfinished clones provide updated information on incomplete sequences. This allows rapid access to new genes of potential interest. Such sequence

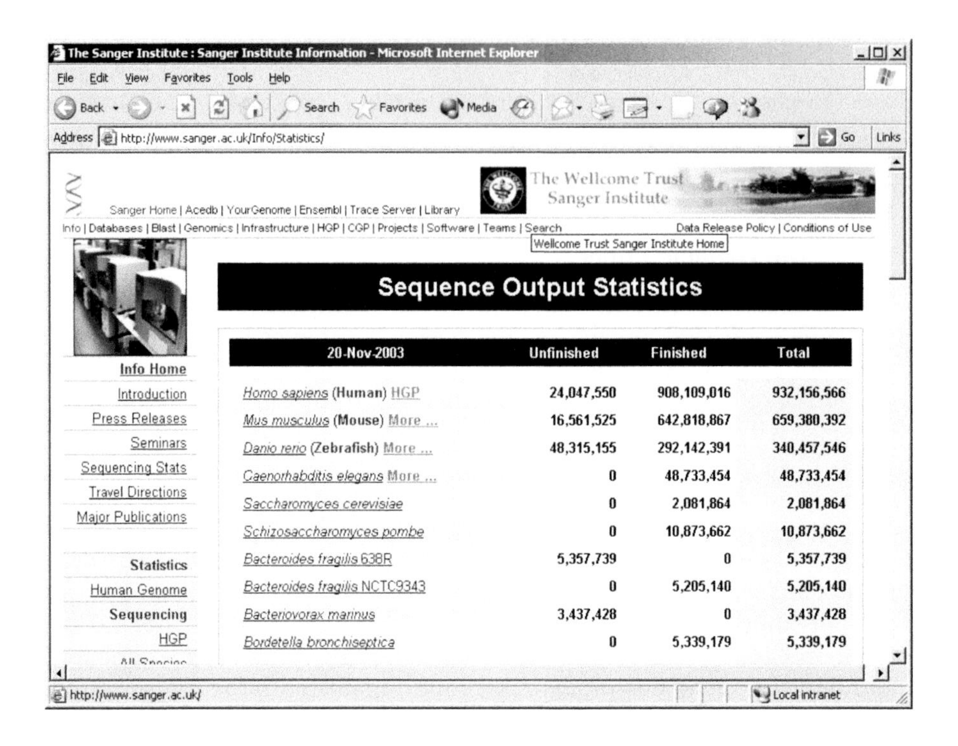

FIGURE 3.6
Genome sequence output statistics at the Sanger Institute.

information must be dealt with cautiously, as it may contain erroneous sequences and, therefore, must be considered unpublished. It is important to note that the clone information pertains only to those obtained at the Sanger Center and does not reflect the total number of sequences for any given organism. Finished clones are annotated and submitted to GenBank, EMBL, and DDBJ (DNA Database of Japan), unfinished ones are not.

Another excellent place to follow the progress in sequencing of whole genomes and from a functional genomics perspective is the *EBI Cogent Database* at the European Bioinformatics Institute (Figure 3.7). It stores complete genome sequences and produces a complete collection of predicted amino acid sequences of proteins as translated from nucleotide sequences of open reading frames of the corresponding genes. While many sequences of proteins are previously known, an average of 40% of all open reading frames code for undescribed proteins. Genomes are only included if complete sequence and taxonomic information on the organism is available to provide a full identifier of the protein sequences included in the database. The identifier includes information about the organism, strain, version of the genome, and a number for the protein sequence.

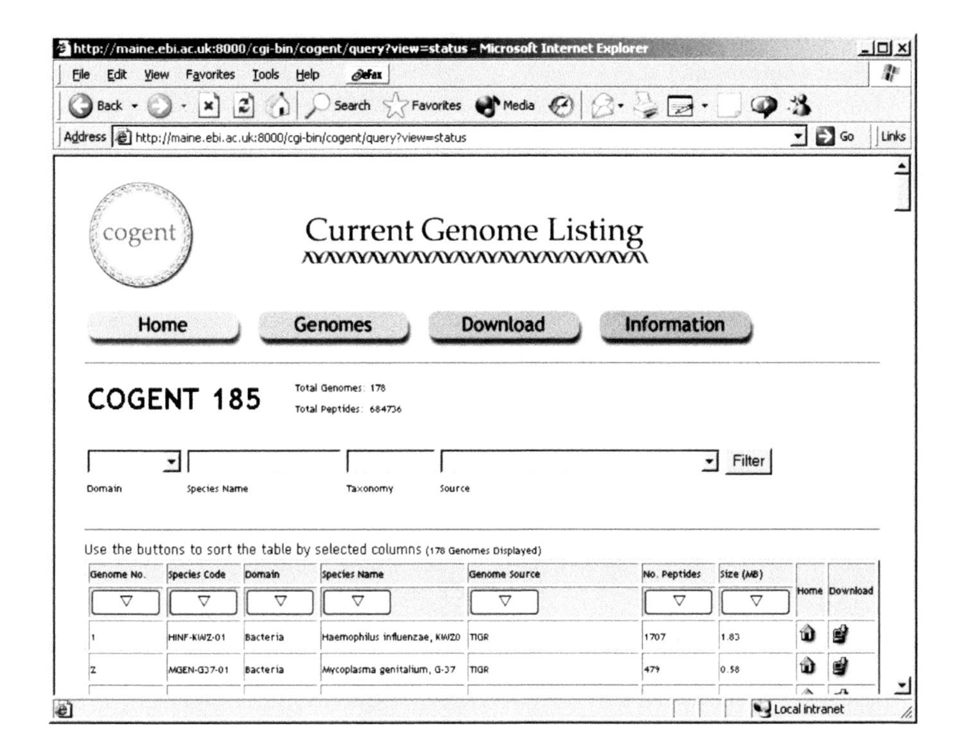

FIGURE 3.7
EBI cogent database cataloging peptides (proteins) from whole genomes. Complete Genome Tracking Database. A database of complete genomes and their protein sequences. http://maine.ebi.ac.uk:8000/services/cogent/. (Janssen, P. et al., *Bioinformatics*, 19 (11), 1451–1452.)

For instance, for the bacterium *H. influenzae*, Strain *KW2*, Version *01*, Peptide *342* the identifier is encoded as:

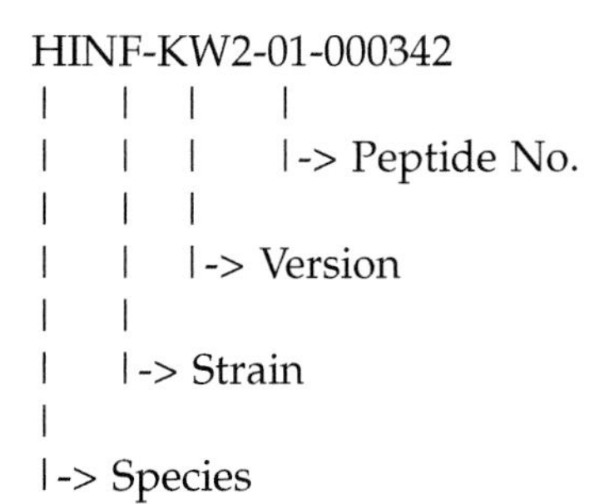

The cogent database at EBI is a curated database and intuitive from a biological point of view. It has been structured with evolutionary information in mind. As only whole genomes are included, the proteins contained in the database allow a direct analysis of orthology (comparing conserved genes among different organisms). More about orthology can be found in

Section 3.2 (Comparative Genomics) on clusters of orthologous groups (COGs) of proteins.

Cogent defines common names of genomes and proteins and is structured using two tables only, one for genome information, the other for protein information. The genome table includes the following fields:

Rel_order (order of publication)
genome_id (timing of sequencing and submission)
fullname (e.g., *Haemophilus influenzae*, Strain KW2)
source (sequencing center)
date_sequenced
species_code (species; strain; order of sequence submitted)
total_genes (number of open reading frames identified)
tax_class (taxonomic classification; e.g., bacteria, proteobacteria …)
source_url (link to genome center or database where genome is located)
size_mb (size or number of nucleotide in sequence)
curator (person who added and annotated the sequence to Cogent)
data_added

The second protein identification table includes the following information:

protein_id (e.g., HINF-KW2-01-000342; see above)
genome_id (see table 1)
old_name
length (in amino acids)
sequence (actual amino acid sequence)
annotation (name of enzyme; similar to Gene Ontology)

An important aspect of Cogent is its use of common naming convention, systematic numbering of genomes and proteins therein. The database is fully curated and offers links to reference databases that contain the original data such as GenBank, Swiss-Prot, or KEGG.

3.1.2.1 How Many Genes are in a Genome?

A recurring challenge stemming from the genome projects is to identify a DNA sequence representing or containing a gene. A gene is a functional unit in the genome of an organism and includes regulatory sequences and a *reading frame* between a *start* and a *stop* codon defining the coding sequence corresponding to the amino acid sequence of a protein or nucleotide sequence of a structural RNA (tRNA, rRNA). The structures of genes come in two major types: (1) those with a continuous reading frame, and (2) those with an interrupted reading frame (exons and introns; all exons together represent the reading frame with the introns being enzymatically cut out at the mRNA level, so-called RNA splicing). The latter are typical for higher organisms (eukaryotes) but can be found in bacterial and archaea at low frequency.

When the human genome project reached the "finished" draft stage in 2001, its most widely known impact was the reevaluation of the number of human genes. Interestingly, estimates since the late 1970s put this number anywhere from 30,000 to 140,000 based on extrapolation data obtained from mRNA analysis, EST sampling, and CpG island density (repetitive sequences with >50% GC content). The lower end of this range estimate appears to be the correct number, as judged from the Human Genome Project that puts the number of genes to less than 25,000 of which some 19,000 are confirmed as functional genes coding for proteins and structural RNAs.[3] This is but twice the number of genes found in the genomes of the fruit fly and roundworm, a finding widely referred to in commentaries on the complexity of the human genome.

Clearly, a central goal of genome sequencing is to build a complete collection of all genes of all known organisms. At NCBI, the *Reference Sequence* (RefSeq; Table 3.6) collection aims at exactly this: to provide a comprehensive, integrated, nonredundant set of sequences, including genomic DNA sequences,

TABLE 3.6

Sample RefSeq Accession Numbers

Accession	Molecule	Method	Note
AP_123456	Protein	Curation	Bacterial proteins. Although protein products may be annotated via prediction, there is some support such as protein homology, identified domains, or more direct experimental support.
NC_123456	Genomic	Curation	Complete genomic molecules including genomes, chromosomes, organelles, plasmids.
NG_123456	Genomic	Curation	Incomplete genomic region; supplied to support the NCBI Genome Annotation pipeline.
NM_123456	mRNA	Curation	Mature RNA (mRNA) protein-coding transcripts.

RefSeq accession numbers can be distinguished from GenBank accessions by their prefix distinct format of [2 characters | underbar]. For example, a RefSeq protein accession is NP_015325.

transcript (RNA) sequences, and protein products, for major research organisms. RefSeq standards serve as the basis for medical and functional studies, as well as studies in evolution and systems biology. The most important aspect of this collection of functional units is to provide a reference for gene identification and characterization, mutation analysis, expression studies, polymorphism discovery, and comparative analyses across species. RefSeqs form the starting point for the functional annotation of some genome sequencing projects, including those of human and mouse. RefSeq Release 5 includes over 977,450 proteins and over 2390 organisms. The release is available by FTP at ftp://ftp.ncbi.nih.gov/refseq/release/. The main features of the RefSeq collection include:

Nonredundancy

Explicitly linked nucleotide and protein sequences

Updates to reflect current knowledge of sequence data and biology

Data validation and format consistency

Distinct accession series

Ongoing curation by NCBI staff and collaborators, with reviewed records indicated

Unlike GenBank entries, which are the original data provided by individual scientists or sequencing centers, RefSeq is a database that is curated and modified as new information is added to public databases referring to the same gene or protein or RNA. GenBank, in contrast, has usually many entries to the same gene or protein provided as raw data and annotation information by different investigators. As a result, because of its integrated content, RefSeq best represents all the available sequence, structural, and functional information from various sources. It is expressly a nonredundant database, i.e., there is one sequence per gene (although alternate transcripts are included if reported in the literature), one sequence per protein, and one structure per protein for each organism where such information exists. Thus orthologs and paralogs (gene families) have as many entries as the family has members. As for any database annotation, RefSeq records are annotated with the level of curation it has undergone; accordingly, records are identified as inferred, predicted, provisional, reviewed, or validated.

But how can we identify a gene? What constitutes an open reading frame (ORF) or an unidentified reading frame (URF)? Identifying such elements based on sequence information alone requires software such as ORF Finder (Figure 3.8) that identifies reading frames by searching for long stretches of sequence between start and stop codons and corresponding splice sites demarcating exon-intron boundaries.

Table 3.7 and Table 3.8 show the most common goals in studying genome structure, the identification of novel genes and their related protein structures, and the major Internet addresses where these programs can be accessed.

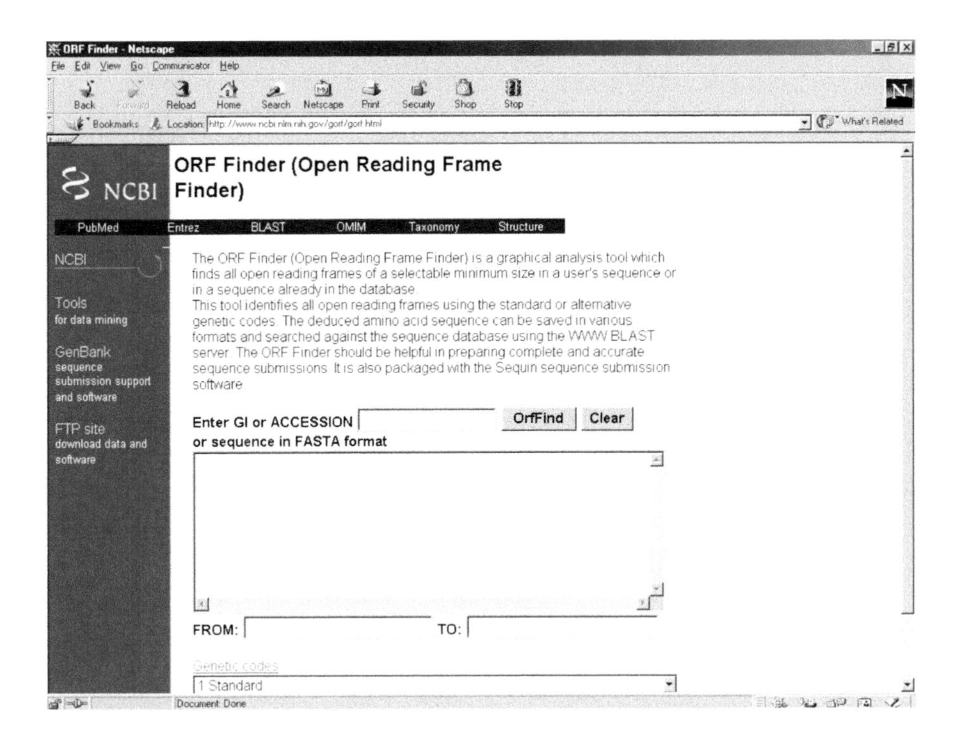

FIGURE 3.8
ORF Finder; entry form. ORF finder Blast version also uses accession number or Fasta format of search string sequence.

Once a reading frame or coding sequence has been identified, what can the sequence tell us about the biological function of a gene? If a genome has been sequenced in the absence of any biological information about much of an organism, no biological function will be associated to most genetic elements. And how do we know how to look for genetic elements? Do we even know what elements to look for? These questions are intrinsic outcomes of genome projects, where long contiguous sequences of DNA have to be analyzed for the presence of genes, coding regions, noncoding regions, recombination hotspots, repetitive elements, and insulators, regulatory elements responsible for large segments of chromosomal activity.

A critical element of coding sequences is their length. The length of the ORF is directly related to the size or molecular weight of the coded protein. Thus studying length distribution of known proteins and their sequences can be a useful indicator for putative reading frames. In eukaryotic genes the signature of splice sites, i.e., the sites delineating exons and introns, are an additional help for identifying a gene. Finally, the presence of transcription consensus sequences close to a start codon need to be found for an ORF to qualify as a functional gene.

Databases provide functional sites to analyze a DNA sequence for the presence of an ORF. They allow the prediction of the corresponding amino

TABLE 3.7

Public Domain Software Analysis Tools for DNA and RNA

Goal	Program	Internet Address
Sequence Similarity	BLASTN TBLASTX BLASTX	www.ncbi.nlm.nih.gov/BLAST/
Finding Open Reading Frames (ORF)	ORF Finder	www.ncbi.nlm.nih.gov/gorf/gorf.html
Finding PCR-Based Sequence Tagged Sites (STSs) in DNA Sequence	Electronic PCR	www.ncbi.nlm.nih.gov/STS/
Translating DNA or RNA → Protein	Translate Transeq	http://au.expasy.org/tools/#translate www.ebi.ac.uk/emboss/transeq/
Comparison of Genomic DNA and Protein Sequence	Wise2	www.ebi.ac.uk/Wise2/index.html
Finding Genes	Gene Recognition and Assembly Internet Link (GRAIL) and PROCRUSTES	http://compbio.ornl.gov/Grail-1.3/ and www-hto.usc.edu/software/procrustes/index.html

acids and potential structural features of the protein. If aligned ORF sequences are found, and if one of them is confirmed as a gene, the comparison of two or more novel sequences is a good predictor for potential biological functions of related open reading frames (inferred function).

TABLE 3.8

Public Domain Software Analysis Tools For Proteins

Goal	Program	Internet Address
Sequence Similarity	BLASTP TBLASTN	www.ncbi.nlm.nih.gov/BLAST/
Automated Structural Modeling	SWISS-MODEL	www.expasy.ch/swissmod/SWISS-MODEL.html
Identification and Characterization	Protein identification and characterization programs	http://au.expasy.org/tools/#proteome
Finding Patterns and Profiles	Pattern and profile search programs	http://au.expasy.org/tools/#pattern
Structure Analysis	Primary structure analysis, secondary and tertiary structure prediction programs	http://au.expasy.org/tools/#primary http://au.expasy.org/tools/#secondary
Sequence Alignment	Sequence alignment programs	http://au.expasy.org/tools/#align
2-D Page Analysis	Melanie II	http://au.expasy.org/melanie/

The ORF Finder is a graphical analysis tool that finds all open reading frames of a selectable minimum size in a user's sequence or in a sequence that is present in the database. This tool identifies all open reading frames using the standard or alternative genetic codes. The deduced amino acid sequence can be saved in various formats and searched against the sequence database using the WWW BLAST server. The ORF Finder should be helpful in preparing complete and accurate sequence submissions. It is also packaged with the Sequin sequence submission software (from www.ncbi.nlm.nih.gov/gorf/gorf.html).

To ensure that novel genes are predicted correctly, one has to take care of applying the appropriate translation code of the organism at hand. NCBI is also providing a genetic code database (http://www.ncbi.nlm.nih.gov/Taxonomy/taxonomyhome.html). The database includes the standard code for all eukaryotic organisms or taxonomic branches with the most common exceptions given below:

The Vertebrate Mitochondrial Code

The Yeast Mitochondrial Code

The Mold, Protozoan, Coelenterate Mitochondrial Code and the Mycoplasma/Spiroplasma Code

The Invertebrate Mitochondrial Code

The Ciliate, Dasycladacean, and Hexamita Nuclear Code

The Echinoderm Mitochondrial Code

The Euplotid Nuclear Code

The Bacterial "Code"

The Alternative Yeast Nuclear Code

The Ascidian Mitochondrial Code

The Flatworm Mitochondrial Code

Blepharisma Nuclear Code

Exons and introns have to be identified based on putative splice sites that conform to consensus sequences.[4] These consensus sequences contain highly conserved dinucleotides at each end of the intron, GT at the 5 end and AG at the 3 end of the intron. Splice site (xx | xx) flanking consensus sequences are MAG | GTRAGT at the 5 splice site and CAG | G at the 3 splice site (where M is A or C, and R is A or G). Many nonconsensus sequences have been described as well. Detailed information can be found at the Danish Center for Biological Sequence Analysis (www.cbs.dtu.dk) and the Alternative Splicing Database (ASD) at EBI (www.ebi.ac.uk/asd).

Some ORF sequences belong to *pseudogenes*, nonfunctional copies of genes that have all or some of the structural features of a gene, but do not express proteins or structural RNA.[5] Pseudogenes can result from faulty duplication and retrotransposition of a gene. The latter inserts cDNA copies of genes into the genome. Such cDNA copies are known as *processed* pseudogenes.

Software for the identification of genes from DNA fragments (sequences) is available at the Baylor College of Medicine (Gene Finder http://searchlauncher.bcm.tmc.edu/seq-search/gene-search.html).

It has become evident from the completed microbial genome projects from bacteria to man that as many as 40% of structural genes are of unknown function. We have no information about them except for the predicted identification of their genomic DNA structure. This means that neither these genes nor their gene products have ever been studied experimentally for their genetic, biochemical, and physiological properties. Therefore, gene identification depends on some rules of gene structures (promoters, start/stop codon, exon/intron structure, protein structure features such as secondary structures, and sites for post-translational modification) through automated analysis. If related sequences or fragments thereof are found in known genes, some biological annotation may be achieved, again through automated prediction. This method of identifying gene structures and predicting the structure and function of gene products provides automated annotation of biological information to novel genes. Some automated annotations, however, increasingly depend on previously automated annotation. As a result, many functions are inferred electronically rather than obtained from experimental evidence. Clearly, the biology behind the sequences, biochemical, genetic, and physiological information about protein structure and function falls far behind. This means that even after the completion of a genome project, years of investigation are needed to understand the full complexity of the organism at the physiological level. One of the first tasks after genomes are completely sequenced is to understand their content and to relate phenotype with genotype. In other words, the task ahead consists of assigning function to sequences based on the organization of genes within genomes and comparing these structures to distantly related genomes.

References

1. Pruitt, K.D., WebWise: Guide to The Sanger Institute's Web Site. *Genome Res.*, 1998, 8(1), 4–8.
2. Janssen, P. et al., COmplete GENome Tracking (COGENT): a flexible data environment for computational genomics. 2003, *Bioinformatics,* 19(11), 1451–1452.
3. Human Genome Sequencing Consortium, I. (2004), Finishing the euchromatic sequence of the human genome. *Nature* 431(7011): 931–945.
4. Zhang, M., Statistical features of human exons and their flanking regions. *Hum. Mol. Genet.*, 1998, 7(5), 919–932.
5. Zhang, Z. and Gerstein, M., Large-scale analysis of pseudogenes in the human genome. 2004, *Curr. Opin. Genet. Dev.*, 14(4), 328–335.

3.1.3 The Human Genome

It was projected in early 1998 that the human genome would be sequenced by the year 2005 based on the available sequencing technology and number

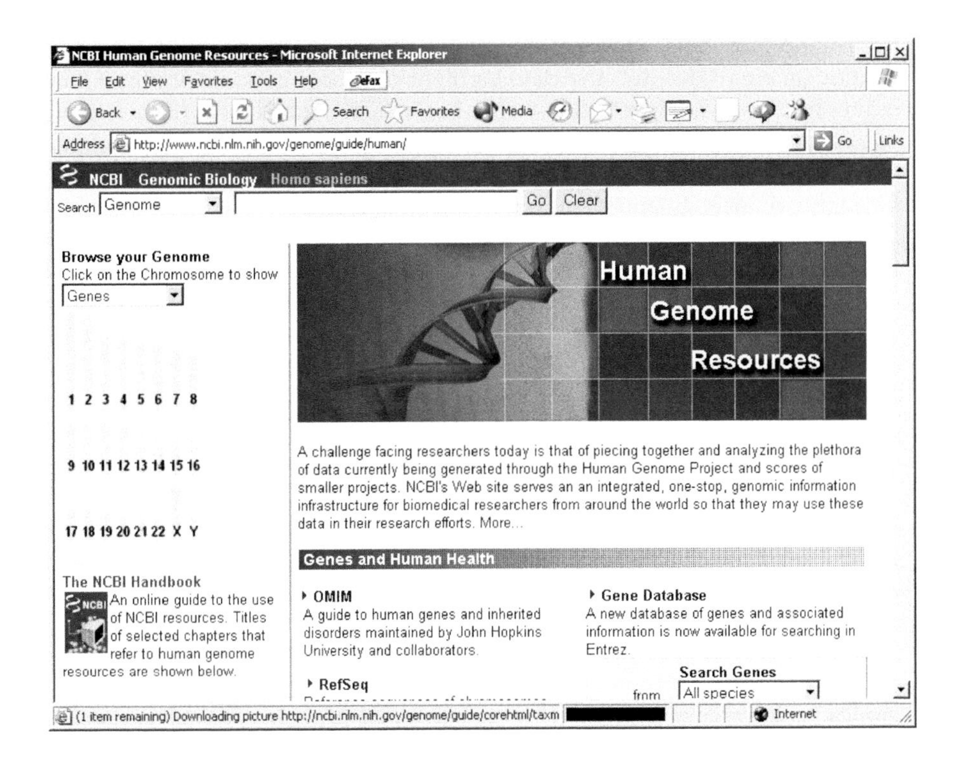

FIGURE 3.9
NCBI's human genome resources.

of base pairs in the entire genome. On June 26, 2000, members of the Human Genome Project announced that they had succeeded in sequencing a "working draft" of the human genome. The DNA sequence of the human genome was made freely accessible to all, for public or private use, from the National Center for Biotechnology Information (NCBI) on February 15, 2001 (Figure 3.9).

The public genome project is the result of the International Radiation Hybrid Mapping Consortium that included many independently operating genome sequencing centers. Some of the larger centers are listed in Table 3.9.

TABLE 3.9
Major Contributors to the International Human Genome Mapping Effort

Genome Center	Country	Internet Address
Généthon	Evrey, France	www.genethon.fr/php/index_us.php
The Sanger Center	Cambridge, UK	www.sanger.ac.uk/
The Stanford Human Genome Center (SHGC)	Stanford, CA	www-shgc.stanford.edu/
The Whitehead Institute for Biomedical Research (WICGR)	Cambridge, MA	www.genome.wi.mit.edu/
The Wellcome Trust Centre for Human Genetics (WTCHG)	Oxford, UK	www.well.ox.ac.uk/

Information on this consortium is available at NCBI at www.ncbi.nlm.nih.govgenemap98/. Yet, the development of whole-genome shotgun techniques by Craig Venter at Celera Genomics, which bypassed the more detailed but cumbersome clone-by-clone approach, combined with the implementation of new hardware and software for sequencing and genome assembly through privately financed means, were critical for the accelerated pace of finishing the human genome. While Venter's effort was performed by Celera and financially backed by PerkinElmer, the public genome project was an international collaboration among academic institutions. Since the finishing of the Human Genome Project, some of the major contributing centers to the public project have shifted their focus. The Stanford Human Genome Center and Généthon are now doing research on post genomics and proteomics projects. Many other centers are still maintaining public databases and are involved in the sequencing of various other genomes.

Since the first draft sequences were announced in 2000 and published in 2001,[1,2] the complete sequences of 11 chromosomes have been published leaving the completion of the Human Genome Project half done. By March 2005, chromosomes 6, 7, 9, 10, 13, 14, 19, 20, 21, 22, Y, and X have been annotated and sequenced with an accuracy of 99.99% or better ("Bermuda Standards"). This means that the sequence data accessible through public databases contain in average less than one wrong base pair in 10,000. A quality assessment performed by the Stanford Human Genome Center in May 2004 concluded that submitted sequences exceed the Bermuda Standards by tenfold, with less than one error per 100,000 base pairs. General rules of what constitute a "finished sequence" can be found at the National Human Genome Research Institute's *genome.gov* web site. They include that

> ... *all regions are double stranded, sequenced with an alternate chemistry, or covered by high quality data, that an attempt was made to resolve all sequencing problems, such as compressions and repeats and that all regions were covered by at least one plasmid subclone or more than one M13 subclone; and the assembly was confirmed by restriction digest* (Source: www.genome.gov/10001812).

The human genome resources at NCBI (Figure 3.9) allow complete access to varied information regarding the human genome as summarized in Table 3.10. These resources include information on genes and human health, the nucleotide genome and clone registry, navigatable maps based and different marker types (physical and linkage maps), transcribed sequences allowing a functional view of the genome, cytogenetics and comparative genomics.

Even at its initial draft stage, the genome sequence has produced some surprises about what we did and did not know about human genetics and genome structure. From this first global perspective, the following aspects about our own genome have emerged:

The genomic landscape is not even, but shows marked variations for the distribution of many genetic elements including genes and gene clusters, transposable elements, GC content, CpG islands, and

TABLE 3.10

NCBI Human Genome Resources

Genome Property	Database Tool
Genes and human health	Omim
	Gene Database
	Locuslink
	Refseq
	Dbsnp
The genome sequence	Blast The Genome
	Clone Registry
Map and markers	Mapviewers / Physical Maps / Genetic Maps (Linkage)
	Unists
	Electronic PCR
Transcribed sequences	Unigene (Transcribed Sequences Ests Into Gene Based Clusters)
	Full Length Cdnas
	GEO
	SAGE Map
Cytogenetics	Human BAC Resources (FISH Mapped Sequence Tagged BAC Clones)
	SKY/CGH (Spectral Karyotyping/ Comparative Genomics Hybridization)
Comparative genomics	Homologene
	Homology Map
	Tax Plot (Three Way View Of Genome Similarities) Human/ Mouse/Rat Or Human/Mouse/Fly

Source: From www.ncbi.nlm.nih.gov/genome/guide/human/. With permission.

recombination rates. 1.1% of the genome DNA contains exons, 24% introns, and 75% is intergenic.

There appear to be about 25,000 genes. Of these, about 19,000 are identified based on corroborating evidence, while 6,000 are identified computationally only.

A richer domain architecture of vertebrates as compared to invertebrates results in a more complex proteome.

About half the genome derives from transposable elements with a marked decline in their activity in the hominid lineage of vertebrates.

Segmental duplications (segment size 10 to 300 kb) are much more frequent in humans than yeast, fruit fly, or worm, particularly in the pericentromeric and subtelomeric regions of chromosomes.

A preferential retention of Alu elements is found in GC-rich regions.

The mutation rate during meiosis is twice as high in males than in females.

The cytogenetic analysis confirms that large GC-poor regions correlate with dark G-bands in karyotyping.

Recombination rates (crossing over) occur at higher frequency in distal regions and short arms of chromosomes and produce in average one recombination per chromosome per meiotic event.

More than 1.4 million SNPs have been identified that will greatly help in measuring linkage disequilibrium mapping of genes (alleles) in human populations.

The human genome has shown an important insight into the distribution and relative size of structural genes and noncoding regions. The latter dominate the human genome, accounting for more than 95% of the sequence. Of this, about a fourth are noncoding regions of genes such as introns and regulatory elements. The rest of the genome has nondescriptive sequences and repeat sequences, of which half belong to one of four transposon-derived elements. These elements are self-replicating units that make copies of themselves randomly, inserting the new copy within the genome. The four transposable elements are long and short interspersed elements (LINEs and SINEs), long terminal repeats (LTR) retrotransposons, and DNA transposons. The LINE machinery is recognized as the most active reverse transposition and responsible for the creation of processed pseudogenes. Unlike pseudogenes, the processed pseudo genes make RNA transcripts but have no translational activity.

The structure of LTR retrotransposons is interesting as it suggests a direct relationship to retroviruses including the human immunodeficiency virus (HIV). These exogenous viruses seem to have arisen from acquiring a cellular envelope gene (env) coding for the coat proteins they use to exit cells and bind to new host cells. Thus transposable elements in form of viruses can breach the cellular envelope barrier, including the occasional species barrier. Accordingly, the existence of this large number of transposable elements in the genome indicates that they may play an important role in the evolution of hominids. Exogenous viruses become agents of horizontal gene transfer mechanisms in the evolution of higher organisms. Such models are consistent with the observation that transposable elements are found in much higher proportions in the euchromatin portion (regions of transcriptional activity) of the humans than other organisms. They also indicate the potential role transposable elements in contributing and maintaining a certain level of genomic complexity needed to explain the observed proteome complexity in humans. Forty-seven genes have been identified that are likely the result of DNA transposon activity, including the recombinase genes RAG1 and RAG2 and the centromere binding protein CENPB. The idea is that transposition events occasionally cause gene duplications, a first step in evolving novel genes from common precursors. Most common transpositions, however, result in repeats of short, nonfunctional sequences.

Although repeats are common, some chromosomal regions are nearly devoid of them. These include the homeobox (HOX) gene cluster (important for developmental programs of body segmentation, e.g., head and abdomen) suggesting that certain regulatory elements do not tolerate random disruptions, even in noncoding regions.[3] The latter are highly conserved in this particular gene cluster, not just the exon sequences. The gene order of HOX genes is highly conserved, and their order seems to reflect the order of expression and order of segments they induce. Thus conserved body segmentation in animals maintains conserved HOX gene order at the level of the genome. Any disruption of HOX gene clusters would thus disrupt body segmentation.

In addition to transposons, 3% of the genome contains simple sequence repeats (SSRs) with short repeat units (1 to 13 nucleotides; micro satellites) or long repeat units (14 to 500 nucleotides; minisatellites). The length distribution of SSRs differs from individual to individual constituting a form of length polymorphism in a human population. Polymorphisms are important tools in human genetics. The dinucleotide SSR $(CA)_n$ has been used as marker in most disease mapping studies. It should be noted here that the 3% figure for SSRs is no small number, as it constitutes a larger fraction of the genome than the amount of protein coding sequences.

Duplications of much larger chromosomal segments are also frequently found. These *segmental duplications* (and deletions) account for about 3.3% of the genome sequence. Such segments range from 10 to 50 kb in length and are thought to contribute to exon shuffling. Since exons often correlate with functional domains at the protein level, segmental duplication may be an evolutionary mechanism to generate more functional complexity by providing for random domain recombination within proteins. Segmental duplications have also become an important measure of evolutionary studies through comparative genomics. They tend to be conserved in closely related organisms and the degree of conservation is of course a measure of divergence from common ancestors.

These mechanisms of genomic complexity associated to the activity of noncoding sequences have been identified as an important source of information to understand the complexity of higher organisms including humans. For instance, the human genome contains more genes, domain and protein families, paralogs, and multi-domain proteins than any other genome so far sequenced. Thus complexity is not just a function of larger size, but increased opportunity for innovation. Copying and shuffling many, albeit small, exons within a large noncoding genome seems just to be such a recipe for complexity.

References

1. Lander, E.S. et al., Initial sequencing and analysis of the human genome. *Nature,* 2001, 409(6822), 860–921.
2. Venter, J.C. et al., The sequence of the human genome. *Science*, 2001, 291(5507), 1304–1351.
3. Seo, H.C. et al., HOX cluster disintegration with persistent anteroposterior order of expression in Oikopleura dioica. *Nature,* 2004, 431(7004), 67–71.

3.2 Comparative Genomics

The function of an open reading frame (coding sequence) in complete genomes is often unknown. The "meaning" of a novel gene can be deduced by looking at its chromosomal position and gene order, or within the context of global gene expression activity. This process of associating gene function to context is similar to learning a new word by inferring its meaning from the context and its position within a sentence. Thus, comparing related sequences for their chromosomal location and order within related organism can often help predict their putative structure and function. With an estimated 30 to 40% of unknown genes in completed genomes, this is an enormous task, not least because of the very large amount of sequence information that has to be analyzed. Comparative genomics examines the number, structure, and function of genetic elements based on the similarity of genomes of various organisms.

Comparing genomes yields a plethora of information about a species because of evolutionary relatedness and multiple conserved and changed features. Besides conserved gene sequences (coding sequences), many non-coding sequences are conserved during evolution such as regulatory elements, exon/intron arrangements, gene clusters, and large chromosomal segments, deletions, duplications, and other rearrangements.

3.2.1 Cluster of Orthologous Groups (COGs)

An important part of comparative genomics is matching the complete set, not necessarily their order, of conserved genes of closely and distantly related organisms. This matching is done by comparing the sequence of every gene or protein of one organisms with every other gene of multiple other organisms and sorting them by best scores. These proteins are called homologs if found in different organisms and paralogs if found in the same organism. Sequence comparison of genes of distantly related organisms fall into three major classes. The first class comprises highly similar sequences coding for proteins with identical structure. They are evolutionarily related and are considered descendants of a common ancestral gene. Some of the largest gene families showing universal conservation (across eukarya, bacteria, and archaea) code for proteins involved in cell replication and information-storage and retrieval processes. The second class includes proteins with similar structure, but dissimilar sequence. Their relationship is inferred from functional studies as they lack significant sequence similarity. They may or may not be related evolutionarily and may be examples of convergent evolution. Examples are gap junction (GJ) channels. In invertebrates these channels are composed of innexin proteins, whereas in vertebrates they are composed from an unrelated family of proteins called connexins. Although these protein families show no sequence homology, they do have a similar predicted topology of four transmembrane domains and intracellular N and C termini.[1] The third class shows no similarity for sequence or structure. Two alternate systems have evolved independently to serve the same cellular function. Well-described examples are the bacterial subtilisin family of serine proteases and animal chymotrypsin family of serine proteases. They share a conserved active site structure (e.g., Asp-His-Ser signature motif). Their overall structures (tertiary structure) and amino acid sequences, however, bear no relation except for their size.

Analysis of each of these three classes requires different approaches, and these differences advance our understanding of protein evolution. First, many protein homologs show little overall sequence similarity; however, they usually contain locally conserved patterns or motifs. Local sequence patterns may code for ligand-binding pockets or transmembrane-spanning segments, and are valuable in assessing evolutionary relatedness. Often, structural features are better conserved than DNA or amino acid sequences of entire genes or proteins and functional domain structures can include amino acids from different locations within the sequence. This relationship conceptually allows us to search for conserved patterns in sequences, and it is those patterns that are significant in establishing evolutionary relationships among genes. Even when the full-length sequence shows little or no similarity to other proteins, functional domains show a high degree of structural conservation. Because only selected stretches of fragments of genes show high similarity, these patterns can point out evolutionary mechanisms like gene duplication and recombination events leading to "chimeric" structures. For instance, nucleotide (ATP)-binding pockets are widespread and highly conserved domains with recognizable signature sequences. Identifying signature sequences often is a first step in functionally annotating novel gene sequences. ATP-binding pockets, for instance, can be found in enzymes and transporters with widely diverse function. ATP-binding domains (sometimes called "cassettes") serve as universal regulatory or energy donor sites, and are often found on domains linked to active site domains of proteins as diverse as ion pumps, ABC transporters, dehydrogenases, and kinases.

Second, the most difficult genes to identify are those for which no sequence or motif similarity can be detected in sequence databases. The genome projects completed thus far have identified approximately 30 to 40% of completely new and unidentified gene sequences. Their coding sequences are referred to as unidentified reading frames (URFs), rather than open reading frames (ORFs), since no biological information can be associated to them. Predictive tools (see Section 4.3) usually fail in the absence of homologous sequences and motifs with known function. Such putative genes thus must be coding for proteins that have never been described biochemically, physiologically, genetically, structurally, or in relationship to a disease.

Secondary structure prediction algorithms are of tremendous help here, but they lack predictive power with regards to protein function. There still exists a substantial gap in knowledge of relating a structure to its function.[2] Many structure prediction methods are statistical methods and rely on information obtained from known structures. The limited sample size (i.e., the number of known structures) limits the accuracy of predicting folds.

Comparative genomics is a new tool that is filling an important need in predictive biology. This bioinformatics tool can be used to explore functions for novel genes by comparing the "behavior" of two or more genomes. In other words, the function of a gene can be inferred from associating it to the functionality of genes colocalized in the same genomic region in one or more organisms. Take as an example the enzymes that link amino acids with a class of small RNA molecules called transfer- or tRNAs.[3] In all mammalian organisms there are at least 20 different enzymes (amino acyl tRNA synthetases), one each for each of the 20 amino acids used to synthesize proteins (a universal mechanism). In the completed genome project of the archaea *Methanococcus jannaschii* (NCBI Entrez genome; NC_000909 complete genome gi|15668172) four of these amino acyl tRNA synthetases have no corresponding gene. Of course, they must exist, since all tRNAs are linked properly with their respective amino acids.

One possibility explaining the lack of genes is to assume a totally new mechanism for amino acyl tRNA synthesis: chemical modification of amino acids linked to tRNA molecules. However, a more mundane explanation is at hand. It has been demonstrated by functional cloning that a novel synthetase responsible for linking the amino acid lysine to its tRNA partner has a *sequence totally unrelated* to any known lysyl-tRNA syntheses and that this enzyme is coded for by one of the unidentified genes in the genome of this archaea. This novel protein represents an entirely new family of lysyl-tRNA synthetases. The conclusion is that two structurally entirely unrelated proteins, as judged on the bases of their DNA sequence, perform the same enzymatic activity.

The existence of evolutionarily unrelated but functionally similar proteins demonstrates that in the complete absence of functional data, the interpretation of what kind of protein DNA sequences code for can be difficult and is often impossible. The use of data from functional genomics efforts is a necessary next step (Section 3.3) to assign a biological function to URFs. Understanding the relationship between structure (sequence) and function of the amino acyl tRNA synthetases requires a great deal of familiarity with the topic.

Third, finding functional relationships between genes across taxonomic groups (orthologs) as well as within a population or the same organism (paralogs) provides the true potential of genome projects. Finding orthologs and paralogs is an effort to use databases to generate new information by linking amino acid sequence information from various *complete* genomes. A cluster of orthologous groups (COG; Figure 3.10) is a phylogenetic classification of *homologous* proteins encoded in complete genomes. Each COG consists of individual proteins or groups of paralogs from at least three lineages and thus corresponds to an ancient conserved domain.

Any two proteins from different lineages that belong to the same COG are orthologs, according to the functional definition used by NCBI to construct

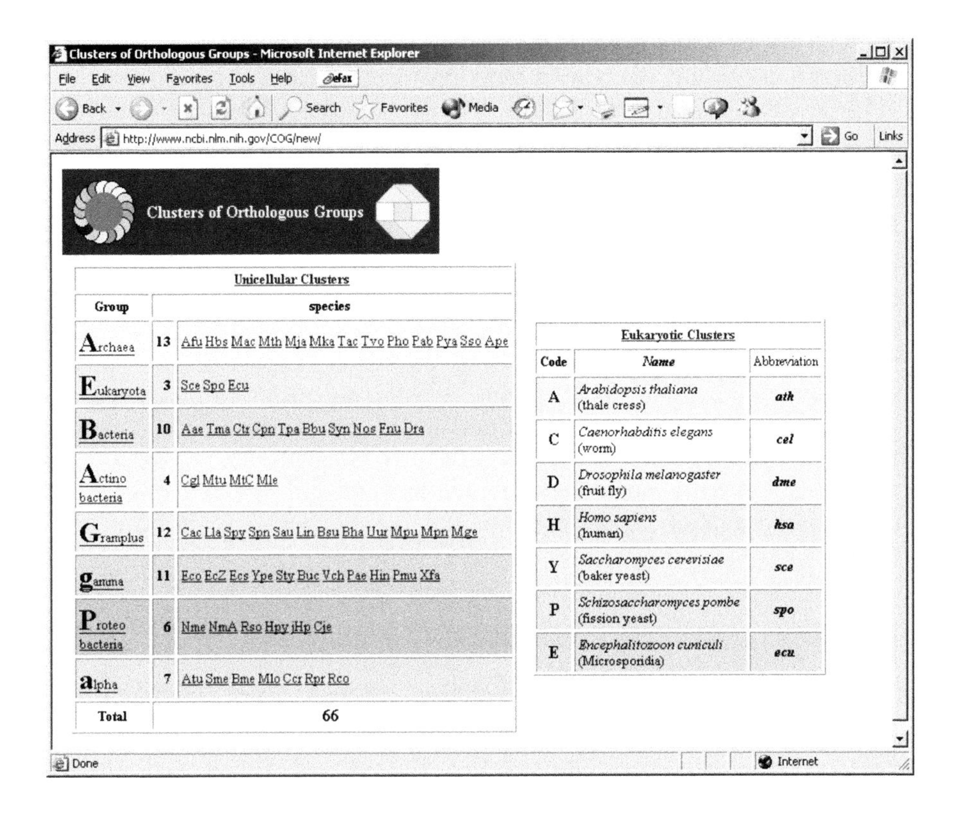

FIGURE 3.10
Cluster of orthologous groups (COGs). A natural system of gene families from complete genomes comparing protein sequences encoded in complete genomes representing seventy phylogenetic lineages (www.ncbi.nlm.nih.gov/COG/new) delineated clusters of orthologous groups (updated version).

COGs, and are assumed to have evolved through speciation. COGs also contain paralogs, which arise through gene duplication events within the same organism.

An initial COG database was built using the first six lineages representing all three domains of life. The current version of the COG database contains information extracted from 70 species. The database is divided into two independent clusters representing unicellular (COG) and multicellular (KOG) organisms, respectively. COG is composed of unicellular organisms including bacteria, archaea, and three eukaryotic single cell species, two fungi and one protozoan. From the 66 species included, a total of 4873 COGs have been identified containing proteins belonging to information storage and processing (groups J, A, K, L, B; 731 COGs), cellular processes (groups D, Y, V, T, M, N, Z, W, U, O; 834 COGs), metabolism (groups C, G, E, F, H, I, P, Q; 1,374 COGs), and predicted or unknown functions (groups R, S). The group of poorly defined proteins (S) comprises a total of 1347 COGs with 702 proteins and domains associated to predicted general function (R). The analysis of COGs allows for the understanding of evolutionary relationships and the identification of related functions across taxonomic divisions.

The eukaryotic cluster includes 4852 KOGs (note the K for euKaryotic cluster of orthologous groups) based on seven species including the three single celled fungi and protozoan from the unicellular group. The eukaryotic clusters allow an interesting analysis of species patterns, i.e., KOGs that include proteins from a subset of the seven species as shown in Figure 3.11. For instance, an animal specific pattern CDH (for species key see Figure 3.11) including the roundworm *C. elegans*, the fruit fly *D. melanogaster*, and *H. sapiens* is composed of 1147 KOGs the largest such cluster. Proteins found in these KOGs are likely animal specific. The second largest pattern contains 928 KOGs represented in all species except the protozoan *Microsporidia*. Patterns that include homologs found in the protozoan tend to have few KOGs and few proteins within each KOG (orthologs). An exception is the "universal" pattern including all seven species with 860 universal KOGs containing a total of 16,377 proteins.

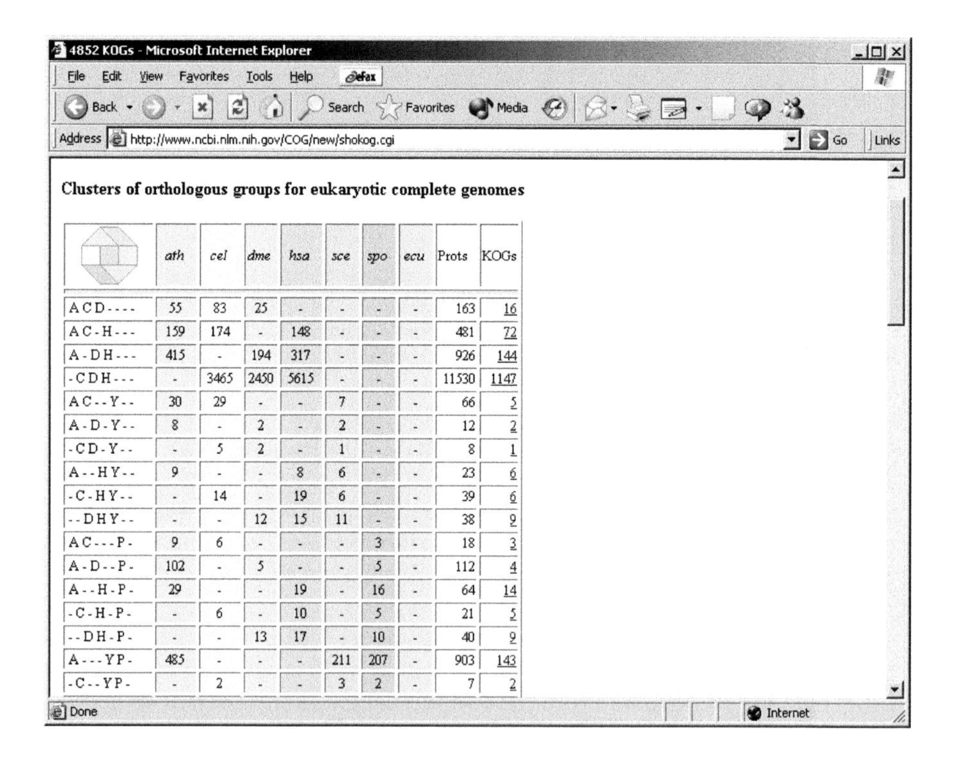

Clusters of orthologous groups for eukaryotic complete genomes

	ath	*cel*	*dme*	*hsa*	*sce*	*spo*	*ecu*	Prots	KOGs
ACD----	55	83	25	-	-	-	-	163	16
AC-H---	159	174	-	148	-	-	-	481	72
A-DH---	415	-	194	317	-	-	-	926	144
-CDH---	-	3465	2450	5615	-	-	-	11530	1147
AC--Y--	30	29	-	-	7	-	-	66	5
A-D-Y--	8	-	2	-	2	-	-	12	2
-CD-Y--	-	5	2	-	1	-	-	8	1
A--HY--	9	-	-	8	6	-	-	23	6
-C-HY--	-	14	-	19	6	-	-	39	6
--DHY--	-	-	12	15	11	-	-	38	9
AC---P-	9	6	-	-	-	3	-	18	3
A-D--P-	102	-	5	-	-	5	-	112	4
A--H-P-	29	-	-	19	-	16	-	64	14
-C-H-P-	-	6	-	10	-	5	-	21	5
--DH-P-	-	-	13	17	-	10	-	40	9
A---YP-	485	-	-	-	211	207	-	903	143
-C--YP-	-	2	-	-	3	2	-	7	2

FIGURE 3.11
Phylogenetic patterns in KOG. Species Key: A *Arabidopsis thaliana* (thale cress) ath; C *Caenorhabditis elegans* (worm) cel; D *Drosophila melanogaster* (fruit fly) dme; H *Homo sapiens* (human) hsa; Y *Saccharomyces cerevisiae* (baker yeast) sce; P *Schizosaccharomyces pombe* (fission yeast) spo; E *Encephalitozoon cuniculi* (Microsporidia) ecu.

This analysis, based on a mere seven species, indicates that the *Microsporidia* is a very different organism and that protozoan are very different from fungi, plants, and animals. It would be interesting to analyze the cluster of orthologous groups combining the unicellular and eukaryotic group, or add a prokaryotic group excluding the eukaryotic unicellular species. Expanding the groups with organisms representing various kingdoms will be uniquely important for evolutionary analysis of the genomic data. For instance, the fourth largest cluster with 484 KOGs in the eukaryotic group belongs to the pattern ACDH including the four multicellular species, but excluding the three unicellular ones. Thus, the proteins included in this pattern are likely important for multicellular functions. However, the small number of species included in the eukaryotic group make these interpretations unreliable due to high percentage of false positive and negatives.

It is important to understand how exactly COGs are calculated. First, sequence comparison is done at the level of amino acid sequences rather than DNA sequences. Second, the decision of including proteins in a COG is based on reciprocal best scores (matches) using Blast. Third, a COG must contain at least one ortholog from three distinct species. The phylogenetic relationship among signal peptidases 1 (COG0691) of distantly related organisms may serve as an example of the structure of the COG database (Figure 3.12). This COG

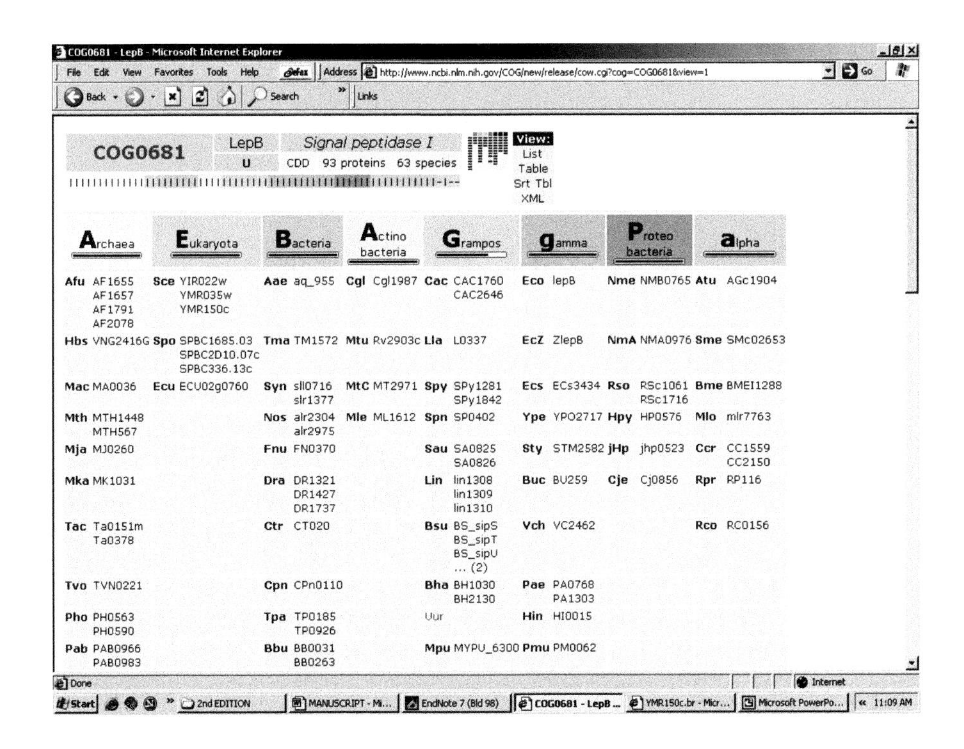

FIGURE 3.12
COG0681 for signal peptidase I.

contains 93 peptidases from 63 different species including 13 archaeal, 3 eukaryotic, and 47 bacterial species. An analysis of best hits for the yeast proteases with bacterial orthologs shows a close relationship of yeast mitochondrial proteases with bacterial orthologs and the relationship of the endoplasmic reticulum (ER) protease paralog with the archaea ortholog of *Methanococcus jannaschii*. This is consistent with the proposed common ancestral single-cell organism of eubacteria and mitochondria (endosymbiotic theory). That the *M. jannaschii* protease shows an orthologous relationship to both yeast and cyanobacteria, stresses the taxonomic classification of archaea as different from both eukaryotes and eubacteria. This COG analysis indicates that of the three yeast peptidases, the two mitochondrial subtypes are true paralogs with a common eubacterial origin, whereas the ER subtype evolved independently or originates from an older ancestral gene that precedes the split into eubacteria and archaea. This split may be the event behind the paralogs in *Synechocystis* sp.

3.2.2 Homologene at NCBI

While COG is a database with protein orthologs from select species with complete genome information, *HomoloGene* is a database of precalculated homologous genes and proteins from the entire Entrez database. Unlike COG, HomoloGene includes proteins from incomplete genomes. HomoloGene thus is a more general approach to identifying related genes/proteins/domains in other organisms grouped by phylogenetic lineages.

HomoloGene is calculated using MegaBLAST for cross-species sequence alignments. It identifies those sequence pairs A and B that share high degrees of nucleotide similarity and uses "reciprocal best matches" as in COG. If sequence A is the query sequence, sequence B must show the best matching score, if sequence B is the query sequence, sequence A must show the highest score. These sequence pairs A B are then used to find cross-species homologies between UniGene clusters. When reciprocal best matches are consistent between three or more organisms, the pair is described as being part of a "consistent triplet." Since the same scoring procedure applies to COG/KOG analysis, a comparison between HomoloGene and COG is in order.

The KOG0223 for the family of water transporters called *aquaporins* is displayed in Figure 3.13 (aquaporins 0 [major intrinsic protein in lens] 1, 2, 4, 6, 8 in kidney, stomach, and a nitrate channel). The KOG table highlights the very different number of paralogs found in each species. Human aquaporin proteins are related to *plasma membrane intrinsic protein,* the prevalent water channel in *Arabidopsis thaliana,* a model plant used in plant physiology and agricultural genomics.[4] The large number of water channel types in *Arabidopsis* comes as no surprise, since plants use osmotic mechanisms for water pumping, cell pressure (turgor) control, and mineral transport and absorption. Regulation of water homeostasis, of course, is important for all cells and aquaporins are found in other organisms including worm, fly, human, and baker's yeast and also prokaryotes. A second KOG for aquaporins (KOG0224) shows a different species distribution. It does not contain

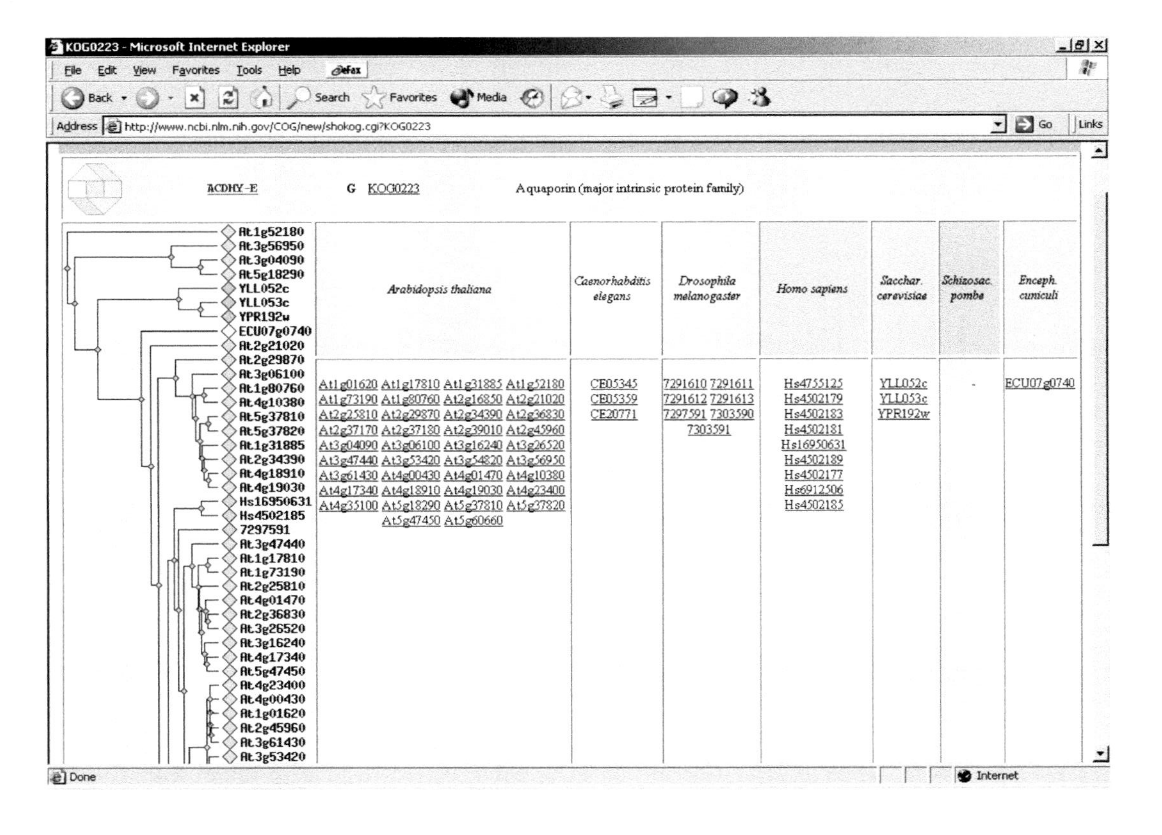

FIGURE 3.13
Aquaporin 4 KOG0223.

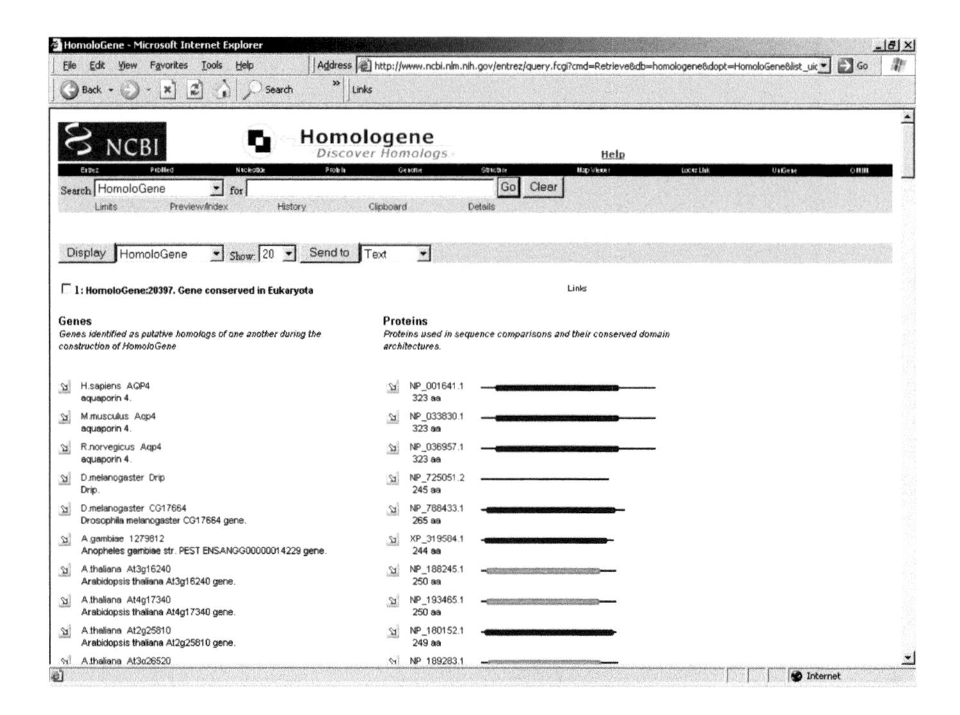

FIGURE 3.14
Homologene page for aquaporins.

any homologs in *Arabidopsis thaliana*. The human paralogs include aquaporins 3, 7, 9, and 10 (lung, adipose, liver, and small intestine). The existence of two KOGs for aquaporins demonstrates a basic subdivision of this family of water channel that requires further investigation.

Interestingly, HomoloGene can provide more information on homologs and paralogs. While KOG includes only species with completed genomes, HomoloGene includes proteins from all available sources, complete or not. In HomoloGene, a report page displays an individual gene, called the "key gene" and shows connections that follow from it. Figure 3.14 shows a HomoloGene display page for human aquaporin 4 (AQP4). This aquaporin is represented by a large family of *A. thaliana* water channels called major intrinsic proteins (MIPs), as seen in KOG0223. Aquaporin 4, the human ortholog, is the major water channel expressed in astrocytes of the human brain (glial cells supporting neurons).

HomoloGene provides links to Entrez Gene (Figure 3.15), which will eventually replace LocusLink and become the centerpiece at NCBI of finding information about a specific gene.

HomoloGene also provides a link for each individual gene entry via the *conserved domain database* (CDD; Figure 3.16), offering access to multiple alignments of both the 3-D structures (insert) and amino acid sequences of all homologs included.

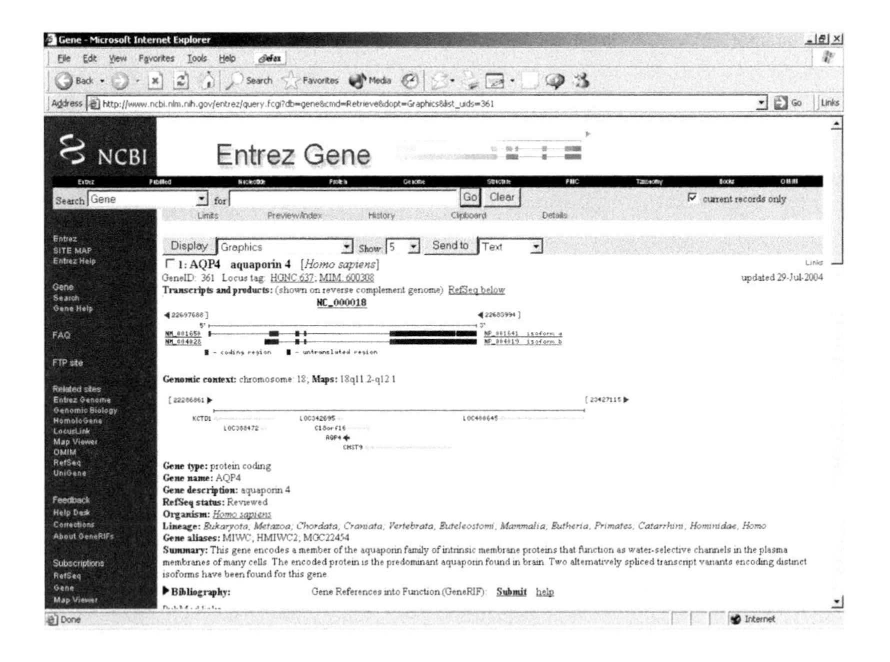

FIGURE 3.15
Entrez Gene for aquaporin 4.

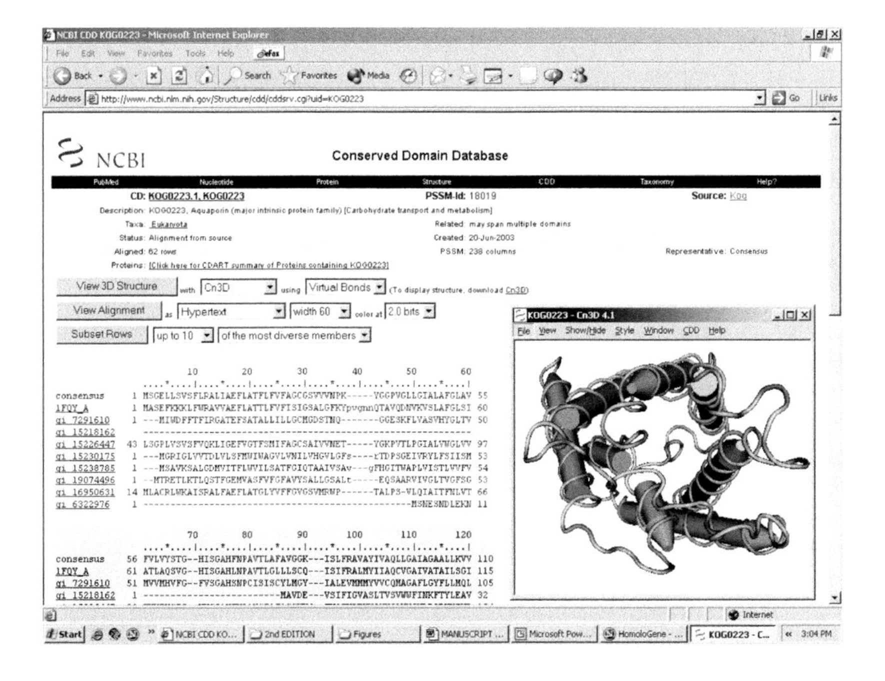

FIGURE 3.16
Conserved domain database (CDD) for aquaporin 4.

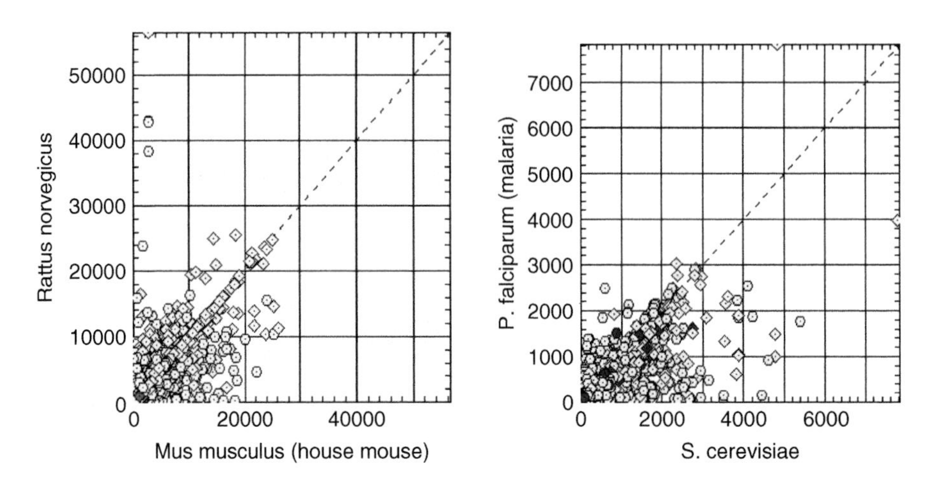

FIGURE 3.17
Tax plot at HomoloGene (NCBI): distribution of human homologs. Plotted by its BLAST scores (*S*) to the highest scoring protein from each of the selected organisms through three pairwise sequence alignments. Symmetrical hits from reciprocal scoring (see also COG) are shown as diamonds.

A third feature to compare proteins among complete genomes is *tax plot*. Tax plot is an interactive function at NCBI's HomoloGene web site and allows a three-way comparison (three pairwise sequence alignments) among completed eukaryotic genomes. Shown in Figure 3.17 are two three-way plots comparing genome wide homologs between human compared to rat and mouse (left) and human compared to *Plasmodium falciparum* (malaria-causing paramecium) and *S. cerevisiae* (baker's yeast). The diagonal in the rodent plot shows 1304 homologs shared between human, mouse, and rat with equal scoring hits (reciprocal best matches). Dots above the diagonal are homologs scoring higher between human and rat representing 13,727 human genes that are more closely related to rat than mouse. Below the diagonal are homologs with higher scores between human and mouse (9538 hits). Humans share only 165 equal scoring homologs with both the plasmodium and yeast, a number much lower than the 1304 human–rodent equal hits, consistent with the larger evolutionary distance between the three organisms representing the animal, fungi, and protozoan kingdoms.

Clicking on a hit result (symbol) extracts multiple homologs with similar scoring values. A link connects each gene to a page with precomputed Blast-2 sequence scores and graph. Figure 3.18 shows such a Blast-2 score for a known pyruvate kinase (gi 6319279; NP_009362) from *S. cerevisiae* and its putative homolog (gi 23612346; NP_703926) in *P. falciparum* strain 3D7.

Both coding and noncoding sequences are ultimately functional elements of the genome, and it is a challenge to understand these functions through comparative genomics. The value of having genome sequences from many

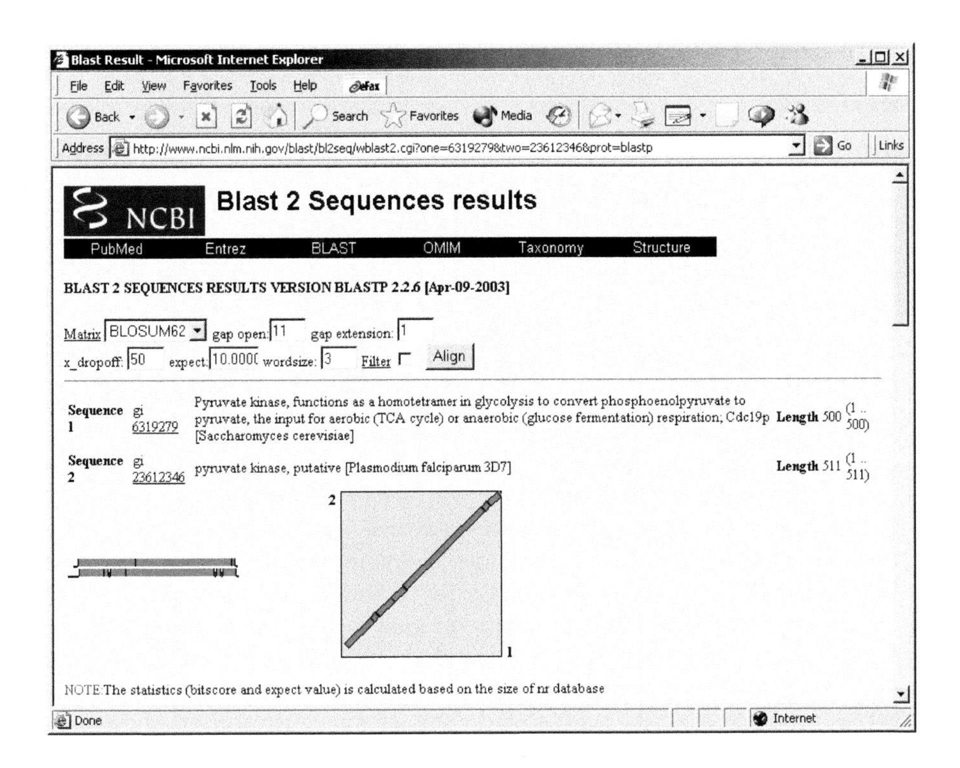

FIGURE 3.18
Blast-2 sequences results (version blastp 2.2.6).

different organisms thus lies in finding genomic features in genomes that are conserved and correlate with evolutionary distance. This information has biological relevance and helps in annotating genetic elements. Consequently, a functional element described in the mouse, rat, and chimpanzee genome is likely to shed light on the function of its human homolog, as it is likely serving the same function in all four mammalian species.

In most cases, the immediate benefit of whole genome comparison lies in high-quality annotation of universally conserved features of chromosomes. Including an increasing number of genomes from all three domains of life will reduce the number of these universal features while favoring the identification of species-specific features. For instance, in the search for the minimal gene set, the number of conserved genes shared by organisms dropped from 240 between two bacterial species (*H. influenzae* and *M. genitalium*; representing two ancient bacterial lineages; Gram-negative and Gram-positive, respectively) to 80 found among 25 whole genomes from all three domains of life.[5] This relatively small number of "universal" genes code for proteins and RNAs involved in protein translation. Incidentally, ribosomal RNA sequences have been used to construct the phylogenetic trees upon which the three domains of life—archaea, bacteria, and eukarya—are based.

References

1. Hua, V.B. et al., Sequence and phylogenetic analyses of 4 TMS junctional proteins of animals: connexins, innexins, claudins and occludins. *J. Membr. Biol.*, 2003, 194(1), 59–76.
2. Buehler, L.K., What's in a Structure. *PharmaGenomics*, 2003, 3(5), 20–21.
3. Doolittle, R.F. and Handy J., Evolutionary anomalies among the aminoacyl-tRNA synthetases. *Curr. Opin. Genet. Dev.*, 1998, 8(6), 630–636.
4. Quigley, F. et al., From genome to function: the *Arabidopsis* aquaporins. *Genome Biol.*, 2002, 3(1), RESEARCH0001. 1–0001.1
5. Koonin, E.V., Comparative genomics, minimal gene-sets and the last universal common ancestor. *Nat. Rev. Microbiol.*, 2003, 1(2), 127–136.

3.2.2.1 Gene Order and Chromosome Rearrangements

The structure of a genome, that is to say the location, arrangement, and codistribution of genes and regulatory elements on chromosomes, are not addressed by COG, HomoloGene, or Tax Plot analysis. Gene *synteny*—the conserved order of genetic elements on genomes—can be identified from whole genome alignments. Comparing two or three genomes will provide reliable identification of large-scale chromosomal regions that are conserved between several species. Such global views of sequence conservation facilitate functional annotation of genes and the identification of species specific as well as universal features.

This method of comparing multiple genomes has been particularly helpful for the identification of regulatory elements. Conserved noncoding regions can be assumed to be functionally important. If a conserved noncoding region is not closely linked to a gene structure, it may not be readily identified from a single genome sequence alone. Thus finding it on genomes in distantly related organisms indicates its importance, such as regulating gene expression, maintaining the structural organization of the genome. They may possibly have other, yet unknown functions. Conserved elements of these binding sites may be better visualized comparing multiple genomic sequences. As has become abundantly clear, genome structure determines the spatial arrangement of chromosome in eukaryotic nuclei, and this spatial arrangement is closely correlated to large-scale genetic activity of the euchromatin portion of chromosomes.[1] Thus where a gene is found on a chromosome strongly affects its activity pattern.

The search for nondescript but otherwise conserved genomic features is called *phylogenetic footprinting*. In other words, no prior knowledge of such elements is necessary. The genome comparison uncovers them in the same way it provides information about novel open reading frames and motifs for proteins with unknown function and structure. For instance, genome comparison has shown that there is an unexpectedly high endosomal transfer of organellar DNA to the nuclear chromosomes.[2] These transfers occurred both recently and earlier in evolutionary history based on comparison between nuclear, organellar, and "ancestral" prokaryotic genomes (bacterial and mitochondrial genomes have common prokaryotic ancestors). *Nu*clear stretches of *mit*ochondrial DNA (Numts) have been found in the grasshopper, primate, and shrimp genome as well as in many plant species. Mechanistically, organellar DNA may be integrated into nuclear DNA by recombination of "escaped" organelle DNA and chromosomes, or via cDNA copies involving some sort of reverse-transposon activity.

3.2.2.2 MapViewer

Direct comparison of genomic sequences can be done at NCBI using MapViewer (Figure 3.19). MapViewer supports comparative views for an increasing number of genomes. Besides searching a genome map for a particular organism, the "Maps & Options" interface can be used to select what other genomes to display relative to a genome of reference, and the maps to display on all.

These homology maps aligning entire genomes map the *synteny* between organisms. Genetic markers that are noncoding regions and whose order is

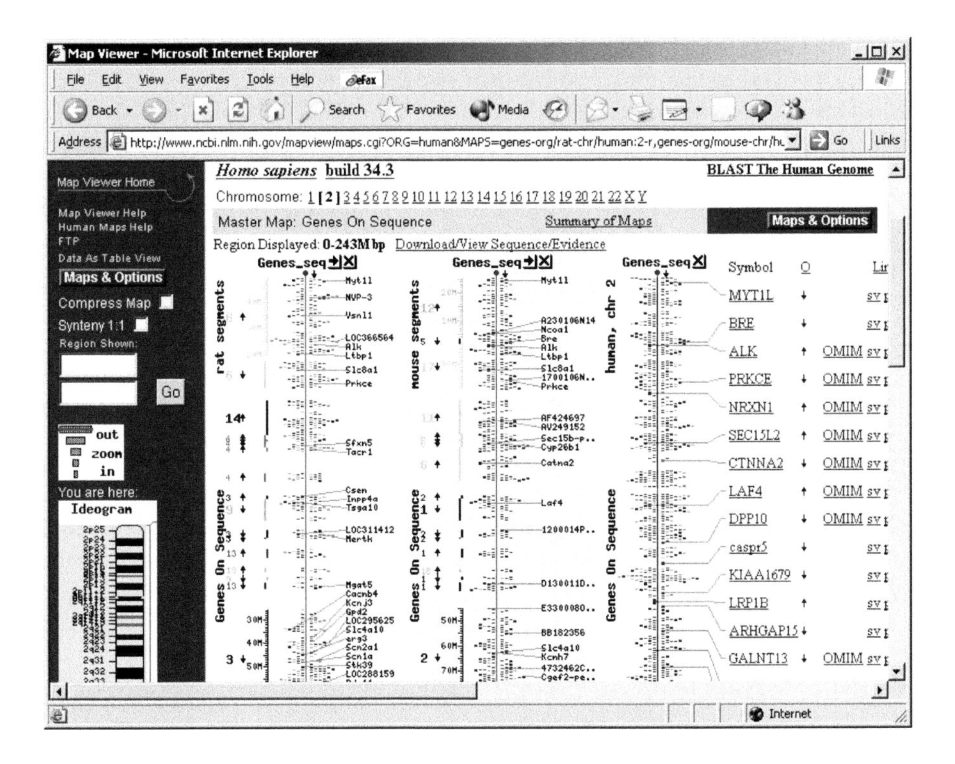

FIGURE 3.19
Human–mouse–rat homology map in Entrez map viewer. Genes found on human chromosome 2 (right-hand sequence) are identified in mouse and rat on various other corresponding chromosomal sequences; e.g., in rat mostly on chromosomes 3, 6, and 9, and on mouse chromosomes 1, 2, 6, 11, 12, and 17 with smaller fragments from other assorted chromosomes.

TABLE 3.11

The Minimum Genome Rearrangements Separating Humans From Other Mammals[3]

Mammal	Rearrangements
Cat	13
Cow	27
Pig	28
Rat	109
Mouse	180

valuable to establish synteny include SNPs and microsatellites, also called short tandem repeats (STRs). Whole genome sequencing has confirmed that gene order in chromosomes is conserved over wide evolutionary distances. Such conserved stretches, however, do not cover large segments or entire chromosomes. In particular the spacing between mapped genes is not always proportional, even though gene order is not affected.

Thus, comparative maps of higher organisms record the history of chromosome rearrangements that have occurred during the evolution of plants and animals. Using synteny allows for quantitative analyses of the patterns of segment conservation, chromosome evolution, and rearrangements. In the human–mouse homology map there are an estimated total of ~180 rearrangements. On average, rearrangements occur at a rate of ~0.8 disruptions per million years of evolution since the separation of the lineages of humans and mice. Comparative genetic assessment expands the utility of these maps in gene discovery, in functional genomics, and in tracking the evolutionary forces that sculpted the genome organization of modern mammalian species. Synteny between humans and mammals shows very high conservancy. This is apparent for the number of rearrangements found between human and cat, cow, and pig. Interestingly, and for unknown reasons, certain groups show accelerated rearrangement rates as occurred in mouse and rat (see Table 3.11). Comparisons of zebra fish and mammalian gene maps have revealed extensive conservation of synteny relationships among vertebrates.

From the analysis of genomes of modern species and based on comparison of novel genes, lost genes, modified genes, and reordered genes, some rules are emerging about chromosome structure conservation. First, rearrangements of chromosome segments occur rarely during evolution. Second, in some lineages, rearrangements happen more frequently. Third, genome *contraction* or *expansion* events occur more frequently than rearrangements.

MapViewer includes 23 eukaryotic genomes (July 2004):

Animals

Homo sapiens (human)

Mus musculus (mouse)

Rattus norvegicus (rat)

Bos taurus (cow)

Canis familiaris (dog)
Sus scrofa (pig)
Danio rerio (zebra fish)
Anopheles gambiae (mosquito)
Caenorhabditis elegans (nematode)
Drosophila melanogaster (fruit fly)

Plants

Arabidopsis thaliana (thale cress)
Avena sativa (oat)
Glycine max (soybean)
Hordeum vulgare (barley)
Lycopersicon esculentum (tomato)
Oryza sativa (rice)
Triticum aestivum (wheat)
Zea mays (corn)

Fungi

Saccharomyces cerevisiae (baker's yeast)
Schizosaccharomyces pombe (fission yeast)
Magnaporthe grisea (rice blast fungus)
Neurospora crassa (orange bread mold)

Protozoa

Plasmodium falciparum (malaria parasite)

References

1. Hernandez-Verdun, D. et al., Emerging concepts of nucleolar assembly. *J. Cell Sci.*, 2002, 115(11), 2265–2270.
2. Timmis, J.N. et al., Endosymbiotic gene transfer: organelle genomes forge eukaryotic chromosomes. *Nat. Rev. Genet.*, 2004, 5(2), 123–135.
3. O'Brien, S.J. et al., The promise of comparative genomics in mammals. *Science*, 1999, 286(5439), 458–481.

3.3 Functional Genomics

3.3.1 The Transcriptome

An important step in understanding the relationship between genotype and phenotype of an organism is to look at the behavior of the entire genome. The behavior of a genome is distinguished by the activity of its genes. The activity of genes is assessed by measuring the temporal changes of messenger RNA levels. By describing the changes in transcript levels for many or all genes of an organism, i.e., its *transcriptome*, patterns of gene expression can be established. Novel genes can thus be identified when expressed in relation to a cellular activity for which we have some biologically significant information. The study of expression patterns then allows the characterization of unknown genes based on their spatiotemporal expression.

Identifying and sequencing novel genes has been greatly enhanced by newer molecular biology techniques such as polymerase chain reaction (PCR) and the DNA synthesizing enzyme reverse transcriptase (RT). Reverse transcriptase can make larger numbers of DNA copies from isolated RNA; or DNA can be amplified by PCR. Even minute amounts of RNA or DNA can be measured and characterized (sequenced) this way. These two techniques have in particular facilitated the identification of actively transcribed genes by isolating whole cell RNA and obtaining a workable cDNA copy for cloning and sequencing. Most of the sequence information deposited in GenBank comes from full-length cDNA sequence analysis or the shorter expressed sequence tags (ESTs) rather than genomic sequences. The biological utility of this approach was the sequencing of the transcriptome of an organism or cell type reflecting the actively expressed set of genes. The combined EST sequence information that can be obtained from cellular RNA reflects a different level of genetic activity, namely that of gene expression. The transcriptome thus represents a functional view of a genome that allows addressing biologically interesting questions such as cell differentiation and disease states of cells that are the result of epigenetic control and misregulation of genomic information.

Gene sequencing based on sequencing RNA-derived ESTs and full-length cDNA showed that the number of transcripts sequenced exceeds the number of genes identified at the chromosomal level. This, of course, is the result of alternative splicing of the complex exon/intron architecture of higher genomes (eukaryotes). However, the high level of redundancy found among transcribed sequences deposited in GenBank and common experimental artifacts made it difficult to make use of the data. As the usual short sequences obtained from RNA screens, multiple entries from often unrelated projects were submitted to GenBank, although many such EST sequences belonged to a single genomic gene sequence or locus. This problem has been addressed by UniGene at NCBI, a largely automated analytical system for producing an organized view of the transcriptome. UniGene provides a comprehensive view that links all original sequence entries to a gene, allowing a researcher to trace back the original contribution from DNA or RNA sequencing projects. UniGene also provides homolog and ortholog information, chromosome mapping information (the location of each original clone submitted to GenBank), gene expression information regarding the tissue origin of the sequenced nucleic acid (e.g., brain, lung, retina, etc.), and functional expression data from microarrays or serial analysis of gene expression deposited in gene expression databases (e.g., GEO, SAGE).

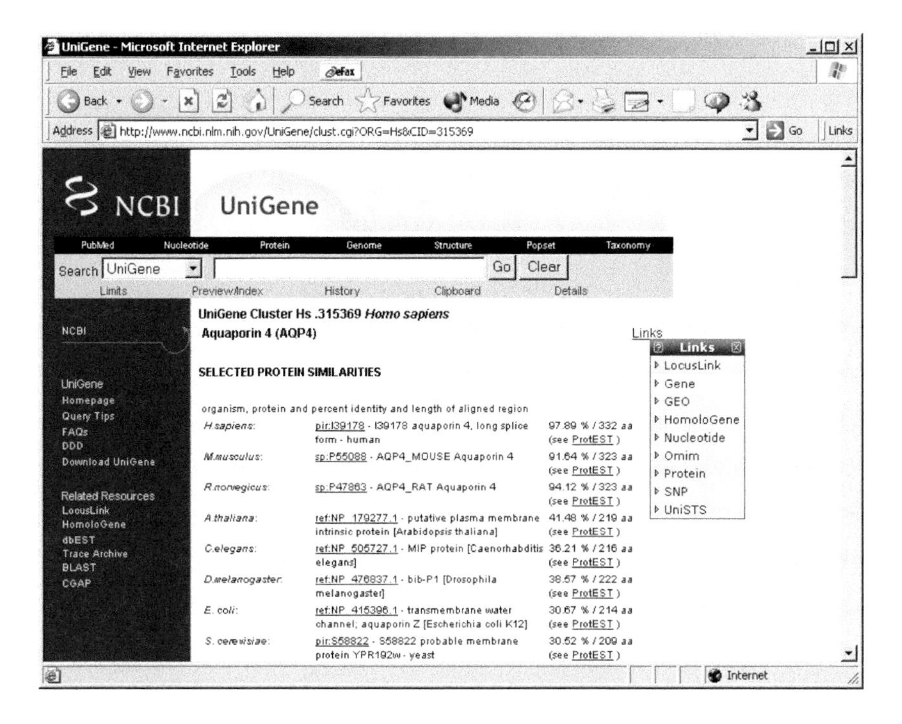

FIGURE 3.20
UniGene Cluster Hs.315369 *Homo sapiens* AQP4: Aquaporin 4. UniGene database automatically partitions GenBank sequences into a nonredundant set of gene-oriented clusters.

Figure 3.20 shows the top part of the UniGene page for Cluster Hs.315369 *Homo sapiens* for aquaporin 4. The UniGene page displays a short list of select homologs from other organisms, chromosome mapping information (18q11.2-q12.1 cM) and sequence-tagged site (STS) accession numbers linked to the aquaporin 4 locus. It also lists functional expression data originating from annotated tags, GEO database, and SAGE tags. Additional information links to cDNA and EST sequences as originally deposited in GenBank. A total of 11 full-length and 168 expressed sequence-tagged sequences are associated to the aquaporin 4 UniGene cluster.

Serial analysis of gene expression, or SAGE and DNA microarrays, are the two major techniques currently employed to generate gene expression data. Both rely on the ability to quantitatively copy and amplify isolated RNA from tissue or cell culture preparation producing relatively short stretches of sequences (20 to 100 nucleotides long) that are nevertheless unique for a particular gene. Gene expression results from DNA microarrays can be accessed at NCBI via the Gene Expression Omnibus (GEO) database at www.ncbi.nlm.nih.gov/geo (Figure 3.21) or ArrayExpress at EBI (www.ebi.ac.uk/arrayexpress).

Both databases contain information on gene expression patterns that can be searched by data set or from a list of experimental conditions. For instance, a study relating defective chloride secretion in the digestive tract

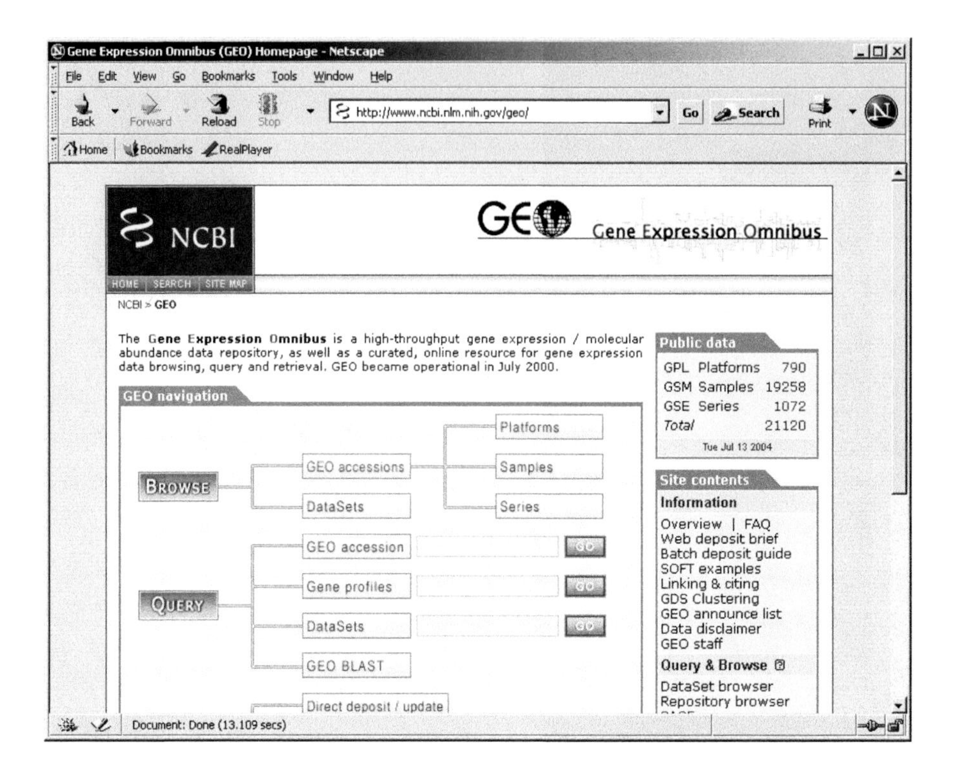

FIGURE 3.21
Gene expression omnibus (www.ncbi.nlm.nih.gov/geo).

to mutations in a chloride transporter, the cystic fibrosis transmembrane regulator (CFTR), provides the following information at GEO:

Cystic fibrosis and small intestine

- GDS588: Examination of small intestine from cystic fibrosis conductance transmembrane regulator null mice (CFTR–/–)
- type: single channel nucleotide
- samples in dataset: 6
- organism: Mus musculus
- GEO platform: GPL81
- `go to full dataset record`

The data summary accessible via the GEO platform link (GPL81) gives detailed information about experimental conditions on test and control group variables such as organism, tissue type, array technology or platform (commercial or custom made), and number of experimental repeats:

GDS588: **Cystic fibrosis and small intestine** [Mus musculus}

Summary: Analysis of small intestine from 40-day-old CFTR–/– and CFTR+/+ mice. The CFTR null mouse has a severe intestinal

phenotype serving as model for cystic fibrosis related growth deficiency, meconium ileus, and distal intestinal obstructive syndrome.
Parent platform: GPL81
Reference series: GSE765
Type: single channel nucleotide count
Subsets: 2 genotype/variation sets
Samples: 6
GSM11902: HK12B5; GSM11903: ZF6B7; GSM11904: HD12E1
GSM11905: HC12E4; GSM11906: HF12E1; GSM11910: HK12B7

Following the link to GEO profiles provides expression data for individual genes from a total of 12,488 tested genes as found in multiple sample conditions tested. A link to a graphical representation of up and down regulation for each gene under the conditions tested comparing CFTR+/+ and CFTR – /– mice is provided. Figure 3.22 shows the expression profiles for the mast cell protease-2 (MMCP-2) mRNA (GenBank accession number J05177) that is not expressed in normal mice, but upregulated in the small intestine of knockout (k.o.) mice (CFTR –/–). Each wild type and k.o. result has been produced three times. Other mast cell-specific genes have also been identified in this study as upregulated in the k.o. animals, indicating that mucosal

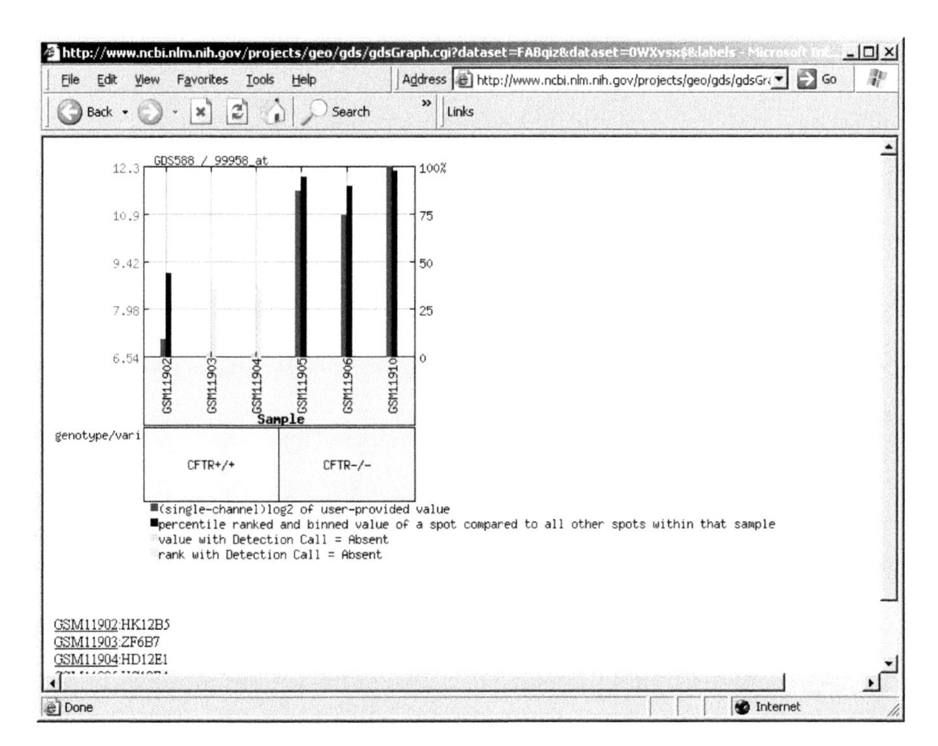

FIGURE 3.22
Entrez GEO profiles. DNA microarray gene expression results showing that a gene is more active in a k.o. mouse (–/–) than in the normal mouse strain (+/+).

inflammation processes play a role in cystic fibrosis. A detailed discussion of microarray technology is presented in the next section.

SAGE is an alternate tool to analyze the transcriptome, and has been used in a wide variety of genomics applications. These include identifying disease-related genes, analyzing the effects of drugs on tissues, and providing insight into disease pathways and the identification of tumor antigens and antiangiogenic targets. SAGE was invented by a group of researchers led by Kenneth Kinzler, Ph.D., of The Johns Hopkins University and Bert Vogelstein, M.D., of The Johns Hopkins University and Howard Hughes Medical Institute, as a powerful technology for the analysis of gene expression. SAGE characterizes a 14-base pair segment of DNA, called a SAGE tag, cut out by restriction enzyme digestion from a defined location in each expressed gene and represents a unique identifier for that gene. Hundreds of unique tags are then ligated (concatenated) to form long DNA strands that can be sequenced. The number of repetitive sequences found for each unique tag sequence in the concatenated DNA is a measure of the transcript or expression level for a specific gene in the tissue under investigation.

Using LongSAGE™ based on 21 base pair markers enables matching of SAGE tags to the genome map and may be valuable in identifying all the genes in the human genome (see Genzyme for more information; http://www.genzyme.com/). SAGE is applicable to any situation where relative levels of gene expression are important. In human disease, SAGE has been used to identify genes that are involved in the disease process. Genes identified by SAGE may be validated by other technologies and used as potential targets for therapeutics or diagnostics. SAGE may also be used once a therapeutic has been identified to determine the mode of action in an experimental animal or cell culture model. As the use of SAGE has increased, it has been possible to combine data from different experiments and sites to produce large-scale analyses of the genes expressed in a particular species. SAGE data is available at a number of sites, including the National Cancer Institute. (See Section 5.1.2 for a SAGE application in cancer research.)

3.4 Microarray and Bioarray Technology

Thorsten Forster and Peter Ghazal

3.4.1 Concept and Use

In genomics, bioinformatics is becoming increasingly important because of the high-dimensional scaling of conventional biochemical and molecular experiments to high-throughput technologies. One of the key technologies in post-genomic research is large-scale expression profiling using a combination of DNA (oligonucleotides/cDNA), Protein (antibody/antigen/peptide) or Aptamer (RNA/DNA) microarray-based chips. These bioarray chips, described in the following summary section, enable measurements of the identity and levels of expression of thousands of biomolecules (DNA, RNA transcripts, protein, and other biomolecules) to be taken in parallel. The profile of such expression levels recorded from different cell types (e.g., cancer vs. normal) under different environmental conditions (e.g., treatment vs. nontreatment) or time-dependent conditions (e.g., embryonic development) enables one to identify characteristics in expression signatures that can uniquely define a cell type or condition. Highly parallel measurements using a microarray platform also allow a data-driven approach to discovery in biology. Because of the volume and type of data produced in these experiments, computational solutions are essential. In this section we will provide a combined empirical and bioinformatics guide to using microarray technology.

3.4.2 Summary of a Typical Experiment Using Microarray Technology

Slides (also referred to as arrays or chips) are manufactured by depositing selected probes onto a solid substrate. Probes are oligonucleotides, cDNA sequences, RNA sequences, antibodies or antigens; common substrates are coated microscopy slides or proprietary formats. The deposition is usually done by robotic printing or *in situ* assembly using lithographic technology.

Target samples are obtained from a biological source, e.g., cancer tissue from mice or cells from a culture. For DNA arrays, total RNA or mRNA is extracted from this sample, reverse-transcribed into cDNA and labeled. Dyes have mostly replaced radioactive labels, in some cases the labeling can be substituted by surface plasmon resonance and potentially other techniques.

For protein arrays, labeling is normally done via a labeled secondary antibody. Labeled target samples are then hybridized to the prepared arrays. In the case of DNA arrays, often two samples—labeled with different dyes—are cohybridized to one array. Sequences, antigens, or antibodies bind with matching probes on the array, and unbound material is washed off. Laser-scanning of the array will excite fluorescent dye molecules (radioactivity is measured by a detector system) and the amount of fluorescence measured for each probe is representative of expression or abundance levels. In addition to this, off-the-shelf arrays exist and often come with proprietary laboratory protocols and basic analysis workflows. Color Figure 3.23 and Figure 3.24 illustrate basic technology and typical experiment workflow.

3.4.3 Microarray Bioinformatics

From a bioinformatics point of view, there are numerous challenges accompanying this technology. They can be broadly categorized into

- Image processing
- Data annotation
- Data analysis
- Data storage
- Data mining

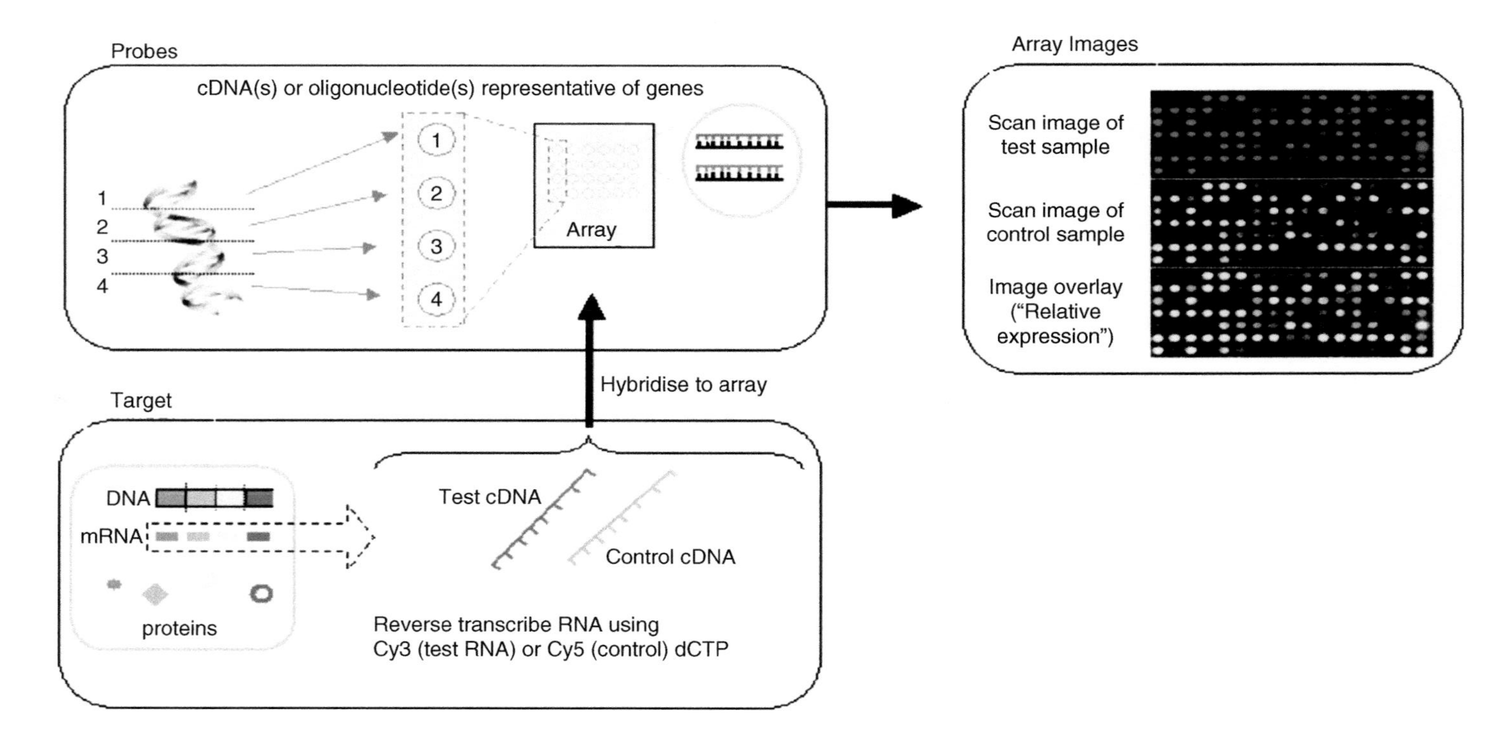

FIGURE 3.23 (See color insert)
Overview of microarray technology as used for DNA readout. The process is very similar for protein arrays, with differences in deposited material and labeling protocols.

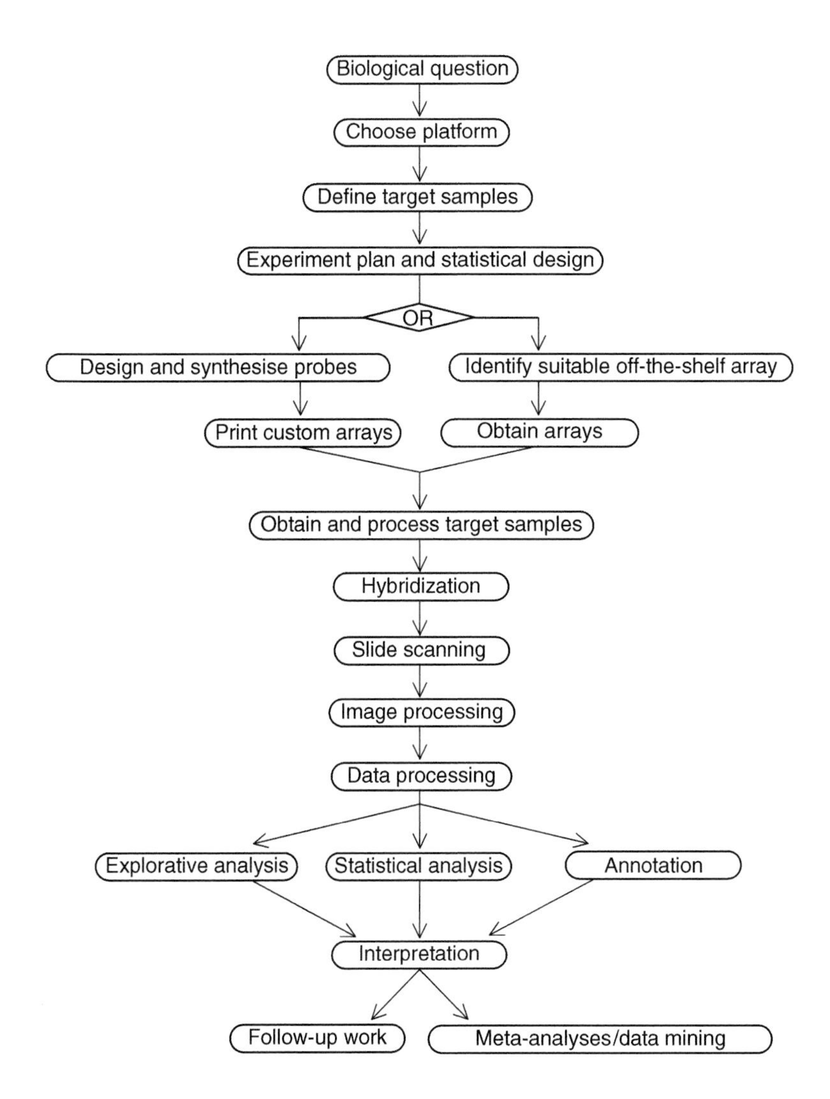

FIGURE 3.24
Outline of a complete microarray experiment workflow. Every step outlined is of similar importance with respect to carrying out a high-quality microarray experiment. The standardization of microarray technology, in particular custom printed arrays, has not yet reached the level where any of these steps can be carried out lightly and without due care and consideration.

All approaches concerning these challenges are also applicable to other technologies in the wider field of functional genomics and systems biology.

In the following sections we will briefly explain these approaches. They are meant to be used as starting points and reference for bioinformaticians starting out in this field.

It is worth noting that some off-the-shelf microarray platforms like Affymetrix have their own standardized approaches, and although there is always room for improvement, they tend to work well within the system.

3.4.4 Image Processing

Microarrays are usually scanned with lasers or other light sources, occasionally radiation detectors, in order to detect the fluorescent (or radioactive) labels attached to the hybridized gene probes. In case of a "dual-target" experiment, where each chip is cohybridized with two differently labeled target samples simultaneously, one scan is performed for each "channel," i.e. at the appropriate excitation-frequency for the labels. For simplicity, they are often referred to as red and green dye-channel. Most scans are carried out at a depth of 16-bit and varying resolution, resulting in image sizes (TIFF format) of several megabytes, depending on the size of the array.

The images are only an intermediate step used for eyeballing and basic quality control, the challenge lies in converting the imaged *fluorescence level* obtained from each gene probe on the array into *numeric values*.

Current image processing software often requires the researcher to manually align an overlay grid with the spotted array, and to select a number of parameter values for the image-processing algorithm. If not carried out with the necessary knowledge and care, this can have a detrimental effect on data quality. It is also expensive in terms of time, and therefore is an obstacle for high throughput rates. Current algorithms are also not fully capable of automatically dealing with noise and quality control issues, e.g., identifying sources of noise and adjusting for them, identifying aberrant spot morphologies and interpolate appropriate signal values. These are processes which then either have to be carried out separately or are ignored due the computational efforts involved. New algorithms taking into account spot morphologies, statistical distributions of pixel values, repeat scans, and the need for automation are currently being developed.[1–5]

Links:

Microarray slide scanners (often provided with image processing software)

ScanArray www.perkinelmer.com/

Affymetrix www.affymetrix.com/

GenePix www.axon.com/

+ many more

Microarray image processing software (standalone or provided with scanner)

TIGR Spotfinder www.tigr.org

Quantarray www.packardbioscience.com

Koadarray www.koada.com

ImaGene www.biodiscovery.com

GenePix www.axon.com

+ many more

References

1. Glasbey, C.A. and Ghazal, P., Combinatorial image analysis of DNA microarray features. *Bioinformatics*, 2003, 19(2), 194–203.
2. Yang, Y.H., Buckley, M.J., Dudoit, S., and Speed, T.P., Comparison of methods for image analysis on cDNA microarray data. *J. Computational Graphical Stat.*, 2002, 11, 108–136.
3. Angulo, J. and Serra, J., Automatic analysis of DNA microarray images using mathematical morphology. *Bioinformatics*, 2003, 19, 553–562.
4. Jain, A.N., Tokuyasu, T.A., Snijders, A.M., Segraves, R., Albertson, D.G., and Pinkel, D. Fully automatic quantification of microarray image data. *Genome Res.*, 2002, 12, 325–332.
5. Ramdas, L., Wang, J., Hu, L., Cogdell, D., Taylor, E., and Zhang, W., Comparative evaluation of laser-based microarray scanners. *Biotechniques*, 2001, 31, 546, 548, 550.

3.4.5 Data Annotation

If the results, be it statistically significant gene expression differences, simple up/down-regulation or expression profiles and cluster memberships (see Section 3.4.6), are to be put into context, it is essential to provide as much information on each gene as possible. Such information can consist of the following:

Have these genes been mentioned in context of diseases or phenotypes elsewhere?

Is there any functional information on these genes?

In which genetic pathways have these genes been shown to participate?

Are these genes associated with known protein expression?

Are these genes in any particular location on the chromosomes?

Answers to all or many of these questions can be found in various databases.[1–7] Using or mining these to best effect is often done by bioinformaticians. Combined and individual efforts from numerous research teams have resulted in a large number of databases, some of which are "standalone" and some of which are interconnected with other databases to aggregate information. They are implemented in a variety of programming languages, interfaces, and database systems, which complicates straightforward retrieval of all possible information by an individual researcher. The ideal situation, and the bioinformatics challenge, is a "one-stop" interface allowing the researcher to enter his gene or protein identifiers in order to automatically obtain all information found on them in journals, protein databases, functional databases, pathway databases, gene expression databases, etc.

Some moves in this direction have been made by interlinking a small subset of databases, but this is not yet of a scale allowing a one-stop approach. The database-linking aspects of Grid technology may enable this in the near future.

Links:

Portals, multi-database sites, individual annotation tools

Entrez (NCBI) www.ncbi.nlm.nih.gov/Entrez/

SRS (EBI, Sanger) http://srs.ebi.ac.uk/

EMBL-EBI assorted databases www.ebi.ac.uk/services/index.html

ENSEMBL (Sanger, EBI) www.ensembl.org/

NetAffX www.affymetrix.com/analysis/index.affx

GoMiner/MatchMiner http://discover.nci.nih.gov

SOURCE http://source.stanford.edu/

Dragon http://pevsnerlab.kennedykrieger.org/dragon.htm

DAVID http://apps1.niaid.nih.gov/David

GenMapp http://www.genmapp.org

ARROGANT http://lethargy.swmed.edu/

Manatee http://manatee.sourceforge.net/

References

1. Bouton, C.M. and Pevsner, J., DRAGON View: information visualization for annotated microarray data. *Bioinformatics*, 2002, 18, 323–324.
2. Brazma, A., Sarkans, U., Robinson, A., Vilo, J., Vingron, M., Hoheisel, J., and Fellenberg, K., Microarray data representation, annotation and storage. *Adv. Biochem. Eng. Biotechnol.*, 2002, 77, 113–139.
3. Brazma, A., Parkinson, H., Sarkans, U., Shojatalab, M., Vilo, J., Abeygunawardena, N., Holloway, E., Kapushesky, M., Kemmeren, P., Lara, G.G., Oezcimen, A., Rocca-Serra, P., and Sansone, S.A., ArrayExpress—a public repository for microarray gene expression data at the EBI. *Nucleic Acids Res.*, 2003, 31, 68–71.
4. Guffanti, A., Reid, J.F., Alcalay, M., and Simon, G., The meaning of it all: web-based resources for large-scale functional annotation and visualization of DNA microarray data. *Trends Genet.*, 2002, 18, 589–592.
5. Diehn, M., Sherlock, G., Binkley, G., Jin, H., Matese, J.C., Hernandez-Boussard, T., Rees, C.A., Cherry, J.M., Botstein, D., Brown, P.O., and Alizadeh, A.A., SOURCE: a unified genomic resource of functional annotations, ontologies, and gene expression data. *Nucleic Acids Res.*, 2003, 31, 219–223.
6. Bouton, C.M. and Pevsner, J., DRAGON View: information visualization for annotated microarray data. *Bioinformatics*, 2002, Feb, 18(2), 323–324.
7. Kanehisa, M., Goto, S., Kawashima, S., and Nakaya, A., The KEGG databases at GenomeNet. Nucleic Acids Res., 2002, 30, 42–46.

3.4.6 Data Analysis

Microarrays or bioarrays are a high parallel throughput technology producing amounts of data that before have only been seen in other research disciplines (e.g., physics and astronomy). Bioinformatics is an essential technique, with proliferative research currently ongoing for all aspects of array technology and data.[1] Together with image processing and database methodology, data analysis is one of the important work disciplines for bioinformaticians working in this field.

The result of analyses for microarray data usually consists of expression fold-changes between treatment conditions, statistical significance of these changes, and information on similarity between biological samples and/or gene expression profiles.

Microarray data analysis has a strict dependence on the following factors.

3.4.6.1 Experiment Design/Plan

Best practice in microarray analysis is to consider experiment design before embarking on an experiment (see Color Figure 3.25). This encompasses consideration of single- or dual-dye experiments, the number of independent biosources (e.g., mice, human tissue samples) that are required to identify expression changes of a given magnitude with a desired level of confidence, the levels of replication in an experiment (multiple instances of the same probe on an array, multiple hybridizations for one sample), or the necessity for positive or negative control probes on an array.

Experimental design is one of the key statistical considerations in bioinformatics, and with large-scale experiments there is a critical need for this.[2–4]

3.4.6.2 Volume of Data

The amount of microarray data can be considered a game of multiplication: If we consider the processed image data (i.e., numeric value for each gene probe) to be the raw data, then an average-sized experiment with a 10,000 gene chip, 2 biological conditions across 5 time-points and 10 mice per time-point/condition and each gene probe printed three times on an array, then the total number of data points is 3 million. Just using more mice, or larger arrays, or repeat hybridizations of one sample to more than one array can easily multiply this number many-fold.

Color Figure 3.25 is an example experimental design, detailing source and processing of biological sample material, labeling and hybridization strategy. It is useful in establishing details for every array hybridization, and clarification of which arrays are biological replicates (different patients), technical replicates (multiple arrays for the same labeled cDNA sample, multiple RNA extractions etc.) or dye-swaps. Interpretation of statistical and explorative analyses depends on the nature and number (per treatment group) of the hybridized samples. For example, an overemphasis on technical replicates without sufficient biological replication would preclude statistically significant results to be also biologically reliable. In this example, the number of biological replicates is only three, i.e. three patients provide the source tissue material. If biological variation at the patient level was a factor or of interest

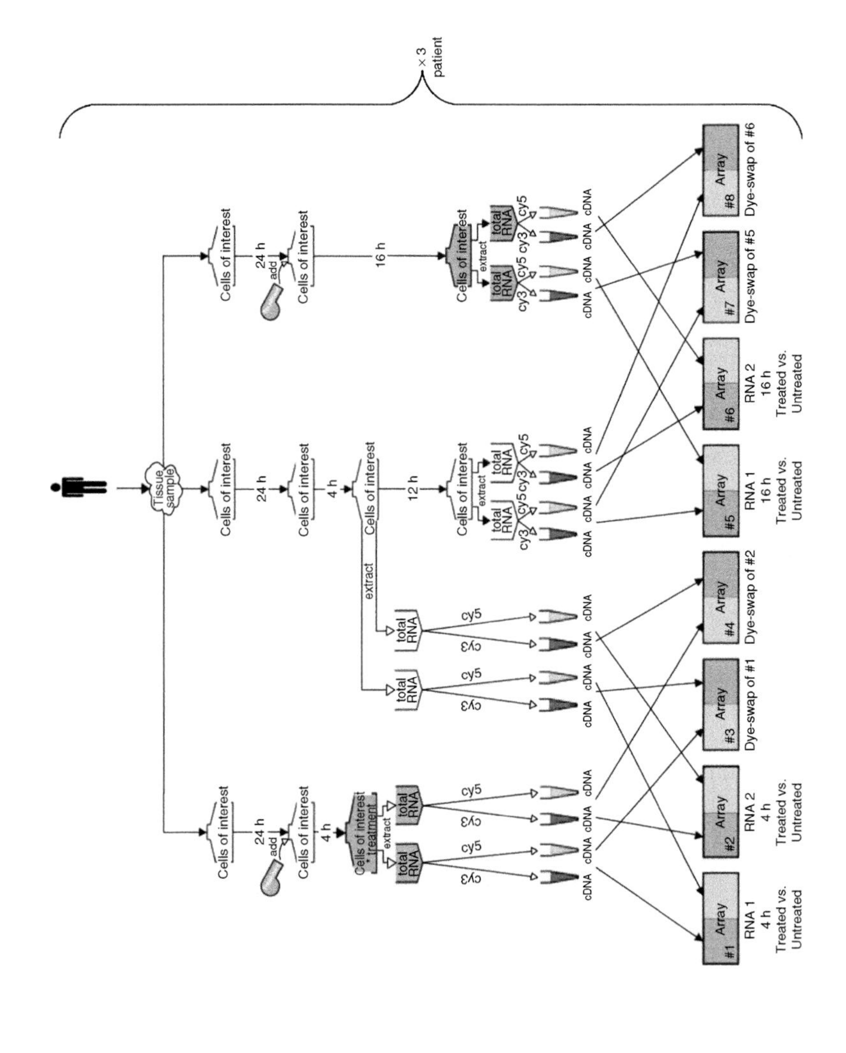

FIGURE 3.25 (See color insert)
Example of an experimental plan.

to the researcher, this particular experiment design would be insufficient despite the total array count of 24.

3.4.6.3 Dimensionality of Data

Data are usually arranged in a matrix consisting of genes down the rows and observed values for those genes across the columns. It may be worth noting the nature of the dimensionality problem for microarray data: Most statistical procedures have been developed to deal with large numbers of observations (e.g., tens or hundreds of patients) and a low number of variables or outcome measures (e.g., blood pressure, height, survival time). This situation is reversed for microarray data, where there are thousands of genes (variables) and only few observations (e.g., mice per time-point). Simultaneous inference testing on a multitude of variables is known as "multiple-testing" problem, and statistical procedures have to be adjusted for this situation.

3.4.6.4 Quality of Data

The above also presents the problem of data variation or noise, or more precisely, the data variation due to nonrelevant sources. Examples would be the scanning and image processing, the dye-labeling and dye-label incorporation, the hybridization conditions, and the chip and gene probe quality. It is difficult to estimate the amount of unwanted variation with only a few observations, and ways have to be found to either eradicate the source of variation by having very precise and consistent experiment plans, or by applying statistical techniques to deal with such sources of variation.

3.4.7 Normalization

Another key feature of data quality is normalization, which in this field refers to a collection of processes necessary to be able to compare data. This is an essential step before performing any kind of analysis. A broad explanation would be that the overall brightness of arrays or dye-label channels (see image processing) is scaled to identical levels, using linear (changing all individual gene probe values on a chip by the same amount) or nonlinear (the same, but increases/decreases vary with level of expression) methods.[5,6]

We will explain in more detail the two main approaches to data analysis, statistical analysis, and explorative analysis. Data mining is covered in a separate section.

3.4.8 Statistical Analysis

First-line statistical analyses[7–9] include group means, expression fold-changes as well as simple group comparisons by means of significance tests such as parametric *t*-test or nonparametric Wilcoxon test. Standard

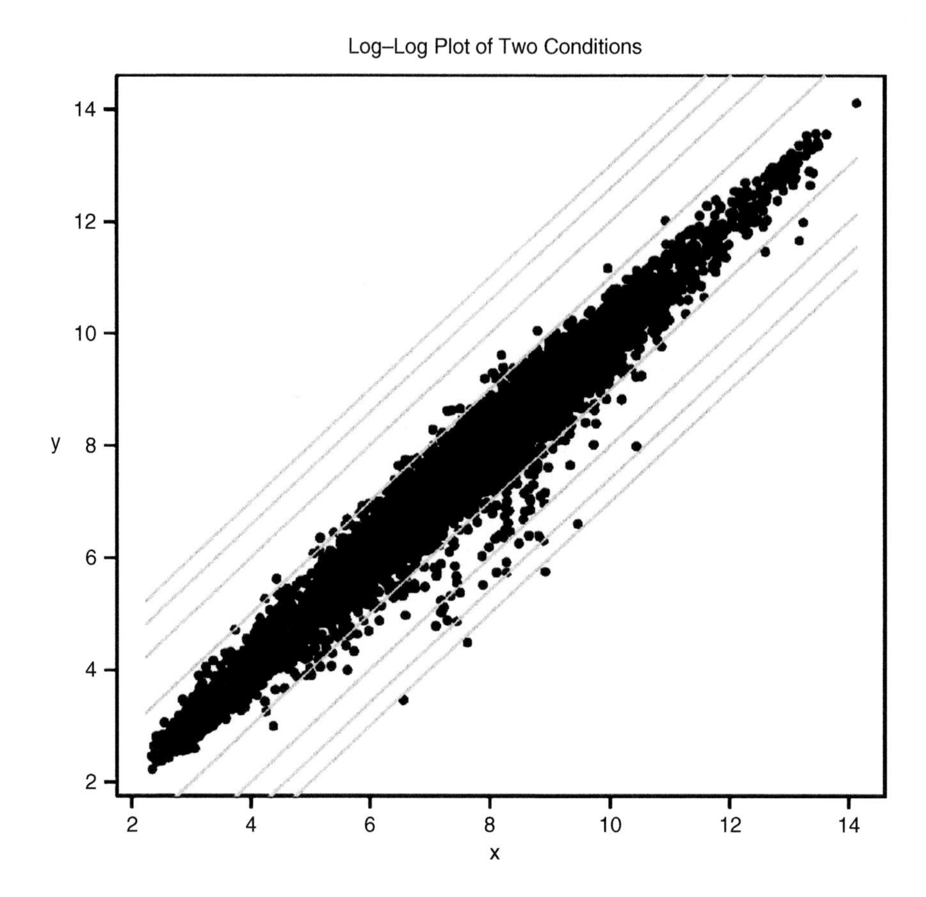

FIGURE 3.26
Log–Log plot. Plotted are the average expression values of two different treatment conditions. Data points away from the central diagonal have a higher fold-change between conditions.

graphs like log–log plots and MA- or RI-plots are also part of the analysis toolbox (see Figure 3.26 and Figure 3.27). Obtaining means and expression fold-changes between any number of conditions poses limited restrictions despite the large number of genes (i.e., variables) for which they are calculated. Of course, the statistical reliability of these may be reduced as a result of the often small numbers of observations.

Calculating confidence intervals for means and fold-changes or calculating p-values for tests of significance can also be performed in this simple way, gene by gene.[10–12] It is noteworthy that obtaining p-values for thousands of genes at the same time is bound to produce false positive results by chance.[13] When using an alpha-level of 5% as significance cutoff, then 5% of all detected significant results may be false positives. This can be corrected for by lowering the alpha-level, or adjusting the p-value based on the number of tests performed.

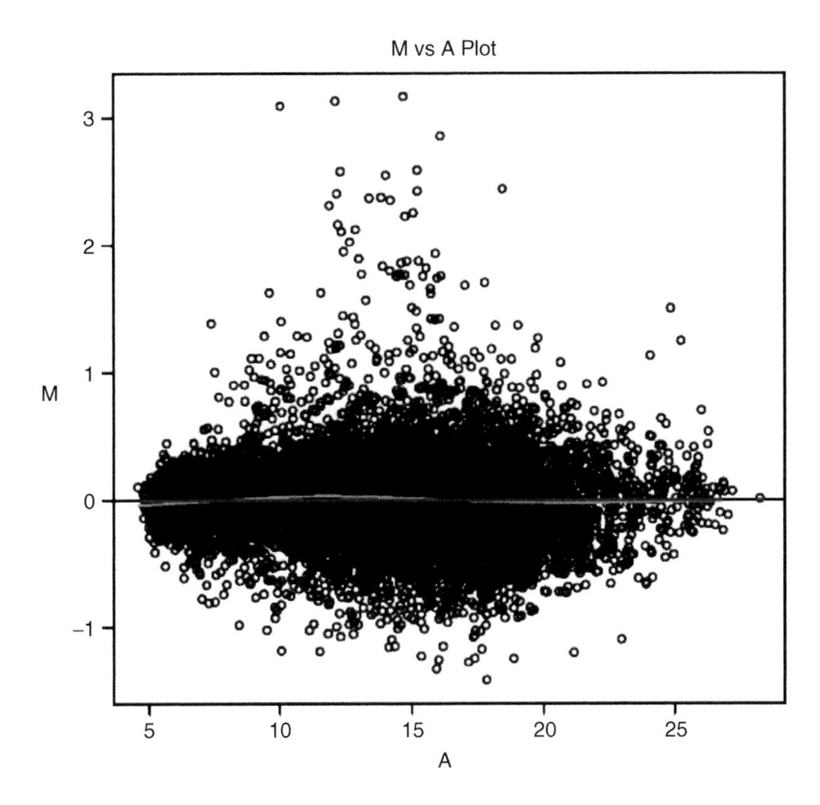

FIGURE 3.27
MA plot (R-I plot). This plot emphasizes any intensity-dependent trends in differential expressions between two arrays or two dye-label channels. For any arrays containing thousands of gene probes, it is expected that the majority of these will be centered around a log-ratio value of zero. If the data cloud deviates from a straight line or shows excessive spread of log-ratios, a nonlinear normalization method and/or a more variance-stabilizing normalization should be applied.

The problem of low numbers of observations per group and the therefore inherent unreliability of any statistics obtained can be alleviated to some degree by using "resampling" or "bootstrapping" techniques.[14,15] These are computationally expensive in that they require the creation of large amounts of virtual data. Basically, this approach consists of repeated random resampling of the original values into the two (or more) statistical groups in order to calculate the same test statistic for each such resampling run. The resulting large number of test statistics is then used to estimate the "true" statistic for the observed data. If one thousand or more of such runs are performed for each individual gene's expression values, then this can turn into a computational problem even for a simple t-test.

A minority of projects profit from custom statistical modeling of data and more involved analyses.[16] This is of course more time-consuming than a first-line analysis and requires the input of statistically/mathematically oriented bioinformaticians. One example objective is to directly infer genetic regulation pathways from microarray data.[17,18] For example, determine by means of Bayesian networks the conditional probability of gene A inhibiting or activating genes B and C, which may in turn also be activated by gene D, etc. Another example objective is to infer clinical relevance and complete biological pathways (genotype to phenotype or genome to proteome) from microarray data that have been combined with clinical or proteomic data.

3.4.9 Explorative Analysis

This type of analysis requires much closer cooperation between biologist and bioinformatician, since algorithm and biological meaning are more intertwined compared to significance analysis. Any chosen algorithm will present lists of interesting expression patterns that require comparison with biological knowledge.

Biological input will allow tweaking algorithms to enhance particular aspects of the data. Explorative analysis does usually not provide measures for statistical significance or probability.

Clustering and the closely related classification are of particular importance for the explorative analysis of microarray data. We therefore give a brief overview of these concepts.

3.4.9.1 Aim of Clustering

Clustering is a relatively straightforward approach for identifying patterns in high-dimensional data. Clustering methods are able to identify and group

- Individual genes that exhibit similar expression profiles across conditions (time-points, treatments, phenotypes, etc.)
- Expression profiles for whole samples across the set of genes present on an array

3.4.9.2 Biological Interpretation of Clustering Results

The biological assumption for genes which are regulated in a similar way across a number of conditions is that they could be coregulated or constitute part of the same genetic pathway. Samples with similar profiles of expression across the array are informative for sample classification, e.g., diagnostic or prognostic arrays.

3.4.9.3 Theory of Clustering

The theory behind clustering is the same for all algorithms. If we take an example gene for which we have one expression value each for three time-points, then we can represent these three values (also referred to as a vector) as a point in a simple XYZ-coordinate system, i.e. 3-D space. Another gene with three values will be somewhere else in this 3-D space. The clustering algorithm now sorts these points into clusters, or groups, of genes depending on how close they are to one another. There are differences in how this distance is measured or how clusters are assembled, but this is the theory behind it. Of course, the same ideas also apply to more than three observed expression values, although the resulting point in n-dimensional space cannot be represented in a 3-D graph any longer. One can try to imagine it as a point inside a sphere.

Most types of clustering require the user to choose at least two parameters: a suitable *distance metric*, e.g., Pearson correlation or Euclidean distance, to determine how distance between points in n-dimensional space is measured, and an *algorithm* to determine similarity/dissimilarity of clusters themselves (i.e., gene profiles), to separate them into distinct clusters.

3.4.9.4 Clustering vs. Classification (Unsupervised vs. Supervised)

Clustering is closely related to classification; however, whereas the former finds clusters in unknown data, the latter is used in context of *predicting* the cluster membership of an unknown gene or sample based *on predefined* genes or samples, i.e., training data sets.

This difference is also often referred to as *unsupervised* (clustering) vs. *supervised* (classification) approaches to machine learning. Broadly speaking, many clustering methods can be adapted to become classifiers and vice versa.

Classifiers are used to characterize the expression profile of known samples (e.g., cancer stage), and to then determine if new, unknown samples can be classified as belonging to the same sample conditions. This could then be considered a diagnostic method. Similarly, unknown genes could be classified by function on the basis of a training set of genes with known function.

In the next section we concentrate on clustering in the context of microarray data, for which certain defaults of graphical representation have been established. A classifier or supervised method can be implemented based on these, if training sets of data are obtained and used by the algorithm in a separate stage prior to running it on the new, unknown data.

3.4.10 Main Types of Clustering

3.4.10.1 Hierarchical Clustering

This is occasionally referred to as *dendrogram* (Color Figure 3.28) or *heat map*.

Hierarchical clustering shows the entirety of the data at a glance, and global sample differences are easily identifiable. Results for clustered genes can be misleading because once the underlying algorithm has assigned a gene to a cluster, this is not reevaluated by further iterative steps. Similar genes can therefore fail to cluster together.

Two main subtypes for hierarchical clustering are *divisive* and *agglomerative*. The former starts off with considering all genes to be part of the same cluster, and then subdivides it into smaller ones. The latter starts off with

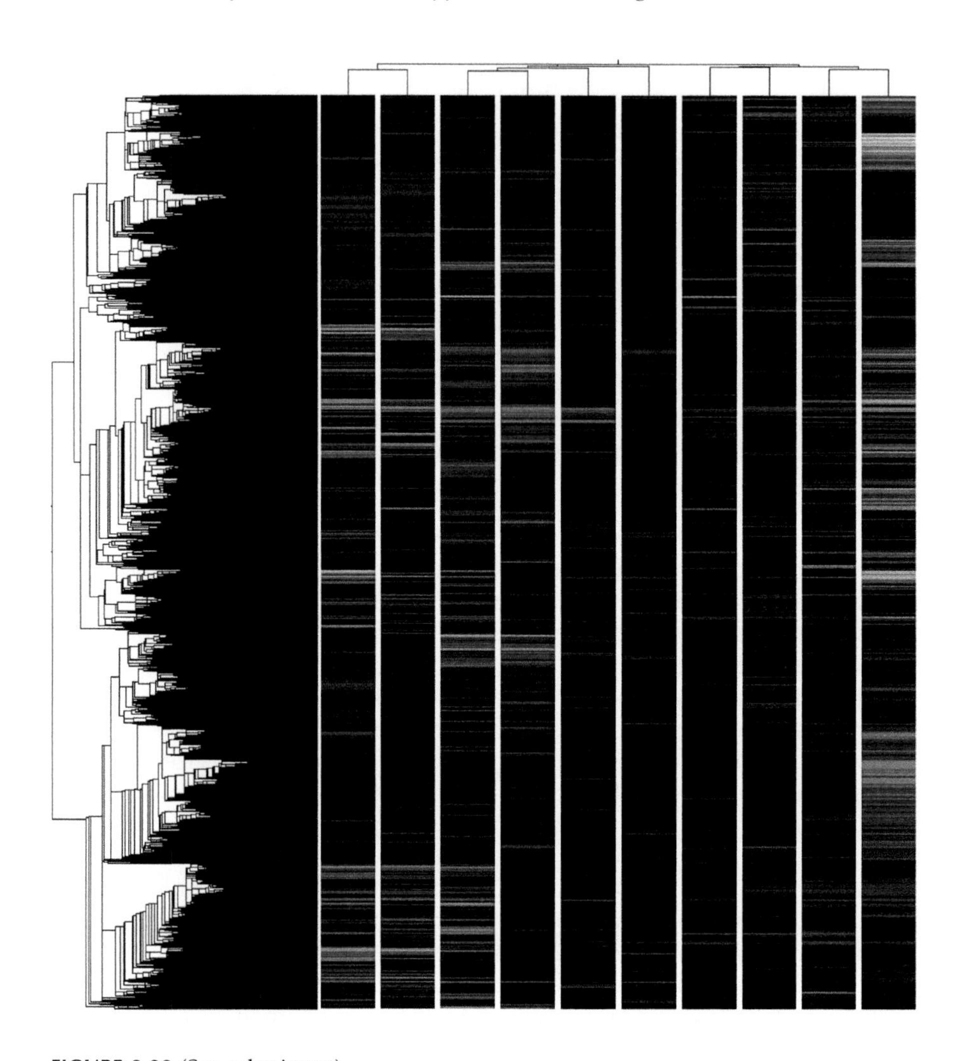

FIGURE 3.28 (See color insert)
Graphical output of a hierarchical clustering. Columns represent arrays; each colored line within the columns is an individual gene. The branches along the top indicate which arrays (i.e. samples) are most similar to one another. The branches along the side indicate which genes are most similar to one another in terms of expression across the four arrays. Interesting genes may be the ones that have high expression (red) in one set of arrays, but low expression (green) in another set.

considering each gene to be its own cluster, and then accumulates them into larger clusters. The results from both are not always very similar.

3.4.10.2 Nonhierarchical Clustering

3.4.10.2.1 K-*means and Self-Organizing-Maps (SOM; see Figure 3.29)*

Expression profiles are directly interpretable. Also, the clustering process is iterative and places genes in the most suitable cluster by means of an iterative process.

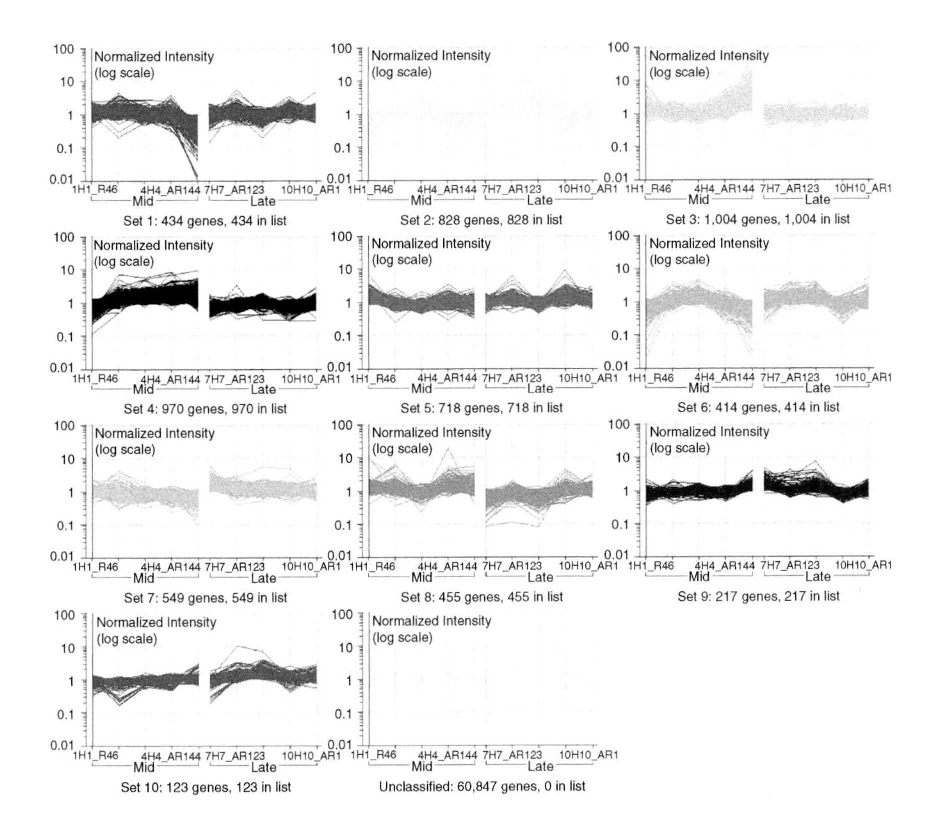

FIGURE 3.29
Graphical output of a *k*-means clustering. Each graph is a separate cluster. Each line within a graph represents the expression profile of a gene across a number of conditions. All genes within a cluster should be similar, but this depends very much on the chosen parameter values. A biologist might conclude that, for this example, cluster number 3 is of biological interest, due to its steady up-regulation in the first condition and no regulation in the second condition.

A disadvantage could be perceived in the need to specify the number of clusters as a parameter value ("*k*") prior to running the algorithm. In cases where there is no expectation as to the number of different types of expression profiles contained within the data, the user has to either experiment with various values for *k*, or use other techniques to gauge the number of clusters that may be present (e.g., the graphical output of a principal components analysis).

K-means is based on *k* randomly generated expression profiles, which are then used as the starting point for the first search for similar genes. After the first run, there follows a user-determined number of iterations to reassign genes based on new cluster means.

3.4.10.3 Other Clustering or Classification Algorithms

Principal Components Analysis (PCA, also used as a data reduction technique)

Support Vector Machines (SVMs)

Neural Networks

Hidden Markov Models (HMMs)

Decision Trees

This is just a small subset—there are many more different types or versions of the above available, but except for PCA and SVM, few of them are used on a regular basis by the microarray community.

3.4.10.4 Advice on Using Clustering

All explorative analyses, including clustering, are strongly influenced by user-selected parameter values. In order to achieve best possible results, it is not good practice to merely select one type of clustering or one set of parameter values for analyzing data. It is more advantageous and reliable to run different types of clustering with different sets of parameter values (although Pearson correlation or euclidian as distance metric could be regarded as a standard for microarray data) with the objective of comparing the resulting gene lists of cluster memberships. Reliable results should show up in a majority of approaches, but grossly different results may also indicate interesting genes.

Keep in mind that the results of cluster analyses rarely come attached with quantitative measures, and the interpretation should be a biological one.

Links:

Microarray-specific software

- Bioconductor packages for R statistical software
- GeneSpring
- Spotfire
- Cluster
- TreeView
- J-Express
- ArrayViewer
- MAExplorer
- ExpressionNTI
- Kensington
- Rosetta
- GeneSight
- Affymetrix

Statistical software

R

S/S-Plus

Matlab

References

1. Nature Genetics special, The Chipping Forecast II. *Nat. Genet.*, 2002, 32(4) (suppl), 461–552.
2. Kerr, K. and Churchill, G., Statistical design and the analysis of gene expression microarray data. *Genet. Res.*, 2001, 77, 123–128.
3. Churchill, G., Fundamentals of experimental design for cDNA microarrays. *Nat. Genet.*, Suppl. 32, 490–495.
4. Yee, H.Y. and Speed, T.P. Design issues for cDNA microarray experiments. *Nat. Rev. Genet.*, 3, 579–588.
5. Schuchhardt, J., Beule, D., Malik, A., Wolski, E., Eickhoff, H., Lehrach, H., and Herzel, H., Normalization strategies for cDNA microarrays. *Nucl. Acid Res.*, 2000, 28, e47.
6. Quackenbush, J., Microarray data normalization and transformation. *Nat. Genet.*, 2001, 32, 496–501.
7. Quackenbush, J., Computational genetics: computational analysis of microarray data. *Nat. Rev. Genet.*, 2001, 2, 418–427.
8. Hegde, P., Qi, R., Abernathy, K., Gay, C., Dharap, S., Gaspard, R., Hughes, J.E., Snesrud, E., Lee, N., and Quackenbush, J. A concise guide to cDNA microarray analysis. *BioTechniques*, 2000, 29, 548–562.
9. Speed, T.P., *Statistical Analysis of Gene Expression Microarray Data*. Boca Raton, FL, CRC Press, 2002.
10. Kerr, K., Martin, M., and Churchill, G., Analysis of variance for gene expression microarray data. *J. Computational Biol.*, 2001, 7, 819–837.
11. Yang, I.V., Chen, E., Hasseman, J.P., Liang, W., Frank, B.C., Wang, S., Sharov, V., Saeed, A.I., White, J., Li, J., Lee, N.H., Yeatman, T.J., and Quackenbush, J., Within the fold: assessing differential expression measures and reproducibility in microarray assays. *Gen. Biol.*, 2002, 3, 0062.1–0062.12.
12. Chen, Y., Kamat, V., Dougherty, E.R., Bittner, M.L., Meltzer, P.S., and Trent, J.M., Ratio statistics of gene expression levels and applications to microarray data analysis. *Bioinformatics*, 2002, 18, 1207–1215.
13. Dudoit, S., Shaffer, J.P., and Boldrick, J.C., Multiple Hypothesis Testing in Microarray Experiments. U.C. Berkeley Division of Biostatistics Working Paper Series 110 (http://www.bepress.com/ucbbiostat/paper110) 2002.
14. Davison, A.C. and Hinkley, D.V., *Bootstrap Methods and Their Application*. Cambridge University Press, 1997.
15. Carpenter, J. and Bithell, J., Bootstrap confidence intervals: when, which, what? A practical guide for medical statisticians. *Statistics in Med.*, 2000, 19, 1141–1164.
16. Moloshok, T.D. et al., Application of Bayesian decomposition for analyzing microarray data. *Bioinformatics*, Apr 2002, 18(4), 566–575.

17. Yeung, M.K.S., Tegnér, J., and Collins, J.J., Reverse engineering gene networks using singular value decomposition and robust regression. *Proc. Natl. Acad. Sci.*, 2002, 99, 6163–6168.
18. de la Fuente, A., Brazhnik, P., and Mendes, P., Linking the genes: inferring quantitative gene networks from microarray data. *Trends Genet.*, Aug 2002, 18(8), 395–398.

3.4.11 Data Storage

As described earlier, microarray experiments are associated with large amounts of data:

Images

Laboratory information, e.g., hybridization protocols

Array designs

Annotation

Raw numeric data

Processed numeric data

Analysis results

None of this information can be discarded, especially in the light of developments in standardization of microarray research. Journals have begun to request gene expression data and experimental documentation to be deposited in public databases on publication.

Experimental documentation is currently defined by the MIAME[1] standard (Minimum Information About Microarray Experiments), which identifies the items that need to be recorded, and ways in which they need to be recorded. However, this covers only parts of all information available, and for microarray printing centers it is important to also have a lab information management system (LIMS) implemented to track other information from start to end.

Submitting gene expression data to public databases currently favors ArrayExpress[2] or GEO. Bioinformaticians are working on simplifying this process by providing better front-ends, universal data formats, and ontologies.

Available tools tend to concentrate on the storage and retrieval aspects, but an important area of work in terms of bioinformatics is the automatic aggregation of data from comparable projects from all potential data sources, allowing for meta-analyses of data and large-scale data mining.

Links:

ArrayExpress: www.ebi.ac.uk/arrayexpress

GEO: www.ncbi.nlm.nih.gov/geo

GeneX: www.ncgr.org/genex/index.html

READ: http://read.gsc.riken.go.jp

SMD: http://genome-www5.stanford.edu

References

1. Brazma, A., Hingamp, P., Quackenbush, J., Sherlock, G., Spellman, P., Stoeckert, C., Aach, J., Ansorge, W., Ball, C.A., Causton, H.C., Gaasterland, T., Glenisson, P., Holstege, F.C.P., Kim, I.F., Markowitz, V., Matese, J.C., Parkinson, H., Robinson, A., Sarkans, U., Schulze-Kremer, S., Stewart, J., Taylor, R., Vilo, J., and Vingron, M., Minimum information about a microarray experiment (MIAME)—toward standards for microarray data. *Nat. Genet.*, 2001, 29, 365–371.
2. Brazma, A., Parkinson, H., Sarkans, U., Shojatalab, M., Vilo, J., Abeygunawardena, N., Holloway, E., Kapushesky, M., Kemmeren, P., Lara, G.G., Oezcimen, A., Rocca-Serra, P., and Sansone, SA., ArrayExpress—a public repository for microarray gene expression data at the EBI. *Nucleic Acids Res.*, 2003, 31, 68–71.

3.4.12 Data Mining

This is a fairly generic term often used for very different problems, and often used synonymously with knowledge discovery. We will consider data mining as a tool that can find patterns and relationships in accumulated data. Are some genes related, given certain conditions? Are some genes very different from other genes under a majority of conditions? Does the presence of some gene always mean the absence of others? Does gene activity relate to protein activity? In this respect, data mining tools could also be regarded as providing answers to questions that have not been asked yet, or at least not in any specific way.

One precondition for useful data mining is the aggregation of data. The more data are available to a good data mining tool, the more potentially reliable conclusions can be drawn. Data are not limited to gene expression data, any annotation information or even literature citations can be used. Data mining algorithms can also include advanced visualization techniques and a degree of interactivity (Color Figure 3.30).

Data mining also includes the basic techniques described in the section on explorative data analysis as well as advanced queries on databases.

The challenges are obvious: integrate data from a variety of different sources, define robust algorithms with visualization front-ends, make them computationally viable, interpret results, and fine-tune algorithms. The scope for improving existing methods or adding novel ones is still great, and other areas of science where this issue has already been researched are currently used as sources of inspiration.

3.4.13 Protein Arrays

Protein arrays serve similar purposes to genomic microarrays, although the measurements obtained refer to the proteome rather than the transcriptome. They, too, allow for classification of samples and changes in protein expression over time or conditions. *Preparation* of probes is different, but follows existing techniques for protein purification. Numerical *analysis* of protein

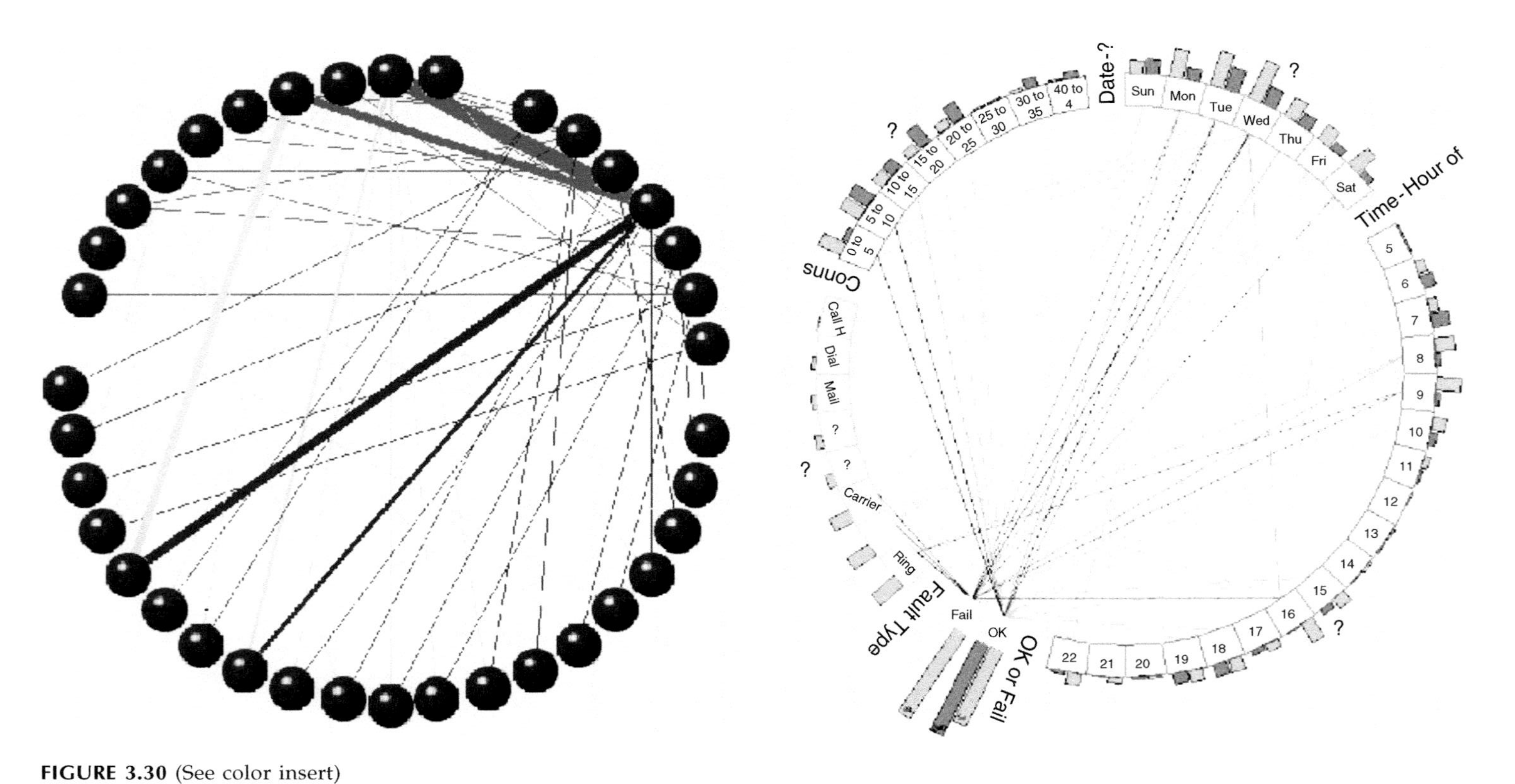

FIGURE 3.30 (See color insert)
Data mining; link analysis (*left*) and Daisy chart (*right*). (Source: http://www.daisy.co.uk/daisy.html. With permission.)

array data is identical to genomic array platforms, with the exception of subsequent annotation and data mining, which may be weighted toward the proteomics domain. The *interpretation* of results is more dependent on the type of analyte (i.e., target sample) in question, comparisons between proteins may be subject to other biological or biochemical factors.

Properties particular to protein arrays:

1. Number of depositable probes. This usually ranges in tens to hundreds rather than thousands. An increase can be expected with accumulation of knowledge and probe sets for the proteome.
2. Arraying and adherence of probes. This is challenging, since tertiary structure requires correct orientation to present a binding site for the target material.
3. Labeling. Approaches in terms of labeling the analyte directly or via another antibody (sandwich method) are broadly equivalent to DNA arrays, but protocols are very different.
4. Dual-target incubations. Competitive binding of target material is often more limited due to the size and number of binding sites of proteins. Current experiments are therefore often carried out with only one target sample per array.
5. Binding properties. The relationship between probes and target protein is more variable in terms of binding strength, whereas for genomic arrays the binding strength is a function of the length of the complementary strands.

3.4.14 Concluding Remarks

Genomics and the application of high-throughput technology such as microarrays are increasingly transforming biology into an informational science. The large amount of numeric data generated from quantitative measurements produced from bioarrays necessitates rigorous statistical approaches and visualization techniques. Accordingly, microarray technology is dependent on an ever-evolving bioinformatics toolbox. We hope this section has provided an initial guide to some of the key issues and challenges in this area.

3.5 Genomes as Gene Networks

How is genetic information used to bring into being and maintain cellular structures and activities? To find answers, it is necessary to probe the temporal interaction between genes within the genome based on gene expression patterns and ensuing protein interactions. Roughly speaking, a gene network is a genotype that defines the phenotype of a cell or organism. The analysis of the transcriptome bestows a functional understanding of the genome revealing in part how the "blueprint" of life is implemented. However, there are thousands of genes expressed at any given moment and one of the central questions in functional genomics are why so many and how the activity of one gene or group of genes (e.g., constituting a metabolic pathway, signaling pathway, cell structure) is controlled by the activity of other genes. There must be a hierarchical structure underlying the organization of interactions among active genes and those that are kept inactive. Studying these interactions requires the following steps:

1. Comprehensive analysis of gene expression
2. Analysis of RNA editing and splice variants
3. Studying functions of nonprotein-coding sequences (RNA interference)
4. Large-scale protein analyses (proteomics)

Gene networks can be analyzed by any of the first three steps, but the impact of the transcriptome must be corroborated at the protein level. Basically, gene networks are functional entities with internally dynamic structures where the number and combination of components (is a gene there or not) and connections (is a gene active or not) varies over time. A gene network includes many noncoding sequences that directly or indirectly affect gene regulation. Gene expression patterns identify which genes are activated or suppressed in coordinate ways and which genes are permanently active (housekeeping genes) or change in response to internal and external cues. But what is the physical state of a gene network? How do genes actually interact and coordinate their activities in a real cellular environment?

A first important step is to identify the components and establish their connectivity within a network. Not every gene is connected with every other gene, even though they may be expressed simultaneously. In addition, connectivties may not be reciprocal, but rather hierarchical (sequential) or cyclical (feedback loops). For instance, the expression of genes coding for enzymes of a specific metabolic or signal transduction pathway must be coordinated by regulatory genes that make proteins (transcription factors), which are not physically part of the pathway per se. Much of what we can extract from gene expression data depends on how much we currently understand about connectivity of pathways, the sharing of an enzyme by two pathways, the channeling of a metabolite through more than one pathway (e.g., is glucose used for respiration, reductive biosynthesis, or stored as glycogen?), or the reversibility of a pathway (e.g., thus a liver cell absorb blood glucose or secrete glucose into the blood serum?). Intuitively we understand that what we call metabolic pathway is the result of small gene networks activated to fulfill a specific physiological need of an organism.

The physical and chemical state of a cell affects protein activity, which in turn controls gene expression. Factors affecting protein and gene expression activity include osmotic pressure and temperature, the post-translational modifications of proteins, methylation states of DNA, localization of proteins and mRNA within the cell (compartmentalization), and ion concentrations (calcium signaling, pH sensitivity, availability of cofactors like Mg^{++} and Zn^{++} etc.), all of which contribute to establish a 3-D, dynamic network of molecular interactions, essentially the internal structure of the cell that defines the actual gene network and how it is implemented in an organism.[1]

One way of understanding gene networks is to ask about the minimal number of genes a cell or organism needs to survive or that differentiates a muscle fiber cell from a skin cell. In other words, what is the smallest gene network or genome that can sustain cellular life in various forms? A partial answer comes from the *minimal gene set* theory. Several interesting biological questions are raised by this theory that relates to the importance of gene networks in general, e.g., cell differentiation or metabolic pathways. Is the number of genes an appropriate measure of complexity? Why must a cell or body (human, bacterial) be so large compared to the size of atoms? Are viruses living things? With the growing number of whole genome sequences in public databases, it has become possible to experimentally address the minimal gene set theory and probe the number of genes an organism needs to be alive and autonomous.

To minimize the complexity of the analysis we restrict the question to prokaryotic organisms. With over one hundred bacterial and archaeal genome sequences now available, the answer to how many genes are needed to make a single cell should not be that hard to find. To summarize the results for bacterial cells, the minimal number of genes for a viable parasitic bacteria or bacteria grown in laboratory settings is estimated at ~300 genes. Experimentally, this number has been arrived at by studying the viability of bacterial cell cultures with single gene knockouts. Any gene eliminated by knockout that does not kill the organism is deemed nonessential, as it is obviously not needed for survival. Therefore, a minimal gene set consists of genes that are absolutely necessary for the survival of a cell. This number is best compared to the 524 genes found in the genome of the parasitic bacterium *Mycoplasma genitalium* and the 230 genes in genome of cytomegalovirus (CMV), a member of the herpes virus family. These two pathogenic organisms represent the lower limit of genes deemed necessary for cellular life and the upper limit of genes found for noncellular, viral life. Both pathogens, of course, depend on the metabolic activity (and thus gene set) of a host cell.

These 300 essential genes identified through genetic screening of mutant strains can be compared to minimal gene sets obtained through bioinformatics approaches based on whole genome comparison. Here, orthologs from complete genomes are considered essential if they are found in all genomes included in the study. Initial comparison of the two bacterial genomes of

M. genitalium (524 genes) and *H. influenzae* (1789 genes) identified 240 genes these prokaryotes have in common. All other genes have no orthologs and thus should be thought of as specific to each bacterium. The question, however, is if these genes can be deemed nonessential. A more inclusive comparison base on 25 bacterial genomes revealed only about 80 orthologous genes common to all species. These genes are dubbed *universal* genes. Incidentally, they all code for proteins and RNAs involved in transcription and translation. Clearly, this approach does not identify a gene set regarding the viability of cellular structures, but identifies evolutionary lineages and early common ancestral genes.

The environment plays a central role in defining a minimal gene set for an organism. The bacteria and virus mentioned earlier are all parasitic/symbiotic organisms and directly benefit from the gene set of the host cell/organism. The defining parameter of the viability of a single-celled organism is the availability of energy sources and metabolic requirements to synthesize any organic building blocks for cellular growth themselves, i.e., amino acids, lipids, monosaccharides, and nucleotides. For free-living archaea and bacteria, the lower number of genes is about 1500, while obligate symbionts and parasitic bacteria require as little as 500 genes.[2]

In general, any given species can only survive in the presence of other species, a fundamental tenet of ecosystems and process of evolution. There are, however, exceptions in the form of autotrophs such as prokaryotes and plants, which have a much reduced dependency on organic food sources from other organisms and can live off inorganic matter and/or sunlight as a source of energy, and have all the necessary enzymes to assimilate single-carbon sources (e.g., carbon dioxide) to synthesize their carbohydrates, lipids, amino and nucleic acids. Autotrophs include chemoautotrophs and photosynthetic organisms. Phototrophs produce, besides bacterial species, some of the largest life forms (trees), containing some of the largest genomes. All higher organisms, however, still depend on nitrogen-fixing bacteria converting the abundant atmospheric N_2 to nitroxides and amino groups as metabolic precursors of nitrogenous groups in lipids, proteins, nucleic acids, and carbohydrates. Chemotrophs have genomes with usually more than 2000 open reading frames, up to four times the minimal gene set found in obligate symbionts and parasites.

For all its flexibility and wide range of interpretation, the minimal gene set theory offers a useful way of thinking about gene networks. Essentially, a genome is a unique gene network coordinating the growth, reproduction, and survival of an organism. Particularly the small gene networks of viral genomes with as little as seven genes demonstrate how a set of genetic elements can coordinate metabolic activity of a host cell to synthesize a new viral particle. In addition, a minimal cell must not be similar to a modern cell with DNA, RNA, protein, and carbohydrate structures. Furthermore, the question of a minimal gene set or genome for a self-replicating organism (a cell) can be reduced to the question of why a cell must be so large compared to the size of atoms. Why must a cell have a certain size? How many genes are needed to form cells of prokaryotic and eukaryotic size (ratio of area per volume)? Is the need for membranous organelles a prerequisite to increase cellular volume from prokaryotic to eukaryotic structures? How is cell size related to complexity of an organism?

Revealing in this respect is the analysis of the upper limit of prokaryotic genomes. Some free-living archaea and bacteria have expanded genomes of up to 9 Mb. While the relative portion of noncoding sequences is independent of bacterial genome size, expanded genomes show an increase in genes coding for regulation, secondary metabolism, and energy conversion pathways. This gives free-living prokaryotes broad metabolic diversity using alternate electron acceptors and a large range of substrates for energy production. Increasing regulatory capacity to switch among several alternate energy generating modes is advantageous.[3] To extend this line of thinking to the process of cell differentiation during development of a multicellular organism, a similarly complex regulation of the available genome to activate or inactive patterns of gene networks creating specific genotypes or cell types should be found from comparative genomics (i.e. comparing transcriptomes), including single- and multicelled eukaryotes.

The question about the relation between size of atoms and size of living cells was first posed by the physicist Erwin Schröedinger in the 1940s. The question remains unanswered, but with our advanced structural knowledge of genes and proteins we can rephrase the question: what makes biological macromolecules special, why do they have the size they have, forming interaction networks in the context of a viable cell? For Schroedinger, cells and macromolecules are aperiodic crystals with a relatively small number of atoms (less than a million). Because of this limited number of atoms in a macromolecule, structural fluctuation around the mean position of individual atoms or group of atoms (noise) is large relative to the size of the macromolecule. Noise starts to determine the behavior of macromolecules, for fluctuations, or conformational changes and binding–unbinding events in interaction networks, are exploited for catalytic and regulatory mechanisms. Fluctuations are also seen in gene expression data that is often considered noise. However, such noise may reflect stochastic fluctuations that underlie molecular "decision making" in biological processes. For instance, the high fidelity simulations of small networks provide evidence that deterministic models do not fit some biological processes as well as stochastic models.[4] Thus pathway regulation within gene networks may well operate at a stochastic level.

Finally, the analysis of gene networks will also touch on the discussion among evolutionary biologists whether populations, individuals or genes are the object of natural selection. Gene networks highlight the interrelatedness of genes within a genome and explicitly lead away from a gene-centric (selfish gene) view of selection by studying how the behavior of one gene is influenced by the behavior of many others.

References

1. Le, P.P. et al., Using prior knowledge to improve genetic network reconstruction from microarray data. *Silico Biol.*, 2004, 4(2), 0027.
2. Waters, E. et al., The genome of Nanoarchaeum equitans: Insights into early archaeal evolution and derived parasitism. *Proc. Natl. Acad. Sci.*, 2003, 100(22), 12984–12988.
3. Konstantinidis, K.T. and Tiedje, J.M., Trends between gene content and genome size in prokaryotic species with larger genomes. *Proc. Natl. Acad. Sci.*, 2004, 101(9), 3160–3165.
4. Schaefer, C.F., Pathway databases. *Ann. N.Y. Acad. Sci.*, 2004, 1020(1), 77–91.

4

Proteome Analysis

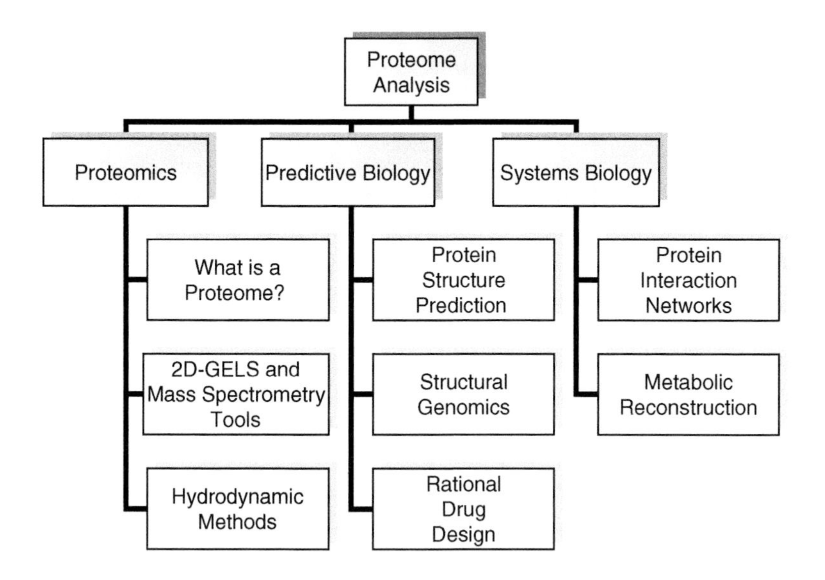

FIGURE 4.1
Chapter overview.

4.1 Proteomics

4.1.1 What is a Proteome?

Most databases are built from sequences of genes, genomes, and structures of proteins. However, less is known about which macromolecules—nucleic acids, proteins, lipids, and carbohydrates—interact how and when to give structure and function to a cell, be they single-cell organisms like bacteria and yeasts, or multicellular organisms like plants and animals. Furthermore, because mRNAs are the molecular intermediate during protein biosynthesis,

TABLE 4.1

Proteomics Analysis of Protein Content

Goal	Methods
Protein composition	2-D gel electrophoresis and mass spectrometry
Protein conformation	Hydrodynamics
Protein interaction	Protein arrays, immune precipitation, 2 hybrid system

transcript levels are often interpreted regarding their biological relevance as if they correlate with actual protein levels.

However, cellular levels of mRNA are not necessarily reliable markers for protein levels. It is therefore of paramount importance to establish the protein profile of a cell independently of transcript analysis. This will not only ensure a more accurate assessment of the relative number of proteins in a cell and potential interactions, but also reveal their conformation and in which chemical form they exist, i.e., post-translational modification such as glycosylation, acylation, ubiquitination, phosphorylation, or proteolytic processing. These modifications are used to control the activity and location of a protein in the cell. As Table 4.1 shows, various techniques are commonly used to study protein content, conformation, and interaction in cells. The techniques include two-dimensional (2-D) gel electrophoresis and mass spectrometry, hydrodynamic methods, protein arrays, immunoprecipitation, and the yeast two-hybrid system.

To make matters more complex, the protein composition of a cell and the posttranslational modifications of proteins can vary rapidly (minutes to hours) such as during cell growth and development using cues from cell–cell interaction or in response to metabolic and environmental stress. Unregulated changes in protein content and activity can cause disease. Tumors, for example, often show altered expression and activity patterns of key proteins as compared to healthy tissue. Since these proteins are related to growth control, carcinogenesis can generally be viewed as some lack of control in growth, or the unhindered multiplication of cells that should no longer divide or are programmed to die (apoptosis).

However, it is very important to realize that not all macromolecules and cellular structures are synthesized off a linear template. In fact, nucleic acids and proteins are the only molecular species directly encoded for by DNA. Everything else—and this includes the posttranslational protein modifications—is guided by molecular interactions, sequential synthesis, and spatial separation known as compartmentalization. Good examples of biological macromolecules lacking a gene template are polysaccharides and protein and lipid glycosylation. Polysaccharides (carbohydrates) exist as linear, as well as branched, multimers, and although polymer sequences are consistently reproduced by the cellular machinery (proteins catalyze carbohydrate synthesis), these sequences are not encoded by other linear, molecular templates as DNA codes for protein synthesis. Instead, polysaccharide synthesis is a sequential catalytic activity performed by the spatial arrangement of enzymes within the cell. Individual genes code for each enzyme that takes part in a pathway. Groups of enzymes that synthesize polysaccharides are therefore not independent, since the lack of an intermediate step in the pathway causes a defect and thus a disease.

It is therefore important to understand the structure of enzymatic pathways. Comparing not only individual genes across species, but entire pathways yields additional information about newly discovered DNA sequences. Are pathways identical across species? Are all enzymes of the same pathway homologs expressing similar degrees of identity? Are certain enzymes in pathways more important or more conserved than others? Do some species have alternative pathways to generic ones, while others do not? Finding the answers to such questions is the true challenge of biology in the 21st century. The Internet (or any equivalent form of public communication) will be instrumental in this discovery process. It will provide the databases necessary for comparing the protein composition of a cell or an organism as a function of metabolic activity and disease from the period of conception to the moment of death.

The study of the entire set of proteins of an organism is called proteomics. Proteomics analyzes protein content and interactions in a cell or an organism using the tools summarized in Table 4.1. Basically, proteomics describes the organizational complexity of the machinery of cells at the level of proteins. The term proteomics refers to the idea that all proteins of any given organism are necessarily linked in their fate with each other.

4.1.1.1 2-D Gels and Mass Spectrometry Tools

To identify protein content of a cell or tissue, proteomics makes use of a biochemical technique invented in the early 1970s[1] where proteins are separated in a gel matrix by electrophoresis in two dimensions using molecular weight for the first dimension and electrical charge as a function of pH for the second dimension (Figure 4.2). 2-D gel electrophoresis allows both analytical and preparative separation of a protein mixture to identify the protein content for differences in composition, quantities, and post-translational modifications. Gels can be used to compare sets of proteins at various conditions to correlate protein expression patterns to cellular activity and environmental influences.

2-D gel electrophoresis separates proteins based on size and charge. The calculated molecular weight of a protein based on its DNA sequence often does not exactly match the experimentally determined value from gel electrophoresis. While size is largely determined by the genetic information (amino acid sequence), it may actually differ depending on RNA splicing, proteolytic processing, and post-translational modification of amino acids. The exact determination of the actual molecular weight and sequence of a protein at the time it was isolated and forming a separate spot on a gel is the whole purpose of 2-D gel analysis.

Modern analytical automated systems assist in the large-scale identification of these protein spots (Figure 4.3). Proteins of interest can be digested

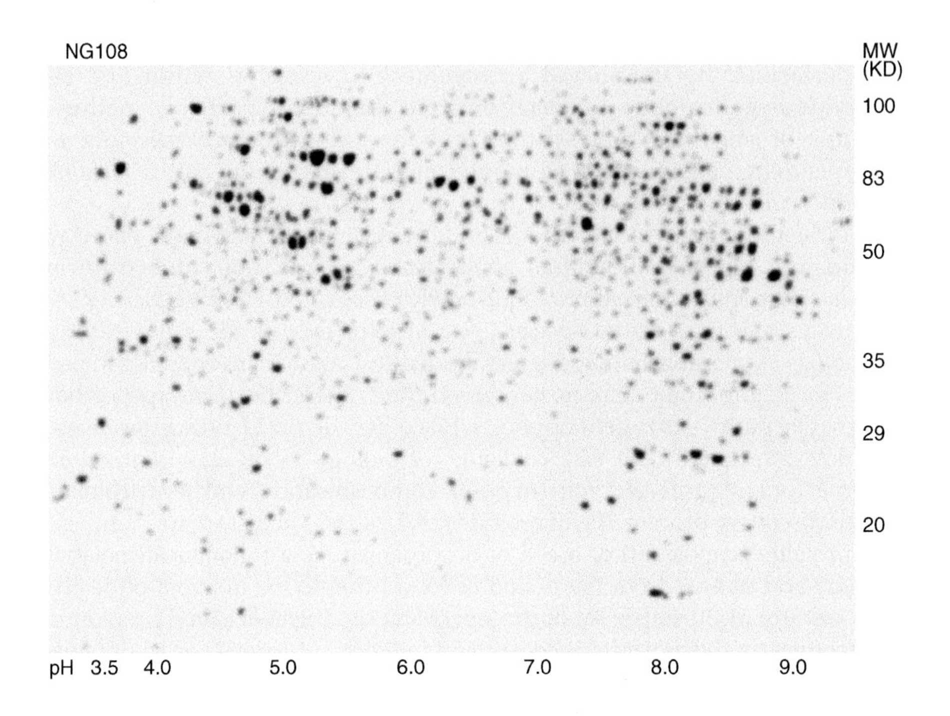

FIGURE 4.2
2-D polyacrylamid gel showing protein contents of eukaryotic cell type. Proteins extracted from the mammalian cell line NG108 are separated according to molecular weight (MW in kilodalton [kDa]) on *Y*-axis and according to charge in a pH gradient ranging from 3.5 to 10 (*X*-axis). Each spot represents an individual protein type. Intensities reflect protein concentrations. Proteins can be extracted from gel matrix for biochemical analysis. (sequencing, mass spectrometry). (From Young Yang, R.W. Johnson Pharmaceutical Research Institute, San Diego, California. With permission.)

within the gel matrix and the resulting peptides are extracted and subjected to high mass accuracy MALDI-MS (matrix-assisted laser desorption/ionization mass spectrometry) analysis. Here, peptide fragments are ionized and their charge/mass ratio is determined. The mass/charge ratio is matched to all possible amino acid sequence combinations. If the matching is ambiguous, the peptide fragment must be micro-sequenced and the sequence subjected to a database search using BLAST algorithms. If all or most fragments from a single 2-D gel spot match the same sequence in the database (e.g., GenBank), the corresponding spot can be assigned a protein including potential isoforms and modifications like multiple phosphorylation or splice variants.

The matching process is often not straightforward because of post-translational modifications. These modifications affect net charge, reactivity, and solubility of proteins. Phosphorylation, for instance, adds negative charges to the protein, thus influencing its mobility during electrophoresis.

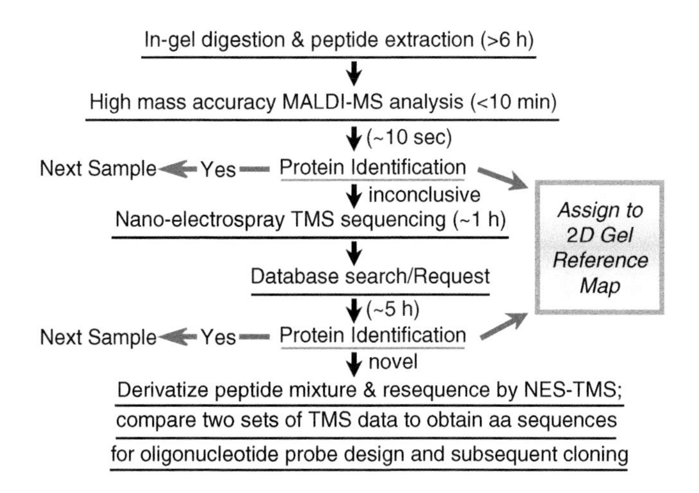

FIGURE 4.3
Strategies for the identification of proteins from 2-D gels. See text for details. (From Young Yang, R.W. Johnson Pharmaceutical Research Institute, San Diego, California. With permission.)

A single negative charge has the equivalent effect of decreasing the molecular weight of a protein by 2 kDa, or roughly 15 to 18 (noncharged) amino acids. Glycosylation also affects the molecular weight of a protein, but not necessarily its pH dependence. Because of the existence of multiple modifications that affect the apparent mobility of a protein on a gel in a similar way, the interpretation of small differences in mobility of proteins on 2-D gels is not always easy and requires careful biochemical analysis.

The entire process of peptide fragment identification has been automated over the last several years. Automated processes require special robotic equipment, as well as customized software. Again, computers play a central role in controlling and analyzing the process. As shown in Figure 4.4, an autosampler collects peptide fractions from a HPLC column chromatography column, which separates peptides according to size. Very small volumes are used in capillary columns and subjected to nano electrospray ionization for mass spectrum analysis. Experimental and predicted mass spectra are used to generate cross-correlation data to identify the sequence of the extracted peptide fragments. If several fragments from a single 2-D gel spot match a single amino acid sequence entry in the database, a protein is identified.

4.1.1.2 2-D PAGE at Expasy (Swiss Bioinformatics Institute)

An important part of proteomic analysis of cellular mechanisms is to compare 2-D gels of cellular extracts obtained after stimulating a cell with an activator (such as insulin on liver cells) with those obtained under metabolic resting conditions.[2] Many public databases include a growing number of such reference gels for preliminary identification of the charge and molecular weights

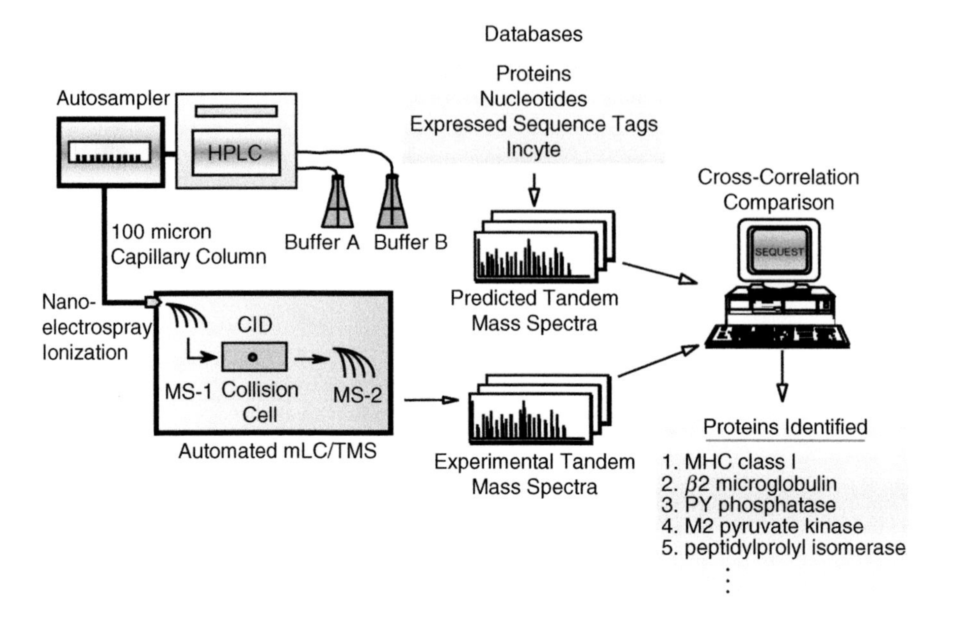

FIGURE 4.4
Fully automated protein identification. (From Young Yang, R.W. Johnson Pharmaceutical Research Institute, San Diego, California. With permission.)

of a novel protein. SWISS-2DPAGE is a public proteome database located at the Geneva University Hospital, Switzerland (http://au.expasy.org/ch2d/) offering proteomics tools useful to study the protein content and interactivity of an organism. The site is organized to access the 2-D database interactively, to provide online help, technical manuals for 2-D gel electrophoresis, and services such as running gels upon submission of samples, training courses, and software packages for the analysis of 2-D gels.

The database (Release 17.1, March 2004) contains 1265 entries in 36 reference maps from human, mouse, *Arabidopsis thaliana, Dictyostelium discoideum, Escherichia coli, Saccharomyces cerevisiae,* and *Staphylococcus aureus* [N315]). The database includes 2-D gels from human and mouse cell types (platelet, red blood cell, macrophage, plasma protein, lymphoma, liver, kidney, two leukemia cell lines, cerebrospinal fluid, and intestinal epithelial cells). Known proteins can be found by searching for an accession number (SWISS-PROT) (Figure 4.5 A and B) or clicking on 2-D gels with marked spots. The putative locations of new proteins can be identified if the amino acid sequence is known. Calculations of the theoretical molecular weight and charge are used to place the protein on the gel. Many proteins yield multiple spots on 2-D gels due to modifications, information that is very valuable to the biochemist in understanding the function of a protein within a cellular environment.

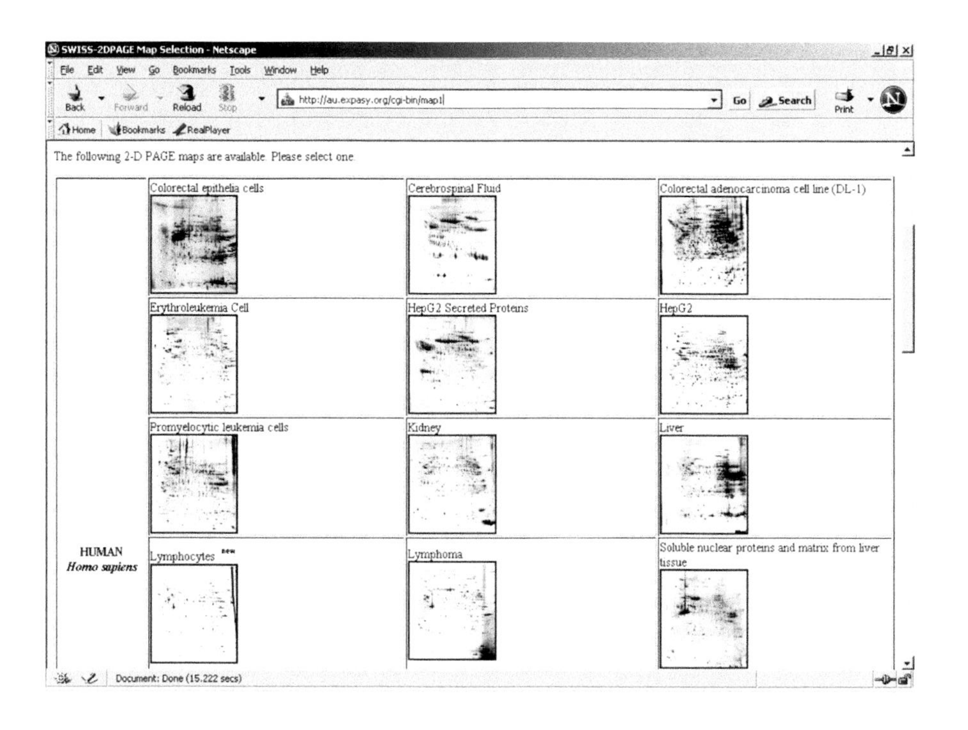

FIGURE 4.5A
SWISS-2-DPAGE Map Selection. (From Swiss Prot; http://au.expasy.org/cgi-bin/map1. With permission.)

A look at some of the gels at SWISS-2DPAGE shows that the majority of spots are not linked to any known proteins. Identification of these spots is a work in progress and will require some time to be completed. New technology is being developed to more quickly identify proteins on 2-D gels. Biochemical analysis using microsequencing of peptide fragments and mass spectrometry of these fragments is the analog approach in sequencing nucleic acid libraries.

Similar to gene expression analysis, the most valuable information from 2-D gel analysis will come from functional comparison. Once a protein is identified in one cell type or organism, its expression level can be compared in other cell types or tissues, potentially revealing different levels of expression and modes of posttranslational modifications. This task of comparing protein expression levels is hardly a trivial one. The way a protein runs on a gel greatly depends on the purification procedure, source, and electrophoresis procedure. Comparison, therefore, requires cautious interpretation regarding relative positioning of spots and their intensities. SWISS-2DPAGE offers an analysis package for rapid image manipulation, complete 2-D analysis, worldwide comparison for referencing, and automated gel matching and comparison (Melanie II 2-D Analysis Software, developed by Denis

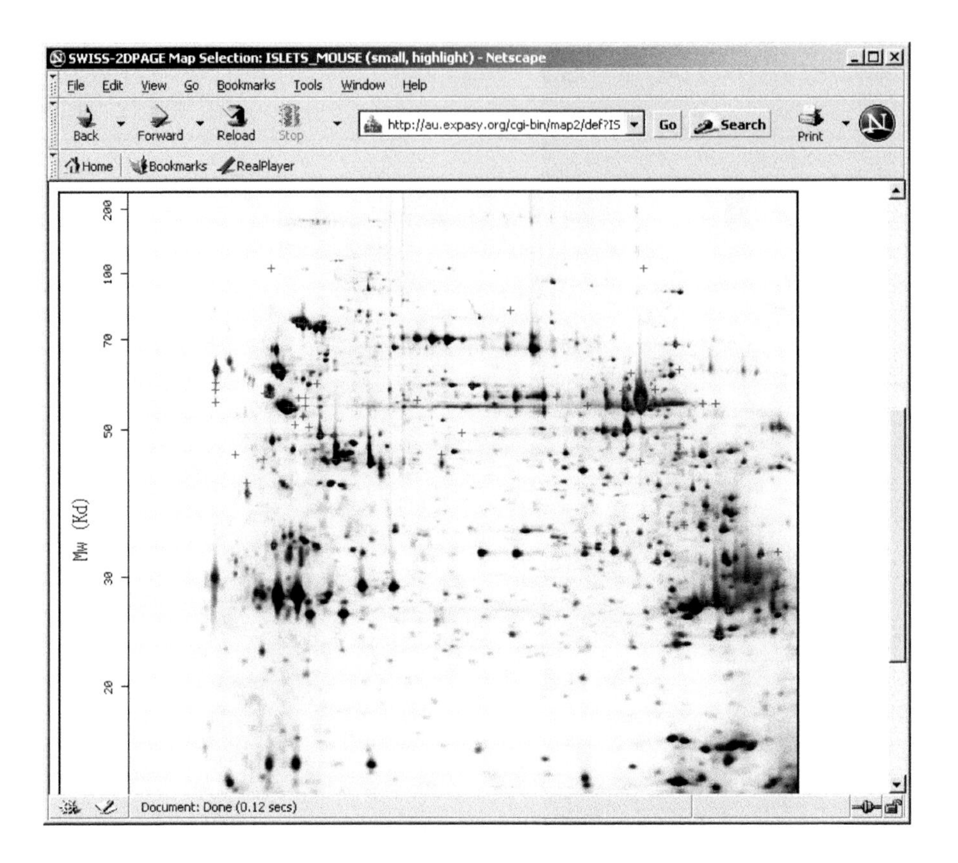

FIGURE 4.5B
Map selection: islets_mouse. Pancreatic islet cells from mouse on Swiss 2-D server; crosses indicate identified proteins and links to Swiss prot data base.

Hochstrasser at Melanie Group in Geneva, http://www.expasy.ch/melanie/MelanieII/description.html). The features of Melanie II include:

Rapid Image Manipulation:

- Zooming
- Filtering (smoothing, contrast enhancement, background subtraction)
- Gel flipping
- Gel stacking for better visualization
- Image stretching

Complete 2-D Analysis:

- Automatic spot identification and analysis
- Gaussian spot modeling
- Gel overlay display

- Point-and-click interface
- Embedded landmarks
- pI/MW setting
- Extended reports
- Histograms
- Statistical data analysis

Worldwide Comparisons:

- Multiple gel display
- Fast, automatic gel comparison and matching
- Reference gels for comparison to all other gels
- Creation of synthetic gels by merging a set of gels
- SWISS-2DPAGE master gels
- Network and online links to biological databases, including SWISS-2DPAGE and SWISS-PROT through Expasy
- WorldWide Web server

Data Import/Export:

- Gel printing
- Image import/export from and to TIFF and PPM
- Data export to Excel and other applications
- Data export as Melanie I format to public statistical and heuristic clustering programs

A variety of additional specialized proteomic databases are tailored with narrower applications in mind. For instance, the Rodent Molecular Effects Database at Oxford's Glycosciences compares expression profiles associated with toxins and xenobiotics. The Danish Centre for Human Genome Research (http://proteomics.cancer.dk/) maintains a keratinocyte database with information on human and mouse skin protein profiles and diseases. Relying on knockout and transgenic animals, the absence or addition of genes should be observable at the protein level by a missing or additional spot on a 2-D gel.

The following section on protein studies is a detailed example of one approach to protein analysis and the relevant computational tools that it employs.

References

1. Klose, J., Protein mapping by combined isoelectric focusing and electrophoresis of mouse tissues. A novel approach to testing for induced point mutations in mammals. *Humangenetik,* 1975, 26(3), 231–243.
2. Wilkins, M.R. et al., Protein identification with N- and C-terminal sequence tags in proteome projects. *J. Mol. Biol.,* 1998, 278(3), 599–608.

4.2 Hydrodynamic Methods

Borries Demeler

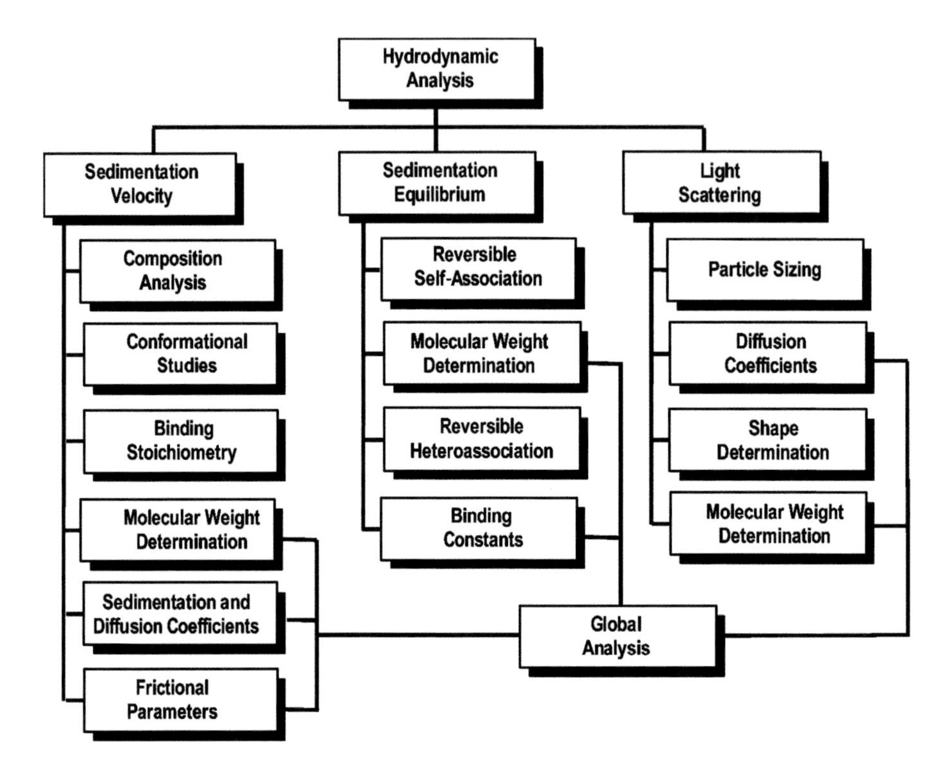

FIGURE 4.6
Section overview.

4.2.1 Introduction

The elucidation of macromolecular structure and function plays an important role in today's research laboratories. With many studies in biochemistry focusing on dynamic interactions among molecules in the solution state, hydrodynamic methods offer a powerful array of tools to study such systems. Modern computers permit sophisticated numerical analysis of the experimental data, which can yield a wealth of information about a wide range of hydrodynamic and thermodynamic properties of the macromolecules under investigation. Examples of parameters that may be studied with hydrodynamic methods include molecular weight, association constants, sedimentation coefficients, diffusion coefficients, virial coefficients, frictional parameters, molecular shape, molecular volume and density. Study of these parameters provides insight into macromolecular organization, oligomerization properties, conformation, binding stoichiometry, and sample composition. These techniques allow the investigator to follow assembly processes of multi-enzyme complexes, to characterize recombinant proteins and assess sample purity before proceeding to NMR or x-ray crystallography experiments. By performing analytical ultracentrifugation (AUC) and light scattering (LS) experiments, it is possible to observe macromolecules and macromolecular assemblies in solution, in a physiological environment, unconstrained by a crystal structure or an electron microscope grid. Hydrodynamic methods also allow the investigator to change the solution environment and then to observe the dynamic change of the molecular properties in response to that environmental change. Among the experimental conditions that may be changed are:

1. Macromolecular concentration: a change in concentration will affect concentration-dependent association and oligomerization behavior, and it provides the basis for the study of solution nonideality.
2. Buffer composition: by changing the buffer composition, the researcher can study the effect of selected ions and ionic strength on the molecule, the effect of drugs such as nucleotides, reductants, inhibitors and other small ligands, and effects of changes in the pH.
3. Temperature: the ability to change temperature allows the researcher to follow folding reactions, provide multiple signals for diffusion measurements and monitor the dependency of binding strength on temperature.

Given the appropriate instrumentation, these methods are applicable for molecules with a wide range of molecular weight and chemical composition. Biological molecules in the range of a few thousand Daltons to multi-megadalton complexes, including proteins (hydrophilic or membrane bound), DNA, polysaccharides, synthetic polymers, and even entire virus particles are all suitable candidates for hydrodynamic analysis. In this chapter the reader is provided with an introduction to hydrodynamic methods. The material is presented from the viewpoint of an experimentalist, but sufficient background information is given to provide an overview of the underlying theory of sedimentation and diffusion transport processes. While the theoretical treatment of the underlying physics have not changed during the past several decades, the experimental equipment and data analysis capabilities have seen significant improvement over just the past few years, allowing the experimentalist to obtain a much more detailed view of molecular structure and function than was previously possible. For a more in-depth treatment of the theory, we refer the reader to the existing literature (see Section 4.2.5), and will instead focus on an overview of the experimental applications and data analysis. Section 4.2.2 discusses analytical ultracentrifugation, Section 4.2.3

discusses light scattering experiments, and Section 4.2.4 discusses global data analysis.

4.2.2 Analytical Ultracentrifugation

4.2.2.1 Experimental Setup and Instrumentation

The instrumentation used for AUC is commonly a Beckman Optima XLA analytical ultracentrifuge. Various optical systems are available to measure samples under multiple conditions. The most common optical systems are the UV/visible absorption system (Figure 4.7) and the Rayleigh interference optics. Other optical systems have been developed as well; they include fluorescence optics, Schlieren optics, and turbidity optics. Multiwavelength, Raman, and light scattering detectors are either under development, or are considered for development. With the UV/visible absorbance system, multiple chromophores can be observed independently and used to measure the sample under a wide range of concentrations. The sample is contained in one of two sector-shaped channels of an epon/charcoal or aluminum centerpiece. The second channel contains the reference solution. The centerpiece is sealed at the top and bottom by a quartz or sapphire window, and assembled in an aluminum cell housing. The cell is placed into a hole in the rotor positioned at a radial distance between 5.8 to 7.2 cm. Rotor speeds between 2000 and 60,000 rpm can be selected to allow accelerations between 50 and 200,000 g, which is suitable for measurement of molecules with a very large range of molecular weights. In the UV/visible absorption optical system, a Xenon flash lamp emits high-intensity light flashes, which have strong emissions in the UV range. The emitted light is passed over a diffraction grating, where a narrow wavelength range can be

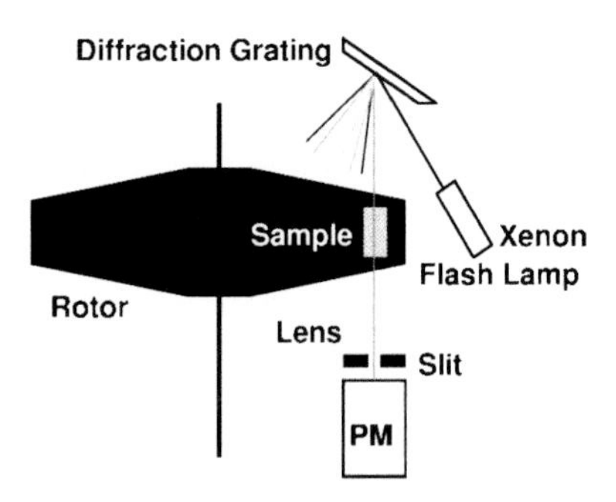

FIGURE 4.7
Schematic view of the AUC experimental setup for a UV/visible optical system. Light from a Xenon flash lamp with emission peaks around 230 and 280 nm is passed through a monochromator. Light of the desired wavelength is transmitted through the sample and the absorbance is recorded below the rotor. A moving slit assembly permits absorbance values to be collected for small radial increments.

selected for analysis. The light is passed through the sample cell, and using a narrow slit assembly, captured below the rotor in a dual-beam spectrophotometer as absorbance or intensity versus radial distance coordinate pairs. Interference systems utilize a laser and CCD arrays for data capture. In the interference optical system the refractive index changes are recorded across the cell instead of absorbance. This requires careful matching of the buffer in the reference channel, since buffer components will redistribute and lead to refractive index gradients. Extensive dialysis of the sample is required, with the dialysate serving as a reference solution. Advantages of the Rayleigh interference or schlieren optical system include cases where strongly absorbing buffer components interfere with the absorbance of the sample.

The current design of the analytical ultracentrifuge also provides higher resolution and faster data collection for the interference optical system. For cases where very tight binding oligomerization processes are studied, fluorescence optics can be employed, since it allows measurement of the sample at very low concentration. The experimental data is collected in digital format on a PC, where it can be analyzed with numerical or graphical methods discussed later in this chapter.

4.2.2.2 Transport Processes in The AUC Cell

In order to better understand the transport processes in the analytical ultracentrifugation cell, a review of the theory of sedimentation and diffusion is in order. Below, each subject is briefly covered. Readers looking for more detailed information on this topic are referred to the reference section in the Appendix.

4.2.2.2.1. Sedimentation

When a solution is accelerated in a rotor, the molecules dissolved in the solution will start sedimenting because several forces are acting on the molecules once the rotor starts spinning. First, the centrifugal force (*Fc*) accelerates the particle towards the bottom of the cell (see Figure 4.8). Immediately, two opposing forces, the frictional drag (*Fd*) and the buoyancy force (*Fb*), balance the centrifugal force and cause the particle to sediment with terminal velocity towards the bottom of the cell. The following relationships hold:

$$Fc - Fb - Fd = 0 \tag{4.1}$$

where:

$$Fc = \omega^2 rm \tag{4.2}$$

$$Fb = \omega^2 rm_0 \tag{4.3}$$

$$Fd = fv \tag{4.4}$$

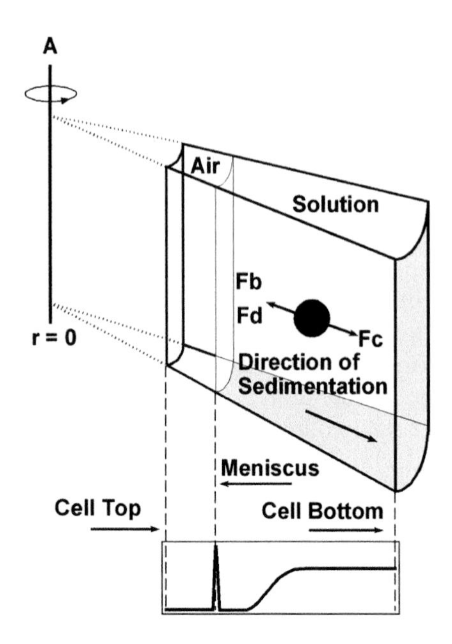

FIGURE 4.8
A schematic view of an analytical ultracentrifuge cell (not drawn to scale, and for simplicity only a single channel is shown). The sector-shaped cell contains the solute in a buffer solution. The air pocket above the solution column gives rise to a meniscus, which generally appears as a spike in the absorbance profile (bottom portion of the figure). Upon rotor acceleration, the particle experiences three forces, the centrifugal force *Fc*, which is balanced by equal and opposing forces *Fb* (the buoyancy force) and the viscous drag, *Fd*. As the solute sediments towards the bottom of the cell, a moving boundary is formed, which can be visualized by the absorbance trace recorded on a PC.

and:

$$\omega = rpm(\pi / 30) \tag{4.5}$$

$$m_0 = m\bar{v}\rho \tag{4.6}$$

(see Equations 4.7 and 4.8)
where ω is the angular velocity of the rotor, r the radius from the center of rotation, m the mass of the particle, and m_0 the mass of the solvent displaced by the particle, $\bar{v}$ is the partial specific volume of the particle and p is the density of the solvent. The buoyancy force is the force required to displace the volume of buffer equal to the solute's volume. The drag experienced by the particle is proportional to the frictional coefficient f and the velocity v with which the particle is moving. When Equations 4.2 to 4.6 are substituted

into Equation 4.1 the equation can be rearranged to obtain:

$$\frac{m(1-\bar{v}\rho)}{f} = \frac{v}{\omega^2 r} \tag{4.7}$$

Multiplying the left side with Avogadro's number (N/N) puts the equation into molar terms:

$$\frac{M(1-\bar{v}\rho)}{Nf} = \frac{v}{\omega^2 r} = s \tag{4.8}$$

The right-hand side of Equation 4.8 is the definition for the sedimentation coefficient, s. This equation illustrates an important relationship: the sedimentation coefficient is proportional to the molecular weight and inversely proportional to the frictional coefficient. Hence, sedimentation experiments provide a means to measure properties of a molecule involving both effects from molecular weight and conformational changes.

4.2.2.2.2 *Diffusion*

In addition to sedimentation, the diffusion of the particle will determine the concentration profile observed in the centrifugation cell. Fick's first law of diffusion states that in an infinite container the transport due to diffusion is proportional to the concentration gradient:

$$L = -D\frac{\partial C}{\partial x} \tag{4.9}$$

where L is the flux, D the diffusion coefficient, and C is the concentration along dimension x. Like the sedimentation coefficient, the diffusion coefficient depends on the frictional coefficient:

$$D = \frac{RT}{Nf} \tag{4.10}$$

Combining Equations 4.8 and 4.10, the molecular weight can be expressed in terms of the sedimentation and diffusion coefficients:

$$\frac{s}{D} = \frac{M(1-\bar{v}\rho)}{RT} \tag{4.11}$$

Equation 4.11 is known as Svedberg's law. Svedberg's law allows us to express the molecular weight in terms of two hydrodynamic parameters, the diffusion and sedimentation coefficient, provided the density of the

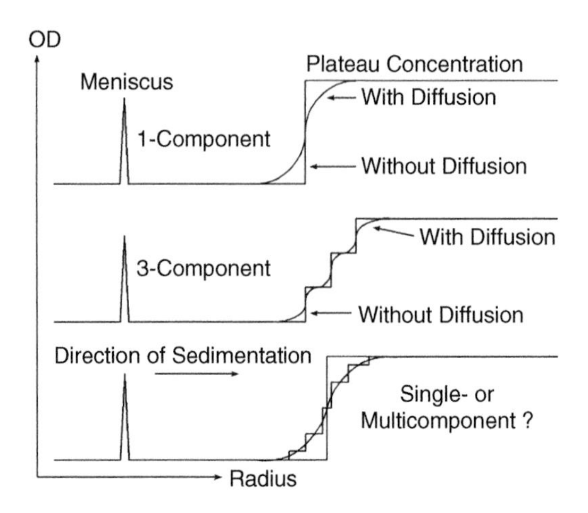

FIGURE 4.9 (See color insert following page 44)
The effect of diffusion on the boundary shape in a sedimentation velocity experiment. Shown in blue is a scan trace for a hypothetical sample with $D = 0$. Shown in red is the actually observed trace with a nonzero diffusion coefficient. As multiple components sediment, the diffusion of each component obliterates the separation due to sedimentation and it becomes increasingly difficult to tell multiple components apart from each other (center panel). In the extreme case, a Gaussian sedimentation coefficient distribution of particles can be indistinguishable from a single component system with a large diffusion coefficient (bottom panel), unless a global method is employed that takes multiple time points (scans) into consideration.

solvent and the partial specific volume of the molecule are known. The density of the buffer solution and the partial specific volume of the substrate are commonly measured with an Anton Paar densitometer. The sedimentation of the solute in the ultracentrifugation cell causes the formation of a moving boundary of solute concentration. Since the moving boundary constitutes a concentration gradient, the boundary shape is also affected by diffusion. The fact that the boundary shape is affected by both diffusion and sedimentation poses a problem when interpreting the boundary shape (see Color Figure 4.9). The analysis has to be able to distinguish between both processes.

The influence diffusion has on the boundary shape starts to become apparent as soon as the boundary forms, causing a sigmoidal boundary spreading, which becomes flatter over time (see Color Figure 4.10). Several approaches exist to deal with this complication, two of them, the van Holde–Weischet method and the finite element solution of the Lamm equation, are discussed in more detail below in the analysis section. Strictly speaking, Equation 4.11 only holds for solutes at zero concentration, since s, D, and f are dependent on concentration, but measurements performed in an analytical ultracentrifuge are often close enough to ideal conditions such that these effects can be neglected. For very large molecules, molecules at high concentration,

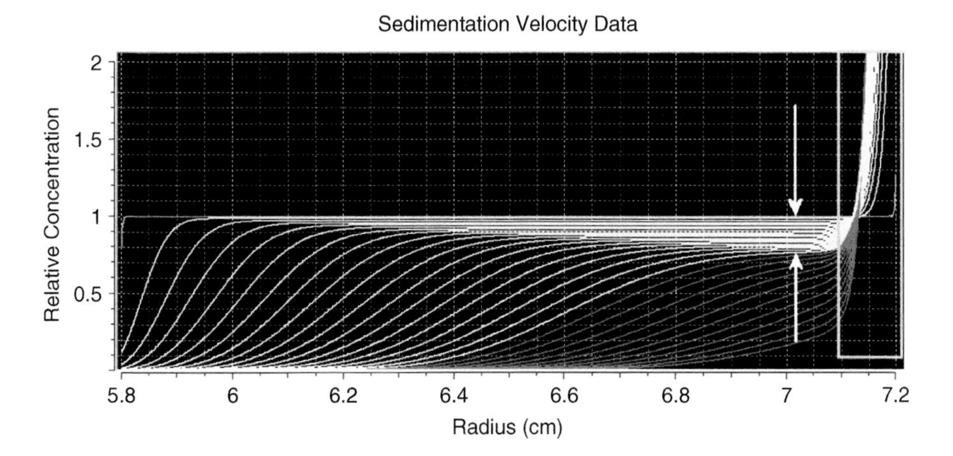

FIGURE 4.10 (See color insert)
Simulated sedimentation velocity experiment showing 30 scans for a 50 kDa solute sedimenting for 8 hours at 35,000 rpm. Each scan represents a snapshot of the concentration profile at a different time during the sedimentation process. The scan depicted in green shows a mostly uniform concentration distribution at the beginning of the experiment, before sedimentation and diffusion have commenced. Later scans are shown in yellow. Note the radial dilution effect, marked as the plateau concentration difference between the two white arrows. The steep concentration gradient between the light blue lines results from sedimented material collecting at the bottom of the cell and back-diffusing into the cell. Scans shown in red are collected at the end of the experiment and are starting to lose their plateau concentration. Most of the material in the red scans has pelleted at the bottom of the cell at this point in the experiment. The longer the sample sediments, the more time the sample has to diffuse. Diffusion causes boundary spreading, which is more pronounced in later scans.

charged or elongated molecules, this approximation may not hold, and correction terms need to be introduced.

4.2.2.2.3 Boundary Conditions

The physical limits of the cell (the meniscus, the bottom, as well as the sector shape), have to be taken into account in models describing the flow in the cell. This presents a number of complications. The first one is related to the sector shape of the cell. Consider a lamella of solute in the sector near the meniscus. As the lamella sediments towards the bottom of the cell, the same lamella will experience an increase in volume due to the fact that the sector opens up and becomes wider at the bottom. This results in the radial dilution effect that can be observed over time (see Color Figure 4.10). The radial dilution effect is described by an exponential function:

$$C_p = C_0 \exp(-2s\omega^2 t) \tag{4.12}$$

where C_p is the scan's plateau concentration, C_0 is the loading concentration, s is the weight average sedimentation coefficient, ω is the angular velocity of the rotor and t is the time of the scan. The second effect is the back-diffusion occurring at the bottom of the cell, which is caused by piling up of sedimented material at the bottom of the cell. This creates a large concentration gradient, which causes the material to diffuse back into the solution column. Color Figure 4.10 highlights the back diffusion effect at the bottom of the cell in the light blue box. A third effect is that of diffusion near the meniscus. At the beginning of the run, the concentration gradient near the meniscus is very large, causing a strong diffusion effect. As sedimentation proceeds, boundary spreading due to heterogeneity in s and due to diffusion proceeds as well, causing the gradient to weaken and hence the diffusion to decrease over time.

4.2.2.3 Analytical Ultracentrifuge (AUC) Experiments

There are two experiments commonly performed in the analytical ultracentrifuge (AUC): A long column, high speed experiment termed "sedimentation velocity" and a short column, lower speed experiment termed "sedimentation equilibrium" (see Figure 4.11). Each experiment provides different, yet complementary information, requiring different modeling approaches. In a velocity experiment, concentration profiles are collected during the sedimentation process, as molecules move towards the bottom of the cell. During the velocity experiment, a net change in concentration occurs as molecules are transported towards the bottom of the cell, and both molecular weight and frictional properties of each sedimenting species contribute to the final concentration distribution. In an equilibrium experiment, the concentration profile at the time of equilibrium is of interest. At that point, sedimentation towards the bottom of the cell, and diffusion away from the cell bottom exactly cancel each other, and net flow of molecules is zero. Because there is no flow, shape information no longer has to be considered, which simplifies the model considerably. Below, each experiment is explained in more detail, including the models used to fit experimental data.

4.2.2.3.1 *Sedimentation Velocity*

Sedimentation velocity experiments can be used to identify differences in shape and size between multiple samples, and for single or two-component ideal systems they can even provide the accurate molecular weights and frictional coefficients of the components. When used with a model-independent analysis method, sedimentation velocity experiments will yield sedimentation coefficient distributions. A typical sedimentation velocity experiment is shown in Color Figure 4.10. Such velocity data can either be modeled by a graphical, model independent approach,[1,2] by direct curve fitting,[3–7] or by fitting time differences.[8] An overview of several representative analysis methods is given below:

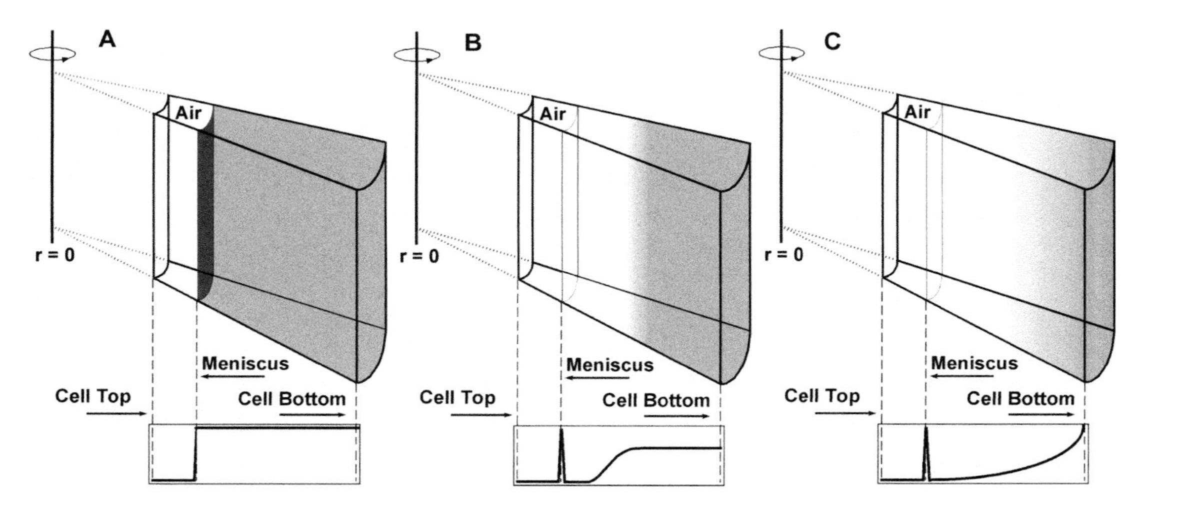

FIGURE 4.11
Concentration distributions for sedimentation velocity and equilibrium experiments. Panel A shows a schematic view of an AUC cell (not drawn to scale) at the beginning of the experiment. All material is equally distributed throughout the cell between the meniscus and the bottom of the cell, and the absorbance is uniform throughout the cell, with a plateau corresponding to the loading concentration. Panel B shows the concentration distribution in a high-speed velocity experiment at some time after the start of the experiment. Some material has sedimented away from the meniscus, causing the concentration near the meniscus to approach zero. A moving boundary is formed, which gives rise to a sigmoidal absorbance trace due to diffusion effects which cause the boundary to broaden. Also notice the drop in the plateau concentration when compared to panel A, which is caused by radial dilution (explained in text). Panel C shows the concentration distribution for a low-speed sedimentation equilibrium experiment after all flow in the cell has ceased and an equilibrium gradient has been established. In this experiment sedimentation and diffusion balance each other and the concentration increases towards the bottom of the cell in an exponential fashion. In practice, columns for equilibrium experiments are often limited to 3 mm in length. Longer columns can require substantially longer run times for equilibrium to be established.

4.2.2.3.1.1 van Holde–Weischet van Holde and Weischet[1] developed a model independent method to analyze sedimentation velocity data from single and paucidisperse solutions which can be used to diagnose a wide variety of experimental conditions.[9] The method benefits from the realization that sedimentation is a transport process proportional to the first power of time, whereas diffusion is a process where material is transported proportional to the square-root of time. This implies that if multiple components could sediment for an infinitely long time in infinitely long cells, all solutes in the solution will separate out. Extrapolating the apparent sedimentation coefficient for a given boundary fraction to infinite time thus provides a method to eliminate the effect diffusion has on the boundary shape. Using the right-hand side of equation (4.8) we can write an expression for an apparent sedimentation coefficient for each position in the boundary by simply measuring the distance the particle has traveled from the meniscus at time t of the scan:

$$\frac{dr}{r} = \omega^2 s dt \tag{4.13}$$

The solution for the definite integral of this differential equation provides the apparent sedimentation coefficient, s_b, at some position in the boundary, r_b, which is given by:

$$S_b = \ln \frac{r_b}{r_a} [\omega^2 (t - t_0)]^{-1} \tag{4.14}$$

where r_a is the meniscus position, and $t - t_0$ the elapsed time since the start of the experiment. A description of the basic algorithm follows:

1. Subdivide each scan into n equally spaced boundary divisions between baseline and plateau.
2. Using Equation 4.14, calculate the apparent sedimentation coefficients for all boundary positions intersecting with the boundary divisions and from the time of the scan.
3. Repeat Step 2 for each scan.
4. Plot the apparent sedimentation coefficients calculated from corresponding boundary positions from each scan against the inverse square root of each scan's time.
5. Fit the apparent sedimentation coefficients from corresponding boundary fractions to a straight line.
6. Extrapolate the line to infinite time (= zero on the inverse scale).

The intercepts on the y-axis will then represent the diffusion corrected sedimentation coefficient distribution of all components in the system. A van

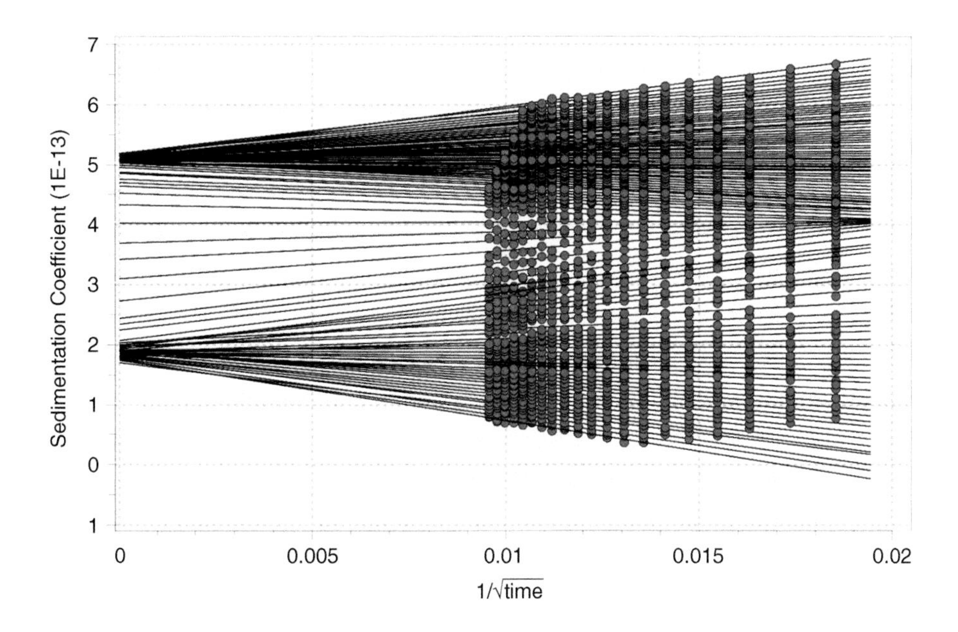

FIGURE 4.12
A van Holde–Weischet extrapolation plot created from the sedimentation velocity data of a two-component, noninteracting mixture of lysozyme and a 208-bp DNA fragment. Apparent sedimentation coefficients calculated from divisions in the boundary are plotted against the inverse square root of the time of the scan and extrapolated to infinite time (zero on the inverse scale). Twenty scans are plotted in this figure, the early scans are on the right, late scans are plotted on the left. Notice the spread of the "fan plot" for the slower component is larger than the spread of the fan for the faster component (DNA). This indicates a larger diffusion coefficient, consistent with a lower molecular weight and smaller sedimentation coefficient of the 1.8 S component (lysozyme). This plot has been generated with the UltraScan software. (From Demeler, B., UltraScan Data Analysis Software for the Analytical Ultracentrifuge. http://www.ultrascan.uthscsa.edu. With permission.)

Holde–Weischet extrapolation plot constructed from the sedimentation velocity data of a two-component system is shown in Figure 4.12. When plotting the boundary fraction against the corrected sedimentation coefficient, an integral distribution plot can be created from the data. Such a plot is shown in Figure 4.13. The strength of the van Holde–Weischet method results from two important properties: (1) the method is model independent and does not require user-biased input to provide a result. (2) The method offers a rigorous approach to remove the ambiguity encountered when trying to determine if boundary spreading is caused by diffusion or by heterogeneity in the sedimentation coefficient distribution. The van Holde–Weischet method is implemented in UltraScan.[10]

4.2.2.3.1.2 Direct Boundary Fitting with Finite Element Solutions of the Lamm Equation Once a model for the experimental data has been suggested by the van Holde–Weischet analysis, the investigator may want to obtain additional detail by using a direct boundary fitting approach. It is possible to

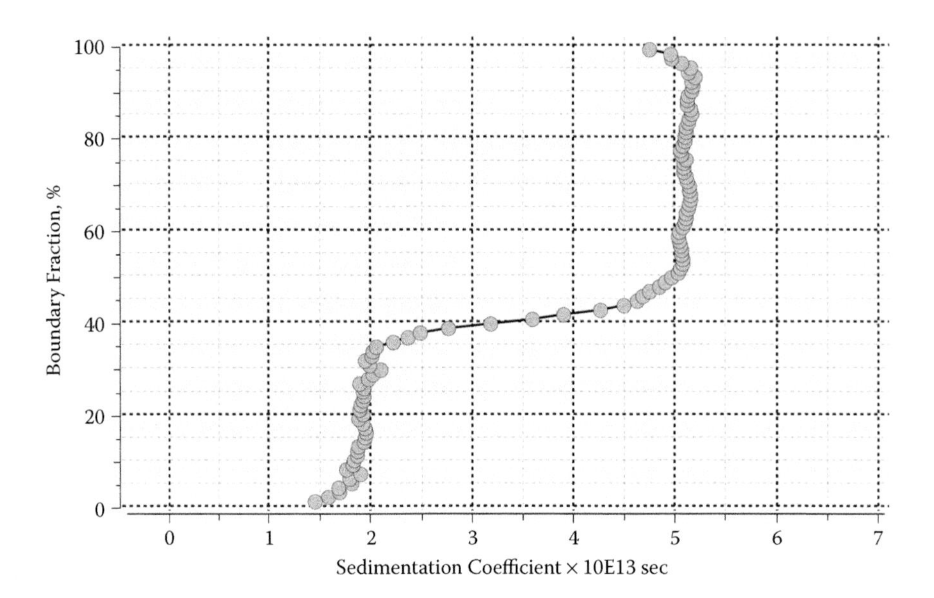

FIGURE 4.13
A van Holde–Weischet integral distribution plot for the two-component system shown in Figure 4.12. Component 1 sediments with approximately 1.8 s, while component 2 sediments with an s-value slightly higher than 5. It is possible to read the partial concentration of each component directly from the y-axis. Component 1 is present with about 45% of the total concentration, while component 2 is present with about 60% of the total OD. Single component systems produce vertical integral distribution plots. Overlaying integral distribution plots in the same plot provides a convenient way to compare sedimentation properties from different samples and to assay for changes, for example between mutants or under different buffer conditions. This plot has been generated with the UltraScan software. (From Demeler, B., UltraScan Data Analysis Software for the Analytical Ultracentrifuge. http://www.ultrascan.uthscsa.edu. With permission.)

directly fit sedimentation velocity concentration profiles C to a model describing the flow of a solute in the sector-shaped ultracentrifugation cell, as long as the number of components in the system is not too large, and interactions, if any, are known and can be described well by an interaction algorithm. For a single ideal solute, the flow is described by the Lamm equation[11]:

$$\left(\frac{\partial C}{\partial t}\right)_r = \frac{-1}{r}\frac{\partial}{\partial r}\left[s\omega^2 r^2 C - Dr\frac{\partial C}{\partial r}\right]_t \tag{4.15}$$

with boundary conditions:

$$C = 0 \quad \textit{for } r < a \quad \textit{and} \quad r > b \tag{4.16}$$

This equation can be conveniently solved with the finite element method.[12,13] For concentration-dependent samples, the diffusion and sedimentation coefficients are not considered constants, but rather functions of concentration:

$$S_k = S_{k,0}(1-\sigma_k C_k) \tag{4.17}$$

and

$$D_k = D_{k,0}(1+\delta_k C_k) \tag{4.18}$$

where σ and δ are parameters describing the concentration dependency in s and D of component k. For multiple, noninteracting solutes, the entire system can be represented by the sum of all solutes k:

$$C_{total} = \sum_{k=1}^{n} a_k C_k \tag{4.19}$$

Direct fitting of sedimentation velocity concentration profiles results in the direct determination of sedimentation and diffusion coefficients as well as partial concentrations for each component. Once the sedimentation and diffusion coefficients are known, the molecular weight and the frictional coefficient f can be determined from Equations 4.10 and 4.11. Fitting results for the system shown in Figure 4.12, when analyzed with a finite element solution of a 2-component noninteracting model of the Lamm equation, are shown in Color Figure 4.14. Directly fitting sedimentation profiles using finite element solutions has a number of advantages. First, the finite element solution includes the boundary conditions (i.e., the meniscus and the bottom of the cell), and hence allows accurate modeling of effects such as back-diffusion and diffusion effects near the meniscus (both are discussed in Section 4.2.2.2.3). Since this method directly measures the diffusion coefficient, the method not only yields molecular weights, but also the frictional coefficient. For reversible, rapidly self-associating systems, this method can also be used to model the reaction boundary of an interacting system, such as a monomer–dimer equilibrium. In such a case, it is possible to obtain equilibrium constants from a velocity experiment. The finite element method can also be used to fit approach to equilibrium data, which cannot be analyzed with any other method. However, as sample complexity increases, a direct measure of the diffusion coefficient is often difficult to obtain and prone to error, and the more parameters are fitted, the lower the confidence in each individual parameter becomes. Therefore, when more than two or three components are present, the method of direct boundary fitting often does not provide satisfactory results. This method is implemented in UltraScan[10] and in SEDFIT.[14]

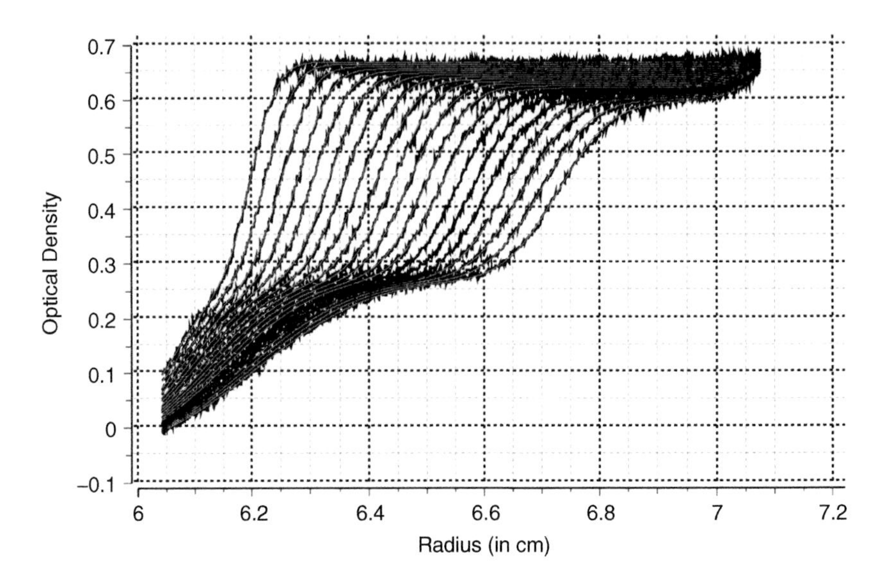

FIGURE 4.14 (See color insert)
Finite element solution for a two-component, noninteracting mixture of lysozyme and a 208-basepair DNA fragment. Blue lines indicate experimental data, the red lines represent the finite element solution obtained in a nonlinear least squares fit. Note that the finite element solution closely traces the experimental data, including border effects such as back-diffusion near the bottom of the cell. Experimental parameters determined from this fit are: s_{Lys}, 1.66×10^{-13}; $D_{Lys} = 1.036 \times 10^{-6}$; $C_{Lys} = 0.324$ OD; $MW_{Lys} = 13.7$ kDa; $f_{Lys} = 3.81 \times 10^{-8}$; $s_{DNA} = 5.25 \times 10^{-13}$; $D_{DNA} = 2.341 \times 10^{-7}$; $C_{DNA} = 0.412$ OD; $MW_{DNA} = 123.0$ kDa; $f_{DNA} = 1.69 \times 10^{-7}$. These data illustrate the detail and accuracy of information that can be obtained from sedimentation velocity experiments. This plot has been generated with the UltraScan software. (From Demeler, B., UltraScan Data Analysis Software for the Analytical Ultracentrifuge. http://www.ultrascan.uthscsa.edu. With permission.)

4.2.2.3.1.3 Direct Boundary Fitting with the dC/dt Method A variation of the finite element direct boundary fitting method has been developed by Stafford (implemented in the SEDANAL software[8]), where concentration differences from finite element solutions are fitted to concentration differences of experimental data. This approach has the advantage of eliminating time invariant noise, which can be substantial for low concentration experiments performed with the interference optical system. The principle of this method is best illustrated by considering an experimental scan, whose total signal $C_{total,i}$ will be composed from three components:

1. The signal arising from the sample's concentration distribution, C_i
2. The signal arising from time invariant contributions, such as finger prints or scratches in the cell window, I
3. The signal due to random noise error, R_i

Subtracting any set of two scans i and $i + n$ from each other results in a time-difference concentration profile that has the time-invariant noise component,

I, eliminated from the data. The penalty is an increase of the random noise error, R_i, by a factor of $\sqrt{2}$:

$$\begin{aligned} C_{total\ I} &= C_I + l + R_I \\ -C_{total\ I+n} &= C_{I+n} + l + R_{I+n} \\ \hline \Delta C_{total} &= \Delta C + (R_I + R_{I+n}) \end{aligned} \tag{4.20}$$

Spacing the time interval between i and $i + n$ sufficiently large provides a reasonable signal to noise ratio to allow fitting to a wide range of custom defined models that can be entered through a model editor. Experimental data differences are then fitted to simulated data differences, where the simulated data is generated with the finite element method.

4.2.2.3.1.4 Direct Boundary Fitting with the C(s) Method Schuck et al.[7] have developed an alternative method for directly fitting sedimentation coefficient distributions to sedimentation velocity experiments. The method is termed *C(s)* and implemented in the SEDFIT software,[14] and is based on fitting the amplitudes a_k for a given sedimentation coefficient distribution of n components using n finite element solutions of Equation 4.19. A nonnegatively constrained linear least squares algorithm is used to find only the positive, nonzero amplitudes a_k for the linear combination shown Equation 4.19. The terms with nonzero amplitudes, when summed, best represent a given velocity experiment. Sedimenting species not present in the given range of sedimentation coefficients are assigned an amplitude of zero, others receive a nonnegative value that is proportional to the partial concentration of the particle with the corresponding sedimentation coefficient. Diffusion coefficients for each component k are estimated from a molecular geometry and from each sedimentation coefficient under consideration. The overall molecular geometry is assumed to be constant (or bimodal) for all components in the system. The measure used to estimate the molecular geometry is the frictional ratio, f/f_0, and can be fitted. This ratio measures the frictional properties of the molecule compared to the frictional coefficient f_0 of a sphere with the same mass and density. A sphere has the lowest frictional coefficient of all possible shapes. Hence, the value for f/f_0 is always ≥ 1.0. Globular proteins have ratios around 1 to 1.3, while extended molecules like DNA can have frictional ratios above 3.0. Various schemes exist to exploit prior knowledge of the frictional ratio, or to perform a nonlinear fit of f/f_0 in the program. To avoid unreasonable oscillations in the coefficients a_k, the solution is smoothed with maximum entropy regularization or Tikhonov regularization, which penalizes large changes in the second derivative of the amplitudes a_k. The results for the same 2-component experiment using the *C(s)* analysis are shown in Figure 4.15. The *C(s)* method works best for sedimentation distributions where overall molecular shape remains constant. Both the van Holde–Weischet method and the *C(s)* method yield sedimentation

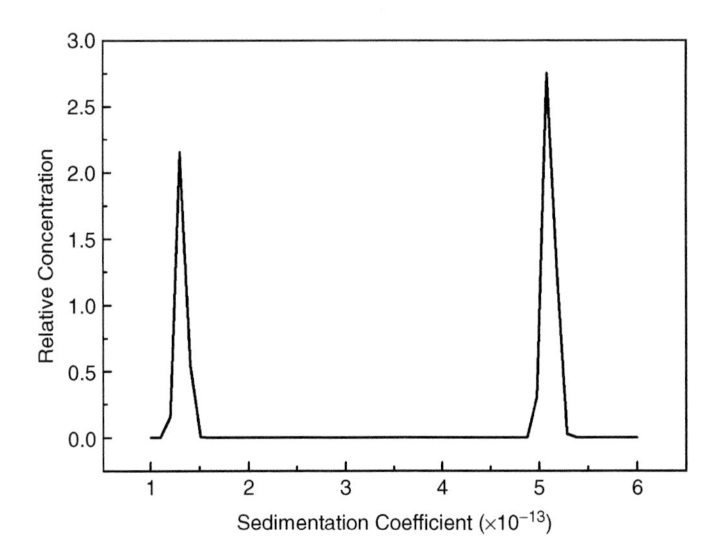

FIGURE 4.15
C(s) sedimentation distribution generated by SEDFIT[14] from the same velocity data for a mixture of lysozyme and a 208-bp DNA fragment shown in Figure 4.12, Figure 4.13, and Color Figure 4.14. Note the two peaks (lysozyme, 1.3 s, and DNA, 5.1 s). While this method achieves nice separation, the method suffers from the constraint requiring a fixed frictional ratio for all components. In this case, the frictional ratio f/f_0 is close to 1.0 for lysozyme, but significantly higher for the DNA component, necessarily introducing error into solution. When frictional ratios remain constant for all species in the system, this method works well and can be used to transform the sedimentation distribution into a molecular weight distribution.

coefficient distributions that can be transformed into molecular weight distributions using appropriate assumptions of molecular shape (see Figure 4.16).

Sedimentation velocity experiments can be used to answer many experimental questions. Table 4.2 identifies the most appropriate method to be selected for several scenarios.

4.2.2.3.2 Sedimentation Equilibrium

Sedimentation equilibrium experiments provide a direct measure of the molecular weight of a solute, and for reversibly self-associating systems, association constants can be determined. At equilibrium, the total flow in the cell ceases, and the sedimentation and diffusion terms in the flow equation cancel:

$$J = s\omega^2 rC - D\frac{\partial C}{\partial r} = 0 \tag{4.21}$$

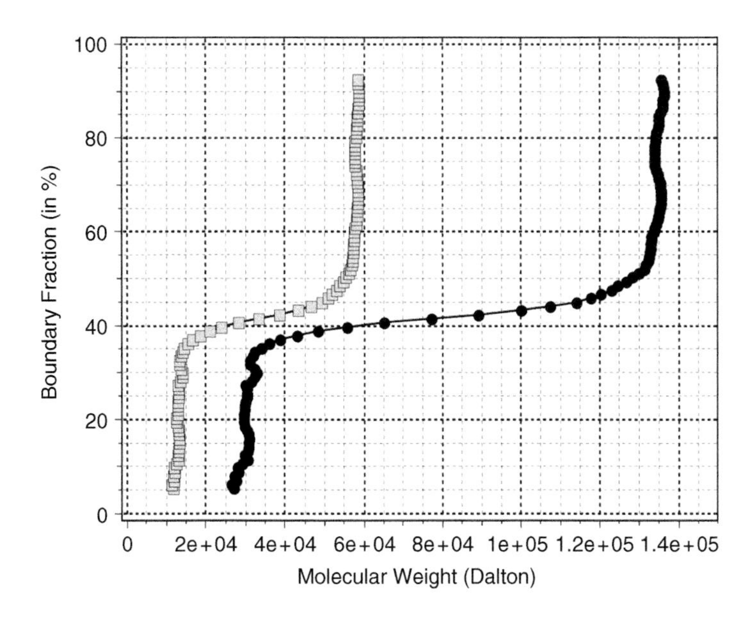

FIGURE 4.16
Molecular weight transformations of the van Holde–Weischet integral distribution plot shown in Figure 4.13. The gray curve represents a transformation based on the partial specific volume of lysozyme, using a frictional ratio of 1.2, the blue curve represents the same distribution transformed using the partial specific volume for DNA and a frictional ratio of 3.2. Using the prior knowledge of partial specific volume and frictional ratio, it is possible to obtain accurate molecular weights for multi-component systems (green curve: ~14 kDa for lysozyme, blue curve: ~130 kDa for DNA, theoretical values: 14.4 kDa for lysozyme, 137.7 kDa for DNA). By plotting the same distribution with multiple parameters it is easy to visualize the possible molecular weight limits for the system. This plot has been generated with the UltraScan software. (From Demeler, B., UltraScan Data Analysis Software for the Analytical Ultracentrifuge. http://www.ultrascan.uthscsa.edu. With permission.)

Solving this differential equation between points r_a and r_b, and by substituting Equation 4.11, an exponential function is obtained that describes the concentration distribution in the AUC cell at equilibrium:

$$C = C_a \exp^{\frac{M\omega^2(1-\bar{v}\rho)}{RT}(r_b^2 - r_a^2)} + C_b \tag{4.22}$$

where C_a is the concentration at reference point r_a and C_b is the baseline concentration, and M is the monomer molecular weight. Just as in sedimentation velocity experiments, multiple components can be described by summing multiple terms of this model, one exponential for each component in the system. The problem with this approach is that each component adds a new molecular weight M and partial concentration C_a to the list of parameters which need to be fitted in a nonlinear least squares fit. In order to improve the confidence in the fitted parameters, it helps to reduce the

TABLE 4.2

Which Analysis Application is Best Used for a Particular Experimental Question

Experimental Question	vH-W	FE	C(s)
Initial characterization of an unknown sample	x		x
Model-independent data analysis	x		
Diffusion-corrected composition analysis (is heterogeneity present?)	x		
Quantitative information for conformational comparisons between samples		x	
Qualitative information for conformational comparisons between samples	x		
Does the sample experience self-association or is it noninteracting? [a]	x	x	x
Is the sample reversibly associating or irreversibly aggregating?	x		
Molecular weights for systems with no more than three discrete species		x	
Diffusion and frictional coefficients for simple single or two-component systems		x	
Qualitative comparisons of diffusion between components	x		
Experimental simulations [b]		x	
Analysis of approach-to-equilibrium experimental data		x	
Determination of binding stoichiometry	x	x	x
Quantification of relative concentration of all components	x	x	x
Does the sample exhibit concentration dependent solution nonideality?	x		
Obtain association constants for self-associating systems		x	

[a] To determine if a sample is reversibly self-associating, the sample should be run at multiple concentrations, spanning at least a 10-fold concentration difference. At the higher concentration, a self-associating system will have a higher proportion of the oligomer present, at lower concentration the monomer will be favored. In the sedimentation velocity experiment, different concentrations will sediment with different rates, which is easily recognized when analyzed with any method. For noninteracting systems, a change in concentration will not affect the sedimentation pattern.

[b] It is very helpful to simulate experiments beforehand to identify optimal run conditions. Components to be analyzed can be simulated for molecular weight, sedimentation and diffusion coefficients, and overall molecular shape (f/f_0). Concentration dependency can also be simulated, and the time required to reach equilibrium can be predicted. Such simulation modules are included in the UltraScan software package.[10]

number of parameters as much as possible. For self-associating systems, the number of fitting parameters can be conveniently reduced by taking advantage of several constraints. For a self-associating monomer–dimer system we have: $M + M \leftrightarrow D$. The equilibrium constant can be expressed as:

$$Ka_2 = \frac{[D]}{[M]^2} \tag{4.23}$$

Using Beer's law, $C = e\,l\,A$, where e is the extinction coefficient, l the path length, and A the measured absorbance, we can substitute:

$$C_{a,D} = C_{a,M}^2 \frac{2Ka_2}{e * l} \tag{4.24}$$

which reduces the equilibrium model to a single molecular weight parameter M (the molecular weight of the monomer) and the concentrations are expressed in terms of the monomer reference concentration C_a at point r_a, and in terms of the monomer–dimer equilibrium constant, Ka_2:

$$C = C_{a,M} \exp^{\frac{M\omega^2(l-\bar{v}_\rho)}{2RT}(r^2 - r_a^2)} + C_{a,M}^2 \frac{2Ka_2}{e*l} \exp^{\frac{2M\omega^2(l-\bar{v}_\rho)}{2RT}(r^2 - r_a^2)} + C_b \quad (4.25)$$

Equation 4.25 can be fitted for C_a, M, Ka_2, and C_b using a nonlinear least squares fitting routine. The equation can easily be adapted to higher order self-associating systems, or even hetero-associating systems. It should be pointed out that fitting sums of exponentials to experimental data is a rather ill-conditioned proposition because equilibrium gradients have very few features to force unique answers for the fit. In addition, experimental uncertainties (for example electronic noise) are contained in the data, and as a result the fitted solution may not be unique. This means that multiple parameter solutions and multiple models can satisfy the least squares condition, and result in the same χ^2 value. In order to alleviate this difficulty, it is advisable to perform multiple experiments under several different conditions. Conditions that can be varied include the loading concentration and the rotor speed. All experiments can then be globally fitted, simultaneously to the same model, since the intrinsic properties of the system (i.e., molecular weight and association constant) do not change under these conditions. Determination of association constants requires information over a large range of concentrations to assure confidence in the results. Taking advantage of variable extinction properties at different wavelengths, protein solutions can therefore be measured at multiple wavelengths, where a good absorbance signal is obtained at a wide range of concentrations. Ideally, the measured concentration range should bracket the equilibrium concentration, to assure that a good signal from all described species is present in the data. An example for a global monomer–dimer equilibrium fit of multiple equilibrium experiments is shown in Figure 4.17. A distribution of the relative concentration for monomer and dimer species is shown in Figure 4.18, and a concentration histogram of all fitted experimental observations is shown in Figure 4.19. Inspecting Figure 4.18 and Figure 4.19, it can be seen that the concentration range measured in this experiment brackets the equilibrium constant on both sides, providing a good signal for both monomer and dimer. Equilibrium analysis can be applied to many other systems besides single ideal systems and monomer–dimer equilibria. For example, equilibrium analysis has been successfully applied to the study of heteroassociating systems, such as receptor/ligand binding and DNA–protein interactions. Global equilibrium analysis software is offered in UltraScan,[10] SEDPHAT,[15] and NONLIN.[16]

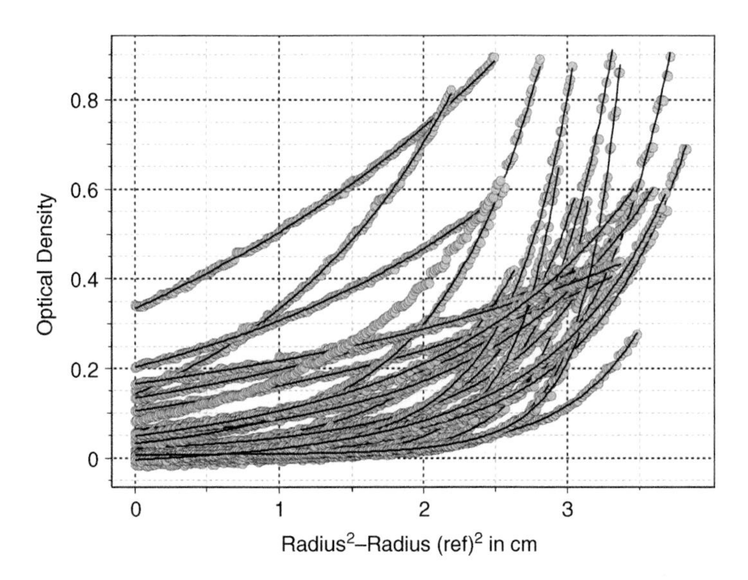

FIGURE 4.17
Global equilibrium fit of experimental data measured under multiple conditions. Loading concentrations were varied between 0.3 and 0.7 absorbance units, and speeds were varied between 11 krpm and 27 krpm. Measurements were made at both 230 nm and 280 nm to exploit a larger concentration range. Experimental data measured under varying conditions have to match the same monomer molecular weight and association constant if the global model is valid. Gray circles represent experimental measurements, black lines represent the fitted monomer–dimer model. This plot has been generated with the UltraScan software. (From Demeler, B., UltraScan Data Analysis Software for the Analytical Ultracentrifuge. http://www.ultrascan.uthscsa.edu. With permission.)

4.2.3 Light Scattering

4.2.3.1 Experimental Setup and Instrumentation

Light scattering experiments rely on the thermal motion of a molecule (Brownian motion) to measure D or molecular weight. In this technique, a laser beam (commonly an argon or helium–neon laser) passes through a temperature-controlled solution of the sample. Molecules diffusing through the laser beam will scatter light. The intensity of the scattered light is dependent on the size and concentration of the molecules. A photomultiplier tube, which is mounted on a goniometer (a turntable that allows rotation of the photomultiplier tube to different angles with respect to the laser beam), is focused on the laser beam and detects scattered light from the molecules passing through a small volume element illuminated by the laser (see Figure 4.20). Since the volume element is very small, the intensity of the scattered light will fluctuate over time, because molecules wander in and out of the volume element, causing concentration changes and thus fluctuations in the intensity of scattered light. The frequency of this intensity fluctuation is proportional to the diffusion coefficient. Small

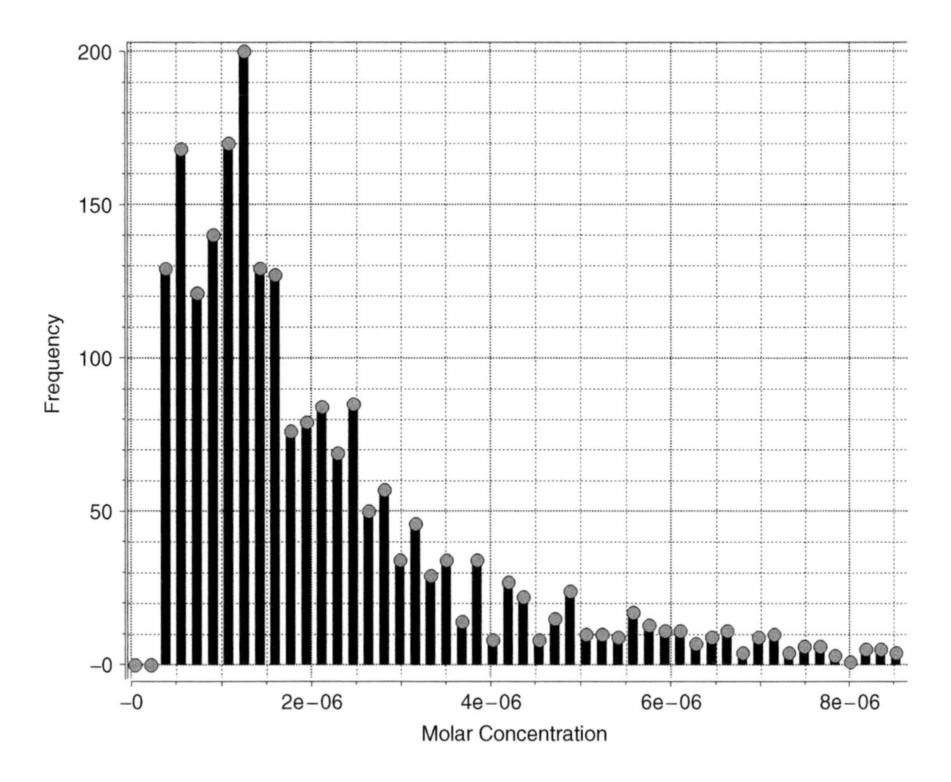

FIGURE 4.18
Concentration histogram obtained by counting the number of data points at a given concentration. Data points are taken from the experiment shown in Figure 4.17. Signal was obtained between 0.3 and 15 micromolar, with most data points in the low micromolar range. This plot has been generated with the UltraScan software. (From Demeler, B., UltraScan Data Analysis Software for the Analytical Ultracentrifuge. http://www.ultrascan.uthscsa.edu. With permission.)

molecules with large diffusion coefficients diffuse rapidly through the volume element, causing rapid changes in the intensity, while large molecules with small diffusion coefficients will diffuse much more slowly, reducing the frequency of this fluctuation. This time-dependent effect is measured in dynamic light scattering experiments. The average scattering over an extended time period is measured in static light scattering experiments, and can be used to determine the weight-average molecular weight of the solution, the radius of gyration, and the second virial coefficient of a solute.

4.2.3.2 Dynamic Light Scattering

The independent determination of the diffusion coefficient can be very useful to provide additional detail about the shape of a macromolecule. Dynamic light scattering (DLS) provides a rapid technique to obtain this information, which is complementary to sedimentation experiments per-

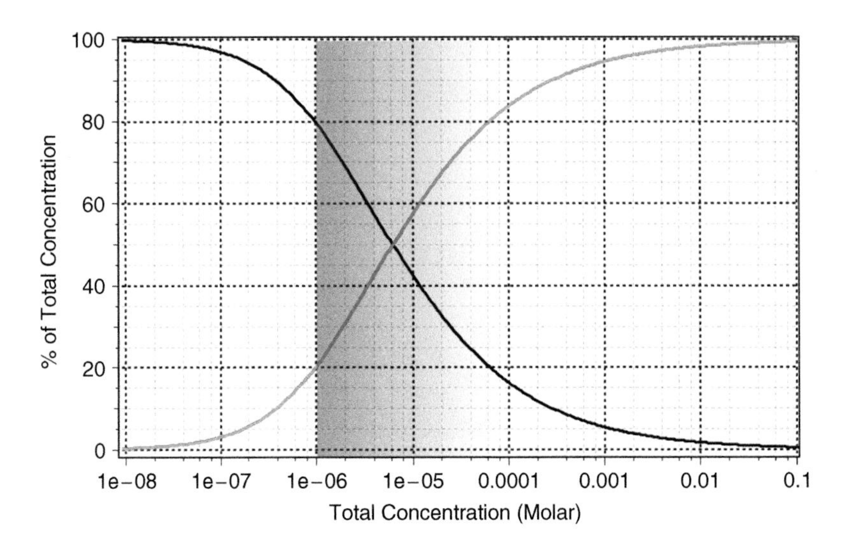

FIGURE 4.19
Self-association profile for monomer–dimer model shown in Figure 4.17. The relative amount of monomer is represented by the black line, the gray line indicates the relative amount of the dimer. According to Le Chatelier's principle, the higher the total concentration, the more monomer associates to form a dimer in a reversible process. The association constant is in the micromolar range. The darkness of the shading indicates approximately the relative signal strength based on the number of data points measured in the given concentration range (see also Figure 4.18). Based on the shading, the information collected was in the appropriate concentration range to assure a reliable measure of the equilibrium constant by AUC. This plot has been generated with the UltraScan software. (From Demeler, B., UltraScan Data Analysis Software for the Analytical Ultracentrifuge. http://www.ultrascan.uthscsa.edu. With permission.)

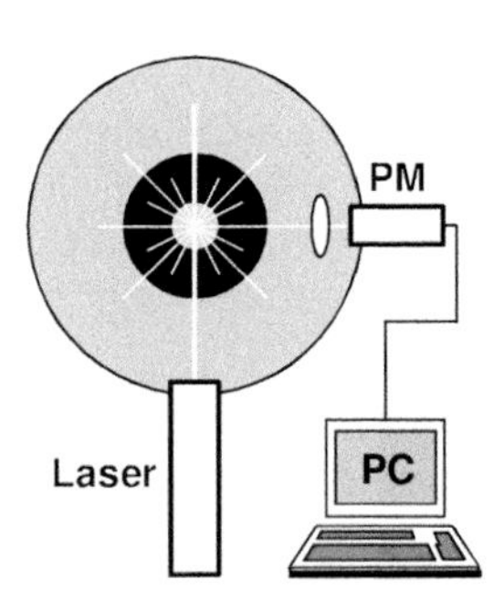

FIGURE 4.20
Schematic view of a light-scattering apparatus. A laser beam passes through the sample solution, which is held in a temperature controlled bath (black circle). Light is scattered by the sample in all directions. A photomultiplier tube (PM) is mounted on an adjustable goniometer so intensity of scattered light can be measured at multiple angles. The signal is fed to a PC equipped with an autocorrelation board.

formed in the analytical ultracentrifuge. An effective method for measuring the periodicity of the fluctuation of intensity is the autocorrelation function A. From the scattered light, an average intensity $\bar{I}$ is determined, and intensities below and above this average are assigned to be of opposite sign:

$$\Delta I(t) = I(t) - \bar{I} \tag{4.26}$$

The autocorrelation function is given by:

$$A(\tau) = \Sigma[\Delta I(t)\Delta I(t+\tau)] \tag{4.27}$$

In the autocorrelation function, products of intensity values separated in time by a small time increment τ, are summed. Correlation is large when the sample time step is short, because intensity measurements spaced close in time are likely to be of the same sign, producing a large sum, while correlation disappears when time steps are far enough apart to be randomly of different sign, so positive and negative contributions to the autocorrelation sum are equally likely and will cancel out (see Color Figure 4.21 for an illustration). Most hardware implementations of the autocorrelation function offer up to 256 channels, each channel performs the autocorrelation for a different sample time, with sample times as short as 5×10^{-8} sec. The length of the sample time for the minimum and maximum channel can be adjusted on the instrument. Most instruments also provide several delayed channels with very long sample times, which provide an estimate for the baseline correlation B. The autocorrelation function has an exponential form, which depends on τ:

$$A(\tau) = A(0)e^{-\tau/\tau_0} + B \tag{4.28}$$

where τ_0 is given by:

$$\tau_0 = \frac{1}{K^2 D} \tag{4.29}$$

and

$$K = \frac{4\pi n}{\lambda_0} \sin\frac{\alpha}{2} \tag{4.30}$$

where n is the refractive index of the solvent, α is the angle of the photomultiplier tube with respect to the laser beam (typically 90°), and λ_0 is the wavelength

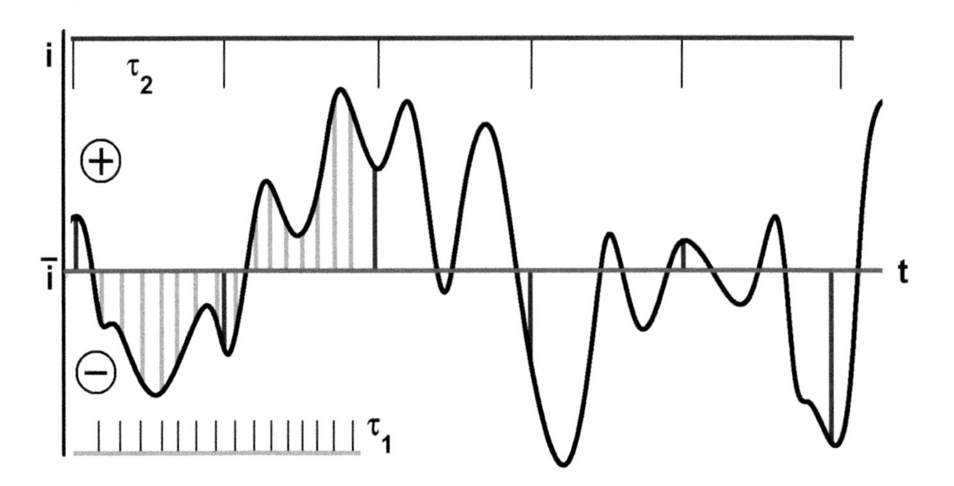

FIGURE 4.21 (See color insert)
Measurement of the autocorrelation function. The intensity fluctuation, *i*, is plotted against time *t*. An average intensity $\bar{i}$ is determined and intensity values above this value are considered positive, values below the line are negative. Shown in green is a short sample time τ_1 leading to high correlation, since adjacent intensity values $i(\tau_1)$ tend to have the same sign and their products are positive, contributing to the autocorrelation sum. A longer sample τ_2 is shown in blue, producing little correlation, since adjacent values have random sign, and positive and negative contributions cancel out. Modern data acquisition hardware allows simultaneous collection of up to 256 different sample times, leading to 256 data points in the autocorrelation function. Sample times as short as a microsecond allow accurate diffusion measurements of even small particles.

of the laser in vacuum. The experimental data can then be fitted to the autocorrelation function using a nonlinear least squares fitting algorithm to directly obtain *D*. For multicomponent systems, a sum of exponentials can be fitted to the autocorrelation data. By adjusting the overall range of the sample time, it is possible to maximize the signal for molecules varying in size and diffusion coefficients. Small sample times emphasize the signal from fast diffusing molecules, while longer sample times better describe slowly diffusing molecules. Multiple measurements performed at different sample time ranges can be combined in a global fit to obtain better fitting statistics. Data analysis software is generally supplied by the manufacturer of the light scattering instrumentation, but can also be found in UltraScan,[10] SEDFIT,[14] and SEDPHAT.[15]

4.2.3.3 Static Light Scattering

Static light scattering, also called Rayleigh scattering, employs the same experimental setup as DLS. Light scattered by macromolecules in solution can be used to measure molecular weight. This information is complementary to results obtained in AUC equilibrium experiments, which also yield molecular weight, and AUC velocity experiments, which provide *s* and *D*, whose ratio is proportional to the molecular weight. Instead of measuring intensity fluctuations with an autocorrelator as is used in the DLS experiment, in static light scattering experiments, the average scattering intensity over time is measured. This can be expressed as the ratio of the averaged scattering intensity $\bar{i}$ over the incident intensity, I_0:

$$R_\theta = \frac{\bar{i}}{I_o} \frac{r^2}{1+\cos^2\theta} \tag{4.31}$$

where R_θ is called the Rayleigh ratio at angle θ, and r is the distance from the scattering origin to the photomultiplier tube. For an ideal solute, the molecular weight is then given by:

$$M = \frac{R_\theta}{KC} \tag{4.32}$$

where C is the concentration of the solute and K is a constant given by:

$$K = \frac{2\pi^2 n_0^2 (dn/dC)^2}{N\lambda^4} \tag{4.33}$$

where n_0 is the refractive index of the solvent, dn/dC is the change in refractive index with respect to concentration, which can be measured with a differential refractometer, N is Avogadro's number, and λ is the wavelength of the laser. For nonideal solutions, the scattering is concentration dependent, and a correction term needs to be included:

$$\frac{KC}{R_\theta} = \frac{1}{M}\left(1 + C\frac{\partial \ln y}{C}\right) \tag{4.34}$$

The first-order expansion of Equation 4.34 yields the second virial coefficient, B, which is a measure of the solution nonideality:

$$\frac{KC}{R_\theta} \approx \frac{1}{M} + 2BC \tag{4.35}$$

Most systems can be well described by this model. Creating a plot of KC/R_θ vs. C yields a slope of B and an intercept of $1/M$. If the solution contains multiple components, the weight-average molecular weight of all components in the system will be observed. This method is an ideal approach to monitor oligomerization and aggregation because experiments require less time than analytical ultracentrifugation experiments. Additional information can be obtained if the molecules are large in comparison to the wavelength of the incident light. In that case even molecular dimensions can be determined, and Equation 4.35 can be expressed in terms of the radius of gyration, R_G:

$$\frac{KC}{R_\theta} \approx \left(1+\frac{16\pi^2 R_G^2}{3\lambda^2}\sin^2\frac{\theta}{2}\right)\left(\frac{1}{M}+2BC\right) \tag{4.36}$$

As can be easily recognized from Equation 4.36, the radius of gyration needs to have sufficient size compared to the wavelength in order that an angular dependence of the light scattering measurement can be observed. The radius of gyration and the molecular weight can be determined by measuring the angular dependence of the scattering at a series of concentration points. The results are graphed in a Zimm plot.[17] In such a plot, light scattering measurements at multiple concentrations are performed at multiple angles between 30 to 90°. The concentration measurements at each angle are extrapolated to zero concentration. The zero concentration points from each angle define a line that, when it is extrapolated to 0°, directly yields the radius of gyration and the molecular weight. At zero concentration, Equation 4.36 reduces to:

$$\frac{KC_0}{R_\theta} = \frac{1}{M}\left(1+\frac{16\pi^2 R_G^2}{3\lambda^2}\sin^2\frac{\theta}{2}\right) \tag{4.37}$$

which is the equation of a line where the intercept provides the molecular weight and the slope provides the radius of gyration. A schematic representation of a Zimm plot is shown in Figure 4.22.

4.2.4 Global Analysis

The information content of the hydrodynamic methods discussed above can be further enhanced when information from different experiments performed on the same system is combined in a global analysis. In such an analysis, complementary information is contributed by multiple experiments and fitted simultaneously to a model that requires all properties and fitting parameters observed by more than one experiment to be identical throughout the fit. For example, the monomer molecular weight has to stay constant for all simultaneously fitted experimental observations, and thus can be used to constrain the fit. All resulting experimental observations are then fitted in a single fit. Since the molecular weight is proportional to the ratio of sedimentation coefficient over diffusion coefficient, sedimentation velocity, sedimentation equilibrium, and dynamic light scattering experiments all produce information that can be used to constrain the fitted parameters of

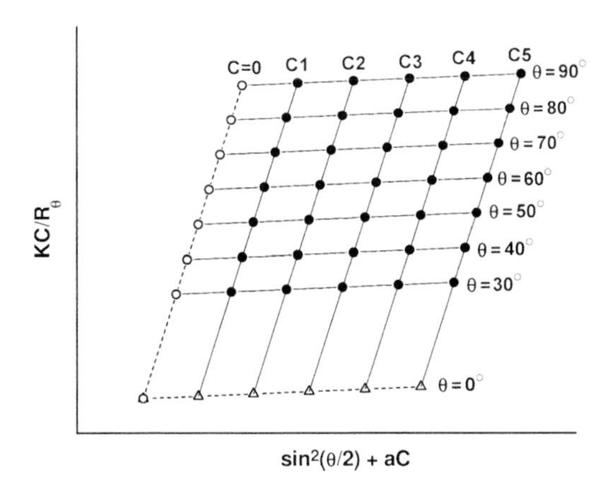

FIGURE 4.22
Schematic view of a Zimm plot. Measurements of KC/R_θ are plotted against $\sin^2(\theta/2) + aC$, where a is an arbitrary constant to provide a convenient scale for the plot. The solid circles represent light scattering measurements performed at five different concentrations (C1 to C5, with C1 being the lowest concentration) and at seven different angles (30 to 90°). The grid defined by these measurements is extrapolated in two directions, first to zero concentration, and then to an angle of zero degrees (extrapolations are represented by the dotted lines). The line of zero concentration extrapolations is represented by a dotted line with open circles, and its slope provides the radius or gyration (see Equation 4.37). The line of zero degree extrapolations is represented by a dotted line with open triangles, and its slope is proportional to the second virial coefficient. The points represented by open circles are the intercepts in Equation 4.35 at zero concentration and can be used to derive the molecular weight of the sample.

any other experiment. Table 4.3 identifies molecular parameters commonly fitted with hydrodynamic methods, and the relative signal strength each experiment provides. Strong signals for parameters can be used to constrain weak signals; for example, the diffusion coefficient fitted in the DLS experiment can be used to constrain the diffusion coefficient in a finite element fit. The molecular weight can be used to constrain the *s/D* ratio in the finite element fit. All three experiments have to agree on all three parameters. As an example, consider global equilibrium analysis, which is often applied by measuring the same system under different conditions, such as different rotor speeds and different loading concentrations. Figure 4.17 shows an example for a global equilibrium fit where multiple experiments performed under different conditions are globally fitted to a single model. Each rotor speed and each loading concentration gives rise to a different concentration profile and curvature in the exponential function, but all experimental observations describe the same system. The benefit of such global analysis is twofold: first, for any fit with the same variance, the confidence limits for each globally fitted parameter is improved, and second, the complexity of the fitted model can be increased by allowing for additional floating parameters, without reducing the confidence limits for the global parameters. Table 4.3 shows a summary of the relative signal strength of each parameter

TABLE 4.3

Approximate Signal Strength of Experimental Parameters Obtained from Hydrodynamic Experiments

Method Parameter	Finite Element	Equilibrium	van Holde-Weischet	Dynamic Light Scattering	Static Light Scattering
Molecular Weight	s/D, 0	+	–	–	WA
Sedimentation Coefficient	+	–	+	–	–
Diffusion Coefficient	0	–	q	+	–
Partial Concentration	+	0	+	0	–
Association Constants	0	+	–	0	–
Radius of Gyration	–	–	–	–	+
Second Virial Coefficient	–	+	–	–	+

Note: (+) indicates a strong signal, the parameter can be determined with high confidence, (0) indicates a weak signal, the parameter can only be determined with a low confidence, and (–) indicates that no information is available. (q) indicates that qualitative information is available, s/D is the ratio of the sedimentation coefficient over the diffusion coefficient, which is proportional to the molecular weight. (WA) indicates that a weight average molecular weight can be determined.

that can be obtained from hydrodynamic experiments. The discussed analysis methods are compared. Global analysis software for hydrodynamic experiments can be found in UltraScan,[10] SEDPHAT,[15] and NONLIN.[16]

4.2.5 Appendix

A summary of literature references is provided here to serve as a starting point for the reader interested in a more in-depth treatment of the subject matter and to identify a selection of publications illustrating the use of these techniques on biological samples. On the topic of analytical ultracentrifugation much of the original theory was developed by Fujita,[18] his somewhat dated publication still is regarded as the "bible" of AUC. van Holde[19] and Cantor and Schimmel[20] provide a good textbook introduction into the experimental techniques. A comprehensive review of the current state of hydrodynamic applications using AUC can be found in Harding et al.[21] An introductory treatment of light scattering can be found in van Holde,[19] a thorough look at light scattering and data analysis of static and dynamic light scattering experiments, including examples, can be found in Harding et al.[22] A very thorough treatment of the hydrodynamic properties of nucleic acids as studied by analytical ultracentrifugation and light scattering is discussed in Bloomfield et al.[23] Reviews of modern AUC analysis methods and software can be found in Lebowitz et al.[24] and in Cole and Hansen,[25] an overview of the application of AUC for the analysis of complex macromolecular systems can be found in Hansen et al.[26] A search of the keywords "Analytical Ultracentrifugation" and "Light Scattering" on PubMed will reveal a large range of biological systems that have recently been analyzed with these techniques.

References

1. van Holde, K.E. and Weischet, W.O., Boundary analysis of sedimentation-velocity experiments with monodisperse and paucidisperse solutes. *Biopolymers*, 1978, 17, 1387–1403.
2. Stafford, W.F. 3rd., Boundary analysis in sedimentation transport experiments: a procedure for obtaining sedimentation coefficient distributions using the time derivative of the concentration profile. *Anal. Biochem.*, 1992, 203(2), 295–301.
3. Demeler, B. and Saber, H., Determination of molecular parameters by fitting sedimentation data to finite element solutions of the Lamm equation. *Biophys. J.*, 1998, 74, 444–454.
4. Philo, J.S., An improved function for fitting sedimentation velocity data for low-molecular-weight solutes. *Biophys. J.*, 1997, 72(1), 435–444.
5. Behlke, J. and Ristau, O., Molecular mass determination by sedimentation velocity experiments and direct fitting of the concentration profiles. *Biophys. J.*, 1997, 72 (1), 428–434.
6. Schuck, P., Sedimentation analysis of noninteracting and self-associating solutes using numerical solutions to the Lamm equation. *Biophys. J.*, 1998, 75(3), 1503–1512.
7. Schuck, P., Perugini, M.A., Gonzales, N.R., Howlett, G.J., and Schubert, D., Size-distribution analysis of proteins by analytical ultracentrifugation: strategies and application to model systems. *Biophys. J.*, 2002, 82(2), 1096–1111.
8. Stafford, W.F. and Sherwood, P.J., Analysis of heterologous interacting systems by sedimentation velocity: curve fitting algorithms for estimation of sedimentation coefficients, equilibrium and kinetic constants. *Biophys. Chem.*, 2004, Mar 1, 108(1–3), 231–243.
9. Demeler, B., Saber, H., and Hansen, J.C., Identification and interpretation of complexity in sedimentation velocity boundaries. *Biophys. J.*, 1997, Jan, 72(1), 397–407.
10. Demeler, B., UltraScan Data Analysis Software for the Analytical Ultracentrifuge. http://www.ultrascan.uthscsa.edu.
11. Lamm, O., Die Differentialgleichung der Ultrazentrifugierung. *Ark. Mat. Astron. Fys.*, 1929, 21B, 1–4.
12. Claverie, J.-M., Dreux, H., and Cohen, R., Sedimentation of generalized systems of interacting particles. I. Solutions of systems of complete Lamm equations. *Biopolymers*. 1975, 14, 1685–1700.
13. Todd, G.P. and Haschemeyer, R.H., General solution to the inverse problem of the differential equation of the ultracentrifuge. *Proc. Natl. Acad. Sci.* 1981, 78(11), 6739–6743.

14. Schuck, P., SEDFIT software and tutorial. http://www.analyticalultracentrifugation.com.
15. Schuck, P., SEDPHAT, global fitting software for sedimentation velocity and equilibrium experiments, and light scattering. http://www.analyticalultracentrifugation.com/sedphat/sedphat.htm.
16. Johnson, M.L., NONLIN, Global Sedimentation Equilibrium fitting software, available from: http://www.bbri.org/rasmb/rasmb.html
17. Zimm, B.H., Development of Zimm's methods for analysis of angular dependence. *J. Chem. Phys.*, 1948, 16, 1093–1099.
18. Fujita, H., *Foundations of Ultracentrifugal Analysis*. John Wiley & Sons, Inc., New York, 1975.
19. van Holde, K.E., *Physical Biochemistry*, 2nd ed., Prentice Hall, Inc., Englewood Cliffs, NJ, 1985.
20. Cantor, C.R. and Schimmel, P.R., *Biophysical Chemistry: Techniques for the Study of Biological Structure and Function (Their Biophysical Chemistry;)* PT. 2, W.H. Freeman & Company, New York, 1980.
21. Harding, S.E., Rowe, A.J., and Horton, J.C., *Analytical Ultracentrifugation in Biochemistry and Polymer Science*. Royal Society of Chemistry, Thomas Graham House, Science Park, Cambridge, 1992.
22. Harding, S.E., Sattelle, D.B., and Bloomfield, V.A., *Laser Light Scattering in Biochemistry*. Royal Society of Chemistry, Thomas Graham House, Science Park, Cambridge, 1992.
23. Bloomfield, V.A., Crothers, D.M., and Tinoco, I. Jr., *Physical Chemistry of Nucleic Acids*. Harper & Row, New York, N.Y., 1974.
24. Lebowitz, J., Lewis, M.S., and Schuck, P., Modern analytical ultracentrifugation in protein science: a tutorial review. *Protein Sci.*, 2002, 11(9), 2067–2079.
25. Cole, J.L. and Hansen, J.C., Analytical ultracentrifugation as a biomolecular research tool. *J. Biomolecular Techniques*, 1999, 10, 163–176.
26. Hansen J.C., Lebowitz, J., and Demeler, B., Analytical ultracentrifugation of complex macromolecular systems. *Biochemistry* 1994, 33(45), 13155–13163.

4.3 Predictive Biology

4.3.1 Protein Structure Prediction

In the early part of the twentieth century, the notion of a molecular understanding of biological systems was considered to be farfetched and something that could be found only in science-fiction books. A few decades later, Avery's experiments showed that DNA rather than protein was the coding molecule (blueprint) in living systems. This evidence created an urgent need to understand every intricate detail of this molecule. This, in turn, attracted many brilliant scientists who continue to contribute to the field's exponential growth and advancement.

TABLE 4.4

PDB Holding List by Technique and Macromolecular Structure

Experimental Technique	Proteins, Peptides, and Viruses	Protein/Nucleic Acid Complexes	Nucleic Acids	Carbo-hydrates
X-ray diffraction and other	21428	1054	747	14
NMR	3246	103	608	4

Note: As of 9/14/04; theoretical models have been removed, effective July 02, 2002.

Modeling and predicting biological macromolecules from basic principles has been used long before advanced computational tools were available, even before the first experimentally determined structures were available. It was first successfully applied to proteins in 1951 by Pauling and Corey who predicted the structure of protein secondary structure,[1] and James Watson and Sir Francis Crick who solved the structure of DNA double helix in 1953.[2] Both predictions depended on crystallographic data that suggested specific symmetries in both macromolecules, e.g., helical conformation. The fruitful interaction of theory and empirical data soon culminated in 1958 with the successful high-resolution analysis of the first globular protein structure, whale sperm myoglobin[3] by a group of pioneering biochemists and crystallographers led by Max Perutz and John Kendrew. In 2004 the number of experimentally determined high-resolution structures deposited in the protein data bank (PDB) passed 27,000. They are mostly protein structures obtained by x-ray crystallography as summarized in Table 4.4. About 10% of all entries are nucleic acids and protein/nucleic acid complexes. There are only 18 carbohydrate structures solved experimentally.

These early experimental breakthroughs validated modeling of the structure of biological macromolecules based on sequence information alone. Linus Pauling is considered the father of molecular modeling and his contributions to this field were instrumental to the work of the scientists who followed in his footsteps. His modeled structure of the alpha helix enabled the identification of helical and pleated-sheet structural patterns, called secondary structures that exist at the molecular level of proteins in all organisms as shown in Color Figure 4.23. This, in turn, helped to recognize additional secondary structures in proteins and enabled initiation of a classification scheme associated with the polypeptide's secondary structures. To date, over 250 types of secondary structures have been classified.[4]

Sir Francis Crick and James Watson were awarded the Nobel Prize for solving the structure of the double-stranded DNA molecule (see Color Figure 1.13a). The structure of DNA was an essential step in solving many of the ambiguities associated with the understanding of the mechanisms of biological inheritance and enabled the development of a new branch of biology that dealt specifically with biological problems at the molecular level: molecular biology. This field is mainly concerned with understanding genetics at the molecular level and the direct and indirect interactions of proteins

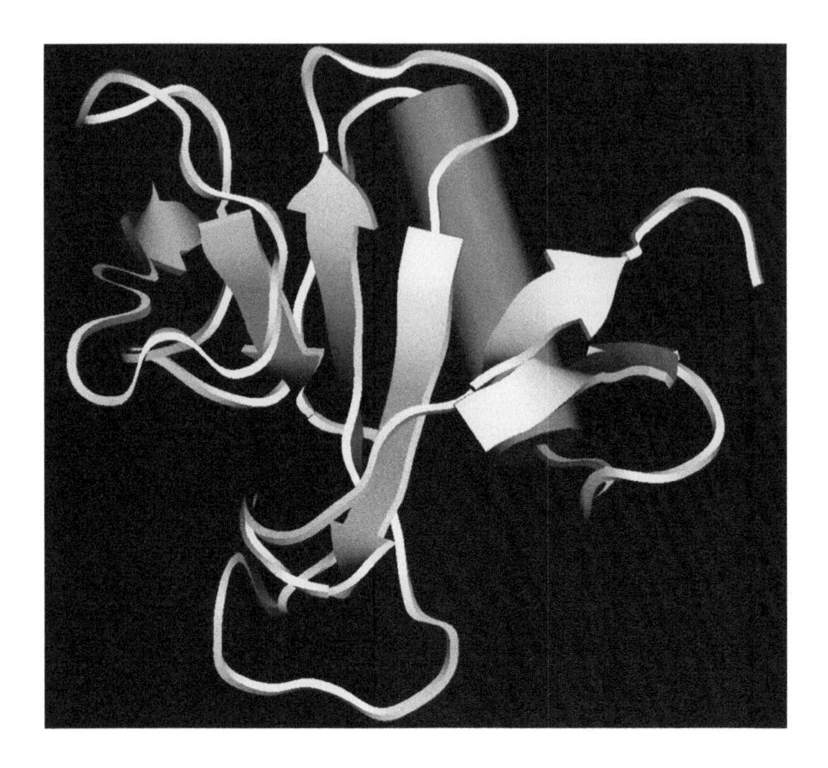

FIGURE 4.23 (See color insert)
Secondary structure representation of a protein. Secondary structure cartoon of penicillin acylase: a small protein with a five-stranded antiparallel beta sheet (yellow ribbons) interacting with a single alpha helix (red extended cylindrical structure) on its "back" side. (From Molecular Simulations, Inc. (MSI). With permission.) Protein motif with beta sheet (golden) and alpha helix (red).

with nucleic acids. The vast amounts of information associated with the structure of the DNA molecule enabled us to gain incredible insight into the phenotypic (physical) and molecular characteristics of many different species.

Today, x-ray crystallography, nuclear magnetic resonance (NMR), and cryo-electron microscopy are the three experimental techniques used in solving the high-resolution structures of biological macromolecules. In x-ray crystallography, the crystallized protein is bombarded with x-rays (electron rays are used in cryo-electron microscopy). The resulting diffraction pattern captured on x-ray sensitive film contains the spatial information used to determine the atomic structure of the molecule, i.e., the electron density within the crystallized macromolecules. In principal, this method is not limited to the size of molecules, yet depends on the successful formation of crystal from millions of units in an orderly fashion. There are, however, many limitations associated with its methodology. First of all, crystallizing the biological macromolecules is not a trivial step, mainly due to the difficulty of obtaining regular crystal lattices of millions of identical units. Many of our essential proteins (e.g., membrane-bound proteins) are still structurally unknown due to the inability to crystallize these polypeptides once they are removed from their membrane environment. Electron diffraction, however, is a good alternative thanks to cryotechniques, where samples are frozen at extremely fast rates (milliseconds) in liquid nitrogen (–180°C).

A second problem is the lack of dynamic information through crystallography. Biological macromolecules (e.g., proteins) are quite dynamic and flexible. Crystallography gives us a rigid picture of the molecule due to the very long time exposure needed to collect the diffraction patterns.

NMR data, on the other hand, yields all possible modes of conformation, since its much higher time resolution is in the millisecond range. Although the use of NMR seems to be advantageous and the most reasonable technique for solving structures, it has a severe limitation in that it can only be used for proteins containing less than 300 amino acids. The use of proton signatures in NMR creates too many overlapping signals in larger molecules. This limits it to smaller macromolecules and prevents the analysis of most proteins and protein complexes. All techniques used for high-resolution structure analysis depend on knowing the amino acid sequence for proper identification of electron densities or chemical shift and magnetic resonance correlation patterns. Thus, the enormous progress in sequencing novel genes dramatically favors the structural analysis of many more proteins.

Nevertheless, the rate at which sequence information is published each day far exceeds that of structural information obtained through crystallography and NMR, hence, the need for and popularity of predictive structural biology. In the past, predictive biology was practiced predominantly by theoreticians with access to super computers. With today's computers, the analysis of a typical protein molecule can take mere days. The time consuming steps in crystallography are protein identification, isolation, and purification, but particularly crystallization.

Understanding the structural aspects of the protein of interest will yield a vast amount of information about its potential function and its relationship to other essential macromolecules. Predicting the high-resolution structure of proteins and their correct folding route, however, is a major problem in biology, because proteins are composed of twenty different alpha amino acids that introduce the essential sequence variability observed in these molecules. Each of these amino acids in turn can hold a variety of conformations with respect to the other residues (the side chain associated with each of the amino acids) in the protein.

The number of different structural possibilities to which the protein can conform with respect to its amino acid sequence are phenomenal in theory, but only one or several closely related active structures are formed in solution. Many different methodologies have been employed to explain why only one preferred conformation is found from these astronomical numbers of conformational possibilities. The most common deductive techniques used are those dealing with energetics (thermodynamics) and those that take advantage of the molecule's relational information to other known homologs. Currently, the consensus in the field is to employ both methods

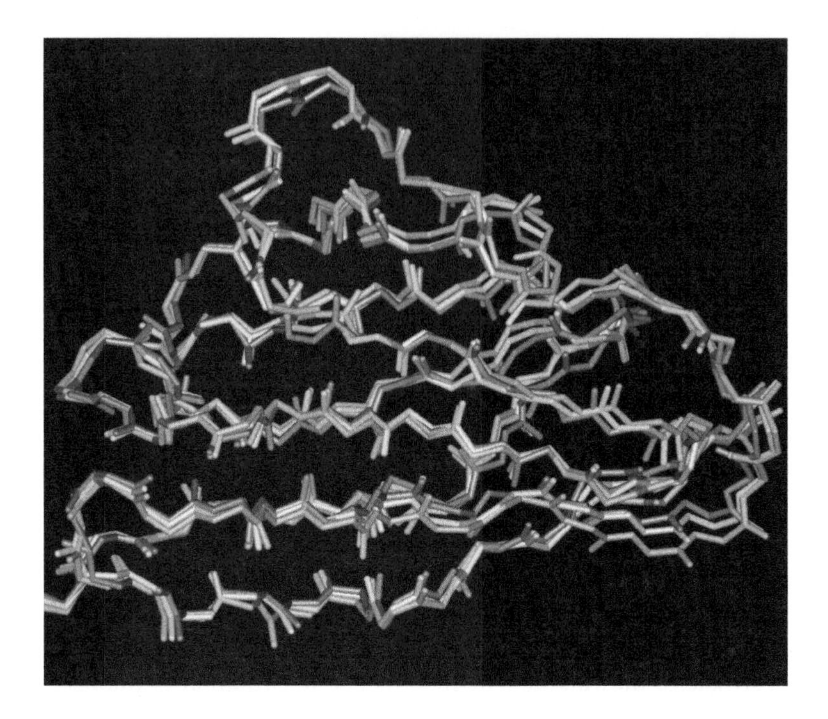

FIGURE 4.24 (See color insert)
Energy minimization of homologous protein structures. The homology model as it undergoes minimization. The structures are colored red, orange, yellow, and green. They represent the results from the starting model, fixing splices, fixing side chain clashes, minimizing whole structure respectively. (From MSI. With permission.)

in an ordered fashion. When possible, homologous molecules would be used to gain an insight into the molecule of interest and the use of energy minimization software would be employed to minimize chemical and physical anomalies of interaction as shown in Color Figure 4.24.

As discussed earlier, the rate of solving structures through x-ray crystallography or NMR is much lower than the high numbers of new DNA and protein sequences that are introduced each day, necessitating a calculated and semi-reliable approach to structural design. The bioinformatics groups are predominantly responsible for identifying key biological macromolecules (e.g., proteins) responsible for pathological events and for proposing potential inhibitors for such macromolecules. In order to understand and, thus propose the possible interactions of the molecule of interest with its substrate and other interacting compounds, the structure of the macromolecule studied must be known. This introduces several problems. The greatest obstacle is the lack of structural data for the molecule of interest. In most cases, the macromolecules studied are structural or regulatory proteins. For proteins that lack NMR or crystallographic structural information, homologous proteins from other species could be used for which a structure is known. Modeling unknown protein structures based on their homologs is better

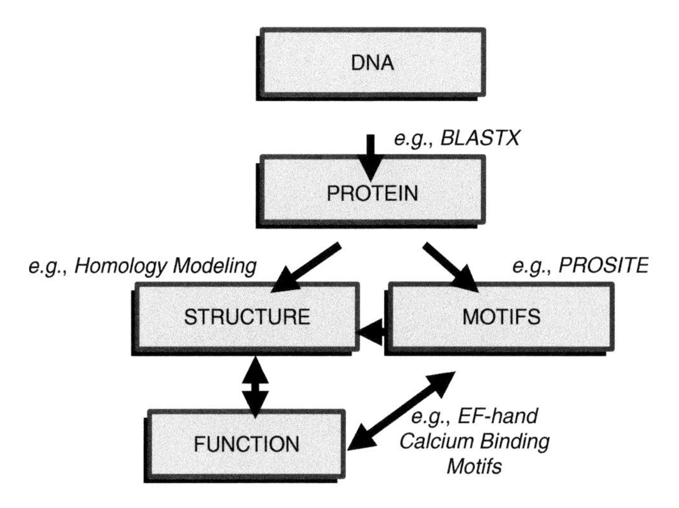

FIGURE 4.25
From DNA sequence to protein function.

known as homology-based structural modeling and the program that best exemplifies such an approach is the homology program from Molecular Simulations Incorporated (MSI) InsightII. In the next several pages we will discuss the details involved in the homology-based modeling approach and point out its most obvious strengths and limitations.

Homologous proteins are generally referred to as polypeptides that share a similar amino acid composition. In most cases, the proteins with relatively high degrees of identity are also structurally and functionally homologous (Figure 4.25). Changes in the protein's amino acid sequence could result in a change of its 3-D structure. It is this relationship between the protein's amino acid sequence and its 3-D structure that allows us to compare proteins that lack x-ray crystallography or NMR-resolved structures with their sequence homologs having known structures.

The sequence homologs with known structures allow us to calculate the structure of the homologous sequences that lack structures by comparative modeling and, thereby, gain insight into the protein's function. In homology-based protein modeling, the experimentally determined structures are generally referred to as the "templates" and the sequence homolog (e.g., a novel string of nucleotides identified from ongoing genome projects) that lacks structural coordinates is called the "target" sequence. The homology-based protein modeling approach entails four sequential steps (Figure 4.26); the first involves the identification of known structures that are related in sequence to our target sequence. This step is typically achieved by using Internet tools such as BLAST to search for the potential templates. In the second step, these potential templates are aligned with our target sequence to identify the closest related template. In the third step, a model of the target sequence is calculated from the most suitable template found in step two.

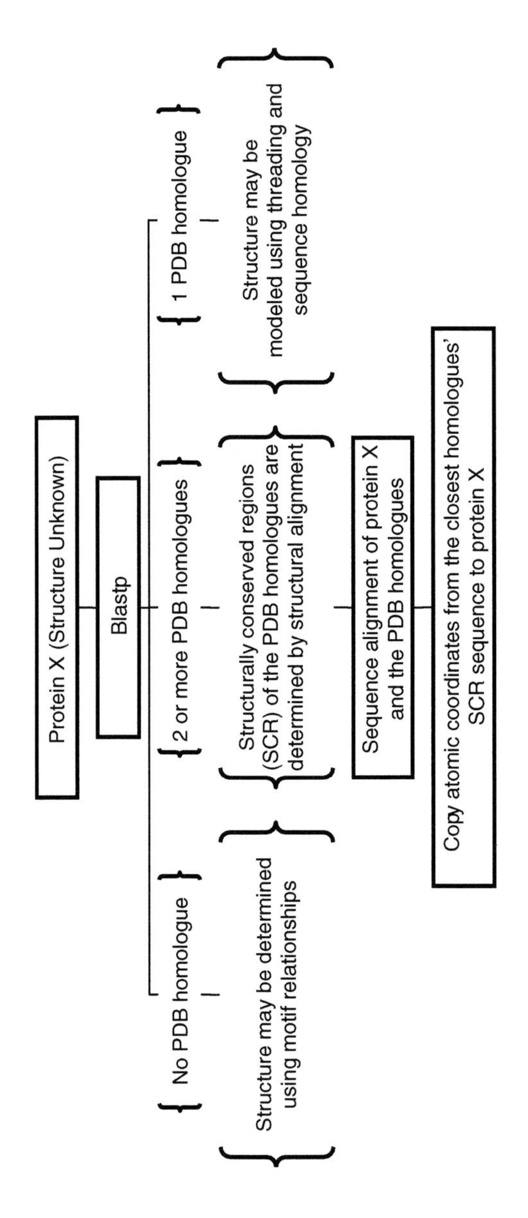

FIGURE 4.26
Homology-based modeling of structures.

The final step involves the evaluation of our modeled target sequence using a variety of criteria (e.g., energetics, etc.).

Ideally, the amino acid composition of the unknown protein and its known structural homologs are quite similar and more than one known structural homolog from a different species is present in the PDB[5] protein database. NCBI's BLAST[6] enables us to find these known homologs. BLAST uses the primary sequence (amino acid sequence) of the target sequence as its input file and conducts a search for homologous protein sequences in a defined database (e.g., SWISS-PROT,[7] PDB, etc.). Upon completion, the closest sequence homologs are displayed. The related proteins are generally those with the highest BLAST scores. By defining the database as PDB, the search would be limited to the sequence homologs that have known structures. The structural coordinates of the highest scoring proteins are then retrieved and used as input files in the protein modeling program of interest (e.g., MSI's homology software). The outcome of such a structural modeling of a consensus structure is shown for serine proteases in Color Figure 4.27. The core structure of four different proteases is highly conserved (yellow) which includes the catalytic center. The surface loop structures are colored red and blue indicating their degree of deviation from the four input structures. The

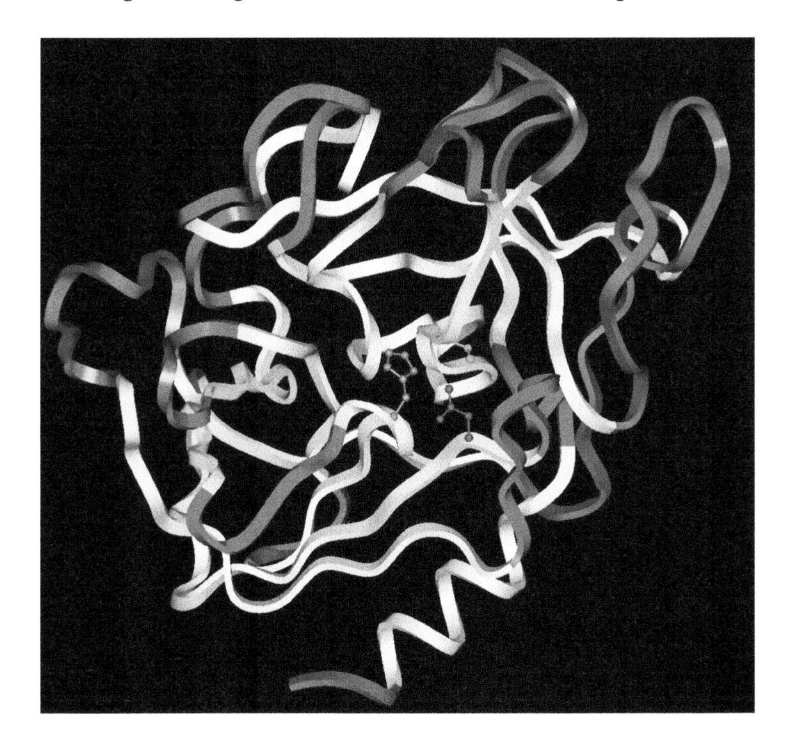

FIGURE 4.27 (See color insert)
Final model structure of serine protease. Structurally conserved regions are shown in yellow, loop regions in blue, and insertions in red. The catalytic triad is also shown. (From MSI. With permission.)

FIGURE 4.28 (See color insert)
Multiple sequence alignment of four serine proteases. Legend: ACB; TGD; EST; RP2. (From MSI. With permission.)

process of obtaining this consensus structure is described in the three subsequent figures: sequence alignment in Color Figure 4.28, loop identification in Color Figure 4.29, and structure overlay in Color Figure 4.30 (not a consensus structure).

The atomic coordinates of the potential templates are then structurally aligned to display the structurally conserved regions (SCRs) of the protein group of interest. The use of SCRs will enable the construction of the evolutionary conserved structural features of the target sequence (structurally unknown protein sequence). Although there seems to be a relationship between the protein's amino acid sequence and its structure, the ambiguities involved in this relationship currently prevent us from constructing models of structurally unknown proteins based solely on their amino acid sequence.

The knowledge of evolutionarily conserved structural features of similar proteins from other species thus enables us with greater confidence to gain insight into the structure of the target sequence. Regions of the protein that are highly homologous in their amino acid sequence do not always correlate to a SCR. In fact, many of the loop regions of homologous proteins are highly similar in their amino acid sequence, but their structural features are quite variable (Color Figure 4.29). In other words, the relationship between the SCRs and their respective amino acid sequence alignment is not as obvious as would be expected. Therefore, the presence of the SCRs is essential in such a comparative approach. After the assignment of the SCRs, the most suitable template must be found. The template is typically the closest amino acid sequence homolog to the target sequence. This

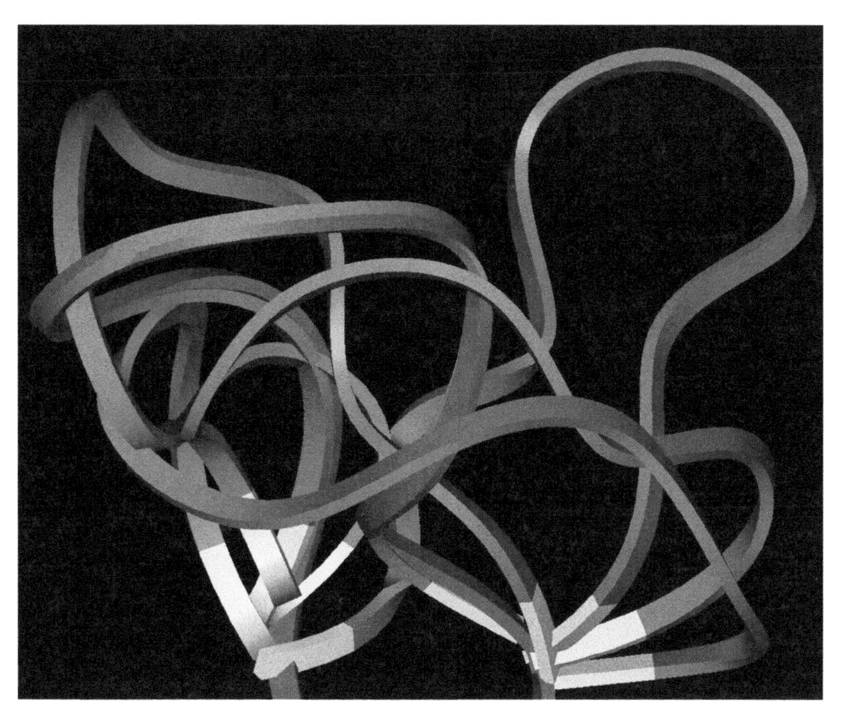

FIGURE 4.29 (See color insert)
Loop conformations located during a database search of the Protein Databank. Chosen conformation is shown in blue. (From MSI. With permission.)

usually happens to be the closest species to our target sequence. If our target sequence is protein X from human, and the potential templates are protein X from rat and Drosophila (fly), then on the basis of phylogeny, the rat protein X is typically selected over Drosophila and is used as the template sequence for modeling the human target sequence (protein X). This is based on the assumption that human and rats are the closer sequence homologs.

The amino acid sequence of the target protein is then aligned with the chosen template sequence (Color Figure 4.28). The atomic coordinates of the amino acids in the SCR regions of the template sequence are copied over to their respective sequence homologs in the target protein. Most of the secondary structures (e.g., alpha helix, beta strand, etc.) are typically covered by the SCRs. The structural features that are typically outside of the labeled SCRs are the loops and the terminal ends of the molecule. After assigning coordinates to the SCR overlapping amino acid sequences, the most suitable structural coordinates for the target protein's loop regions and its terminal ends must then be found.

The following section summarizes a reliable, scientifically sound, and extremely user-friendly set of protein structure prediction-and-analysis software available from Accelrys (formerly MSI). MSI (www.accelrys.com) offers

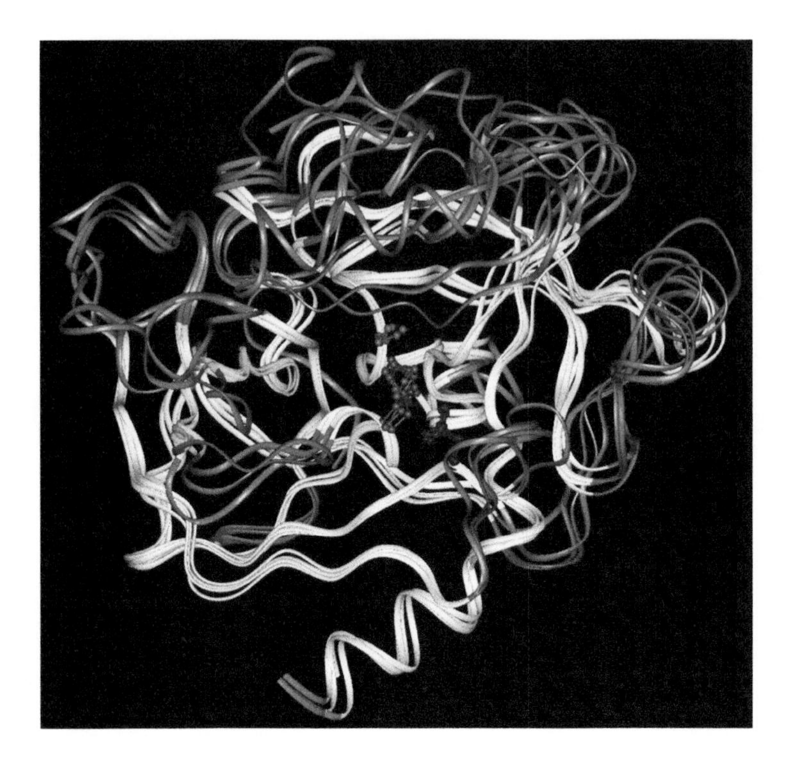

FIGURE 4.30 (See color insert)
Four serine protease structures overlaid. Same proteases as shown in Color Figure 4.28 (sequence alignment). The structurally conserved regions are colored yellow. The catalytic triad is shown in ball and stick (serine: green, histidine: orange, aspartic acid: red). (From MSI. With permission.)

proprietary computational tools that are widely used by academia and are also quite popular in industry.

4.3.1.1 Structure Prediction Software

Structure prediction tools are generally classified into two categories: public domain and proprietary software. Following is a summary of some of the many proprietary software tools exclusively offered through Accelrys:

What is proprietary software (e.g., Accelrys software)?

These are tools that must be purchased by those who maintain the rights to the software. Their popularity is based mainly on their leading-edge approach to predicting and analyzing structural characteristics of the macromolecules and their sequence homologs. These proprietary algorithms are very popular in the pharmaceutical field, and their use has revolutionized the study of drug design by adding an element of prediction to the old empirical approach. One of today's most popular proprietary programs is the InsightII environment offered by Accelrys (it was used to generate Color Figures 4.23, 4.24, 4.27, 4.28, 4.29, and 4.30 shown in this book). MSI is a biotechnology and bioinformatics company based in La Jolla, California specializing in structure prediction and analysis software. Accelrys is predominantly involved in designing and maintaining its own proprietary algorithms and software that meet the needs of a variety of biotechnology and pharmaceutical companies. The company also collaborates with many academicians and constantly strives to sustain its technological leading edge. Accelrys provides a variety of user-friendly yet sophisticated software through the InsightII environment, as well as software specifically designed to perform delicate computational tasks. For instance, the tools in the Affinity software program can be used in rational drug design, while tools such as Homology are utilized to predict the three-dimensional structure of a protein through sequence homologs with known 3-D structures.

What type of hardware platform is required for the Accelrys software?

Currently, Accelrys InsightII environment is only available for Silicon Graphics and IBM RISC system/6000 workstations.

What is the InsightII graphics environment?

Most of the Accelrys software modules require the InsightII 3-D graphics program. Its graphics environment provides a very user-friendly setting (as shown in Color Figure 4.31) that facilitates the use of all other compatible software modules. A variety of interactive tutorials are also provided for many of their sophisticated computational tools to further facilitate the learning process.

Following is a list of some of the Accelrys modules currently utilized in academia and industry (adapted from the Accelrys homepage: www.accelrys.com):

- Biopolymer: this module helps to construct protein, nucleic acid, and carbohydrate models. These models can then be used for a variety of simulation jobs.
- 3-D Profiles: this module enables the user to evaluate the compatibility of the amino acids in their environment to a database of 3-D profiles. The compatibility is a verification measure of the modeled structure and its sequence.[1] This method tries to answer the following question: is the query sequence compatible to a known 3-D profile?
- DelPhi: this module is typically used to calculate solvation energies and electrostatic potentials.
- Homology: this module is used to construct 3-D models of proteins using sequence homologs with known structures as templates (Color Figures 4.24, 4.27, 4.29, 4.30).

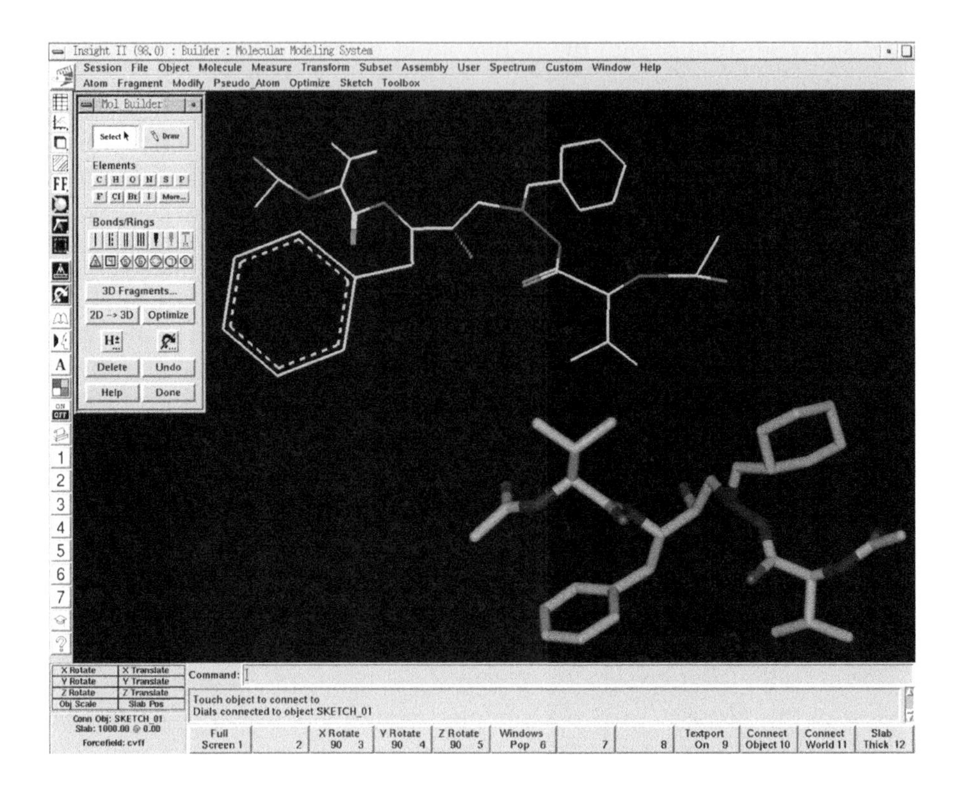

FIGURE 4.31 (See color insert)
Molecular modeling desktop. Sketching a small molecule in 2-D (on the left) that can then be automatically converted into 3-D (on right) (MSI).

- ELF: this module enables the user to calculate the minimum structures and various molecular interactions. It can also be used to calculate solvent-induced effects. These effects include both electrostatic and binding-site interactions.
- Discover: this module is typically used for energy minimization, dynamic simulations, and conformational sampling. It incorporates a variety of validated force fields. Discover enables the user to predict a variety of energetic and structural properties associated with the macromolecule or system of interest (Color Figure 4.24).
- CHARMm: this is a somewhat more advanced module that combines minimization and dynamics features with expert methods such as free energy perturbation (FEP), combined quantum and molecular mechanics (QM/MM), or correlation analysis.
- Affinity: this is an automated docking program. The module is typically used in rational drug design. It docks the appropriate LIGAND into its receptor.

- LUDI:[8] this module allows the investigator to fit the LIGAND or molecule of interest into the active site or binding site of the targeted protein. It accomplishes this by matching the complementary nonpolar-nonpolar and polar–polar substituents.

For a more detailed explanation of Accelrys software and services, please visit their home page at www.accelrys.com. Many of the color figures in this volume were created using MSI software. Special thanks to our colleagues at MSI for providing us with these educational and visually stunning pictures.

References

1. Pauling, L. and Corey, R.B., Configuration of polypeptide chains. *Nature*, 1951, 168(4274), 550–551.
2. Watson, J.D. and Crick, F.H., Molecular structure of nucleic acids; a structure for deoxyribose nucleic acid. *Nature*, 1953, 171(4356), 737–738.
3. Kendrew, J.C. et al., A three-dimensional model of the myoglobin molecule obtained by x-ray analysis. *Nature*, 1958, 181(4610), 662–666.
4. Holm, L. and Sander, C., Mapping the protein universe. *Science*, 1996, 273(5275), 595–603.
5. Sussman, J.L. et al., Protein Data Bank (PDB): database of three-dimensional structural information of biological macromolecules. *Acta. Crystallogr. D. Biol. Crystallogr.*, 1998, 54(1 (Pt 6)), 1078–1084.
6. Altschul, S.F. et al., Basic local alignment search tool. *J. Mol. Biol.*, 1990, 215(3), 403–410.
7. Bairoch, A. and Apweiler, R., The SWISS-PROT protein sequence data bank and its supplement TrEMBL in 1999. *Nucleic Acids Res.*, 1999, 27(1), 49–54.
8. Bohm, H.J., LUDI: rule-based automatic design of new substituents for enzyme inhibitor leads. *J. Comput. Aided Mol. Des.*, 1992, 6(6), 593–606.

4.3.2 Structural Genomics

While Proteomics is the study of the full set of proteins encoded by a genome, Structural Genomics is focused on compiling the structure data of proteins that can be deduced from Genome sequences. It uses predictive proteomics tools, but also experimental techniques for the determination of protein structures.

Knowing the structure of proteins is important in several ways. First, the structure represents a mechanistic view of protein function. This view is most commonly exploited for rational drug design (see next section) where it helps model putative binding modes of novel ligands and substrates of receptors and enzymes. Second, protein structures are more informative about relationship among different proteins than gene sequences alone. This

TABLE 4.5

Protein Families And Sequence Motifs Databases

Database	Content	URL
UniProt	Repository of protein sequence and function created by joining the information contained in SWISS-PROT, TrEMBL, and PIR	www.pir.uniprot.org/index.shtml
PROSITE	Database of protein families and domains	http://au.expasy.org/prosite/
CluSTr	Automatic classification of SWISS-PROT + TrEMBL proteins into groups of related proteins	www.ebi.ac.uk/clustr/
InterPro	Provides an integrated view of the commonly used signature databases	www.ebi.ac.uk/interpro/
Pfam	Protein families of alignments and HMMs and species distribution	www.sanger.ac.uk/Software/Pfam/
PIR	Protein Information Resources maintained by National Biomedical Research Foundation	www-nbrf.georgetown.edu/home.shtml
PRINTS	Protein fingerprint database of groups of conserved motifs used to characterize a protein family	www.bioinf.man.ac.uk/dbbrowser/PRINTS/
ProDom	The protein domain database	http://protein.toulouse.inra.fr/prodom/current/html/home.php
PRF	A peptide structure- function database	www.prf.or.jp/en/
EOL	A project to catalog the complete proteome of every living species in a flexible, powerful reference system including a Genome Annotation Pipeline (iGAP) **to** annotate protein sequences for their putative structure and biological function	http://eol.sdsc.edu/

is due to the observation that the structure of a protein shows a higher degree of conservation than the underlying sequence. The reason are conservative amino acid substitution that change the sequence, but not solubility and folding behavior in solution, and thus do not affect molecular interaction in space. This high degree of conservation is particularly dominant in structural motifs and functional domains. Both are used to classify genes and proteins into families and superfamilies and to identify homologs and paralogs in genomes of diverse organisms. Third, having the structures of all proteins of a genome will facilitate interpretations and predictions regarding the proteome of an organism (see Section 4.3.1).

Table 4.5 summarizes some of the better-known databases emphasizing protein structure, domains, and motif information. These databases are structured according to similarities based on family relationship.

As more and more genome projects are being completed, the structural information for proteins lacks behind. Both predictive and experimental approaches to fill the gap are being pursued. Structural databases containing data from experimentally determined high-resolution structures are listed in Table 4.6. The main depositories are the Protein Data Bank, the Molecular Modeling Database at NCBI (see as an example Figure 3.16) and the Macromolecular Structure Database at EBI. They contain information for all known protein structures, but also include nucleic acid and polysaccharide structures,

TABLE 4.6

Structural Databases

Database	Content	URL
PDB	Repository for the processing and distribution of 3-D biological macromolecular structure data	www.rcsb.org/pdb/
MDB	Molecular Modeling database of macromolecular 3D structures, as well as tools for their visualization and comparative analysis	www.ncbi.nlm.nih.gov/Structure/
MSD	The European project for the collection, management, and distribution of data about macromolecular structures at EBI	www.ebi.ac.uk/msd/index.html
PDBj	Protein Data Bank Japan	www.pdbj.org/
wwPDB	Worldwide Protein Data Bank; a central depository of all available protein structures	www.wwpdb.org/
CATH	Hierarchical classification of protein domain structures	www.biochem.ucl.ac.uk/bsm/cath/
PARTS LIST	Dynamically ranking protein folds based on disparate attributes, including whole-genome expression and interaction information	http://bioinfo.mbb.yale.edu/partslist/
SCOP	3-D fold classifications for folds, families, and super families	http://scop.mrc-lmb.cam.ac.uk/scop/index.html
CSD	Cambridge Structural Database; repository of *small molecule* crystal structures	www.ccdc.cam.ac.uk/products/csd/

where available. While they provide links to gene sequence and genome mapping information, they are not structured or provide whole genome information or classification.

The Protein Data Bank (PDB) currently contains 27,000 entries of known protein and nucleic acid structures. A majority are protein structures, of which some 12,000 are probably unique, while the rest of the entries contain closely related structures, mostly from mutation analysis. The structural databases are biased towards small soluble proteins and protein domains. For instance, while up to 15% of genes in an organism code for membrane proteins, less than 100 unique structures are deposited in the databases. The reason has to do with difficulty of obtaining high amounts of purified and isolated membrane proteins suitable for crystallization. Often, extramembraneous domains are crystallized independently of membrane spanning domains of receptors, adding to the structural information of this class of membrane proteins.

Structural databases offer besides the atomic coordinates of macromolecules interactive 3-D visualization software that makes browsing and studying protein and DNA structures very straightforward. As an example, Figure 4.32 shows the structure of reverse transcriptase (PDB accession number 1REV)

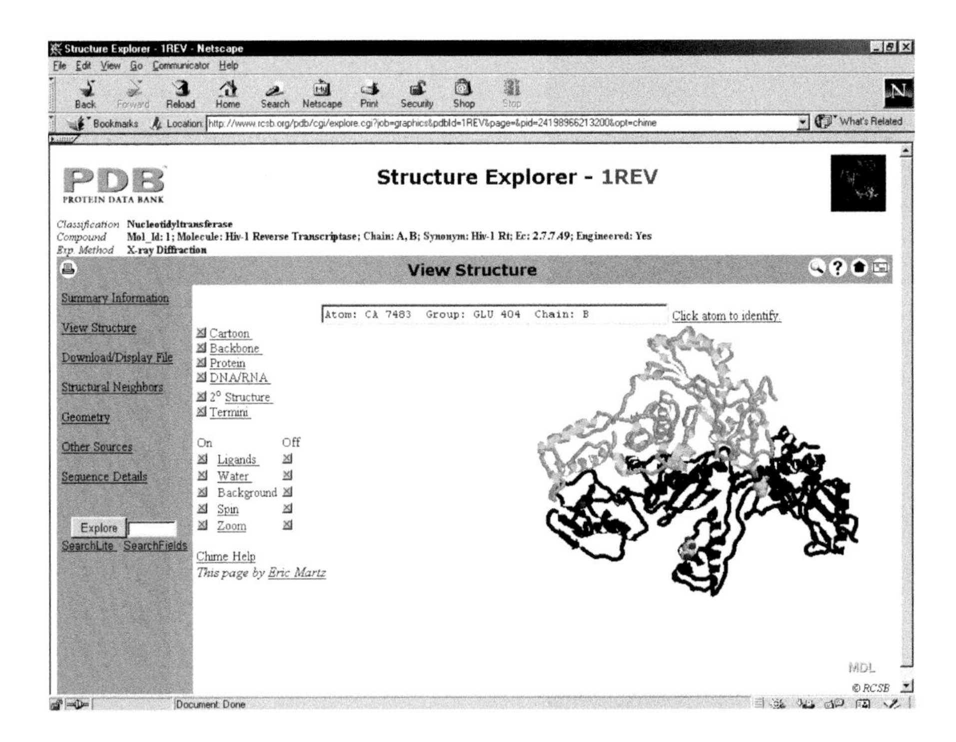

FIGURE 4.32
HIV reverse-transcriptase structure 1REV.PDB viewed with Chime. "View 1REV in 3-D" gives a link to MDL's Chemscape Chime viewer (http://www.mdli.com/chemscape/chime), a software plug-in for easy viewing of the structure in different modes. Alternatively, the PDB structure file can be downloaded onto the local hard drive and viewed by the Rasmol browser plug-in Chime.

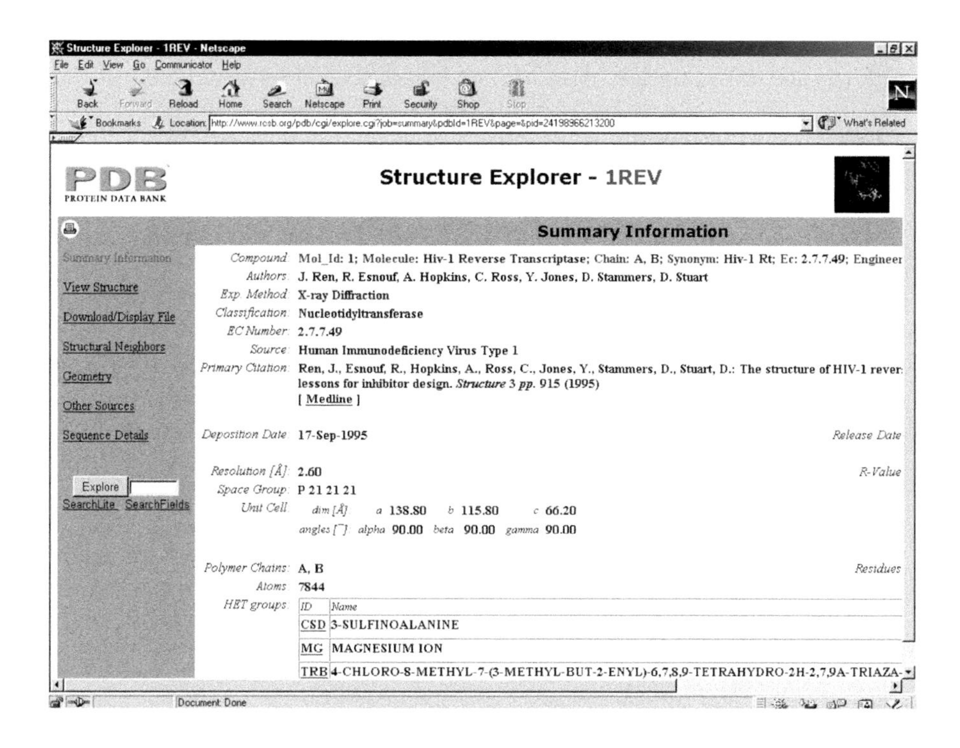

FIGURE 4.33
PDB data sheet for HIV reverse transcriptase.

from the human immunodeficiency virus (HIV). The structure of this enzyme has been solved at 2.6Å resolution describing the spatial coordinates of 7715 atoms, not including all hydrogen atoms. A total of 27,108 reflections have been measured for structure refinement using the program X-PLOR 3.1.[1] Using the accession number 1REV at the query page of the Protein Data Bank (PDB),[2] a summary page indicates date of submission (9/17/1995), title of protein (HIV reverse transcriptase), and the authors (J. Ren et al.).

The PDB datasheet (Figure 4.33) contains classification information and lists the organism/cell type from which the protein originates (it is an engineered, or recombinant, enzyme with the enzyme commission number EC 2.7.7.49, and is an enzyme of the human immunodeficiency virus type 1), and has been expressed in a bacterial system to maximize quantities for crystallization purposes (the expression system of choice is an *E. coli* cell into which the recombinant DNA containing a gene of a virus that can only infect humans and not bacteria, has been inserted in order to over express the protein for purification). Protein crystal parameters (space group), methods, and additional ligands and resolution (2.6 Å) are also indicated. To get a complete understanding of the current structure of this protein, there are several links that will enable the scientist to access the complete coordinates in spreadsheet format and 3-D and 2-D visualization. When following the link (header only) under "data retrieval," additional information and links (to the published article in MEDLINE) and sequence information in GenBank[3] and SWISS-PROT[4] can be found.

It is interesting to look at the history of generating images of biological macromolecules. An historical outline by Eric Martz can be found online at www.umass.edu/microbio/rasmol/history.htm. Computer representation of wireframe models was not possible until 1972 and was pioneered by Levinthal and Katz. Years later, computer generation of models of spacefilled molecules and the rotation of the molecules become possible. In recent years, inexpensive programs have become available for use in classroom settings, which can be operated on Macs or PCs with less than 1 Mb of disk space (however, the data files use hundreds of kilobytes and could easily use up considerable disk space if a personal structural database were built on a home computer). Yet programs like Kinemage[5] and RASMOL[6] demonstrate the ease of looking at protein structures (RASMOL is freeware).

At PDB, several viewer features are available including an interactive immersive ribbon diagram (VRLM), Protein Explorer (Figure 4.34) and QuickPDB (Figure 4.35), a simple java-based applet that allows modifications of sequence and structural features to highlight their location.

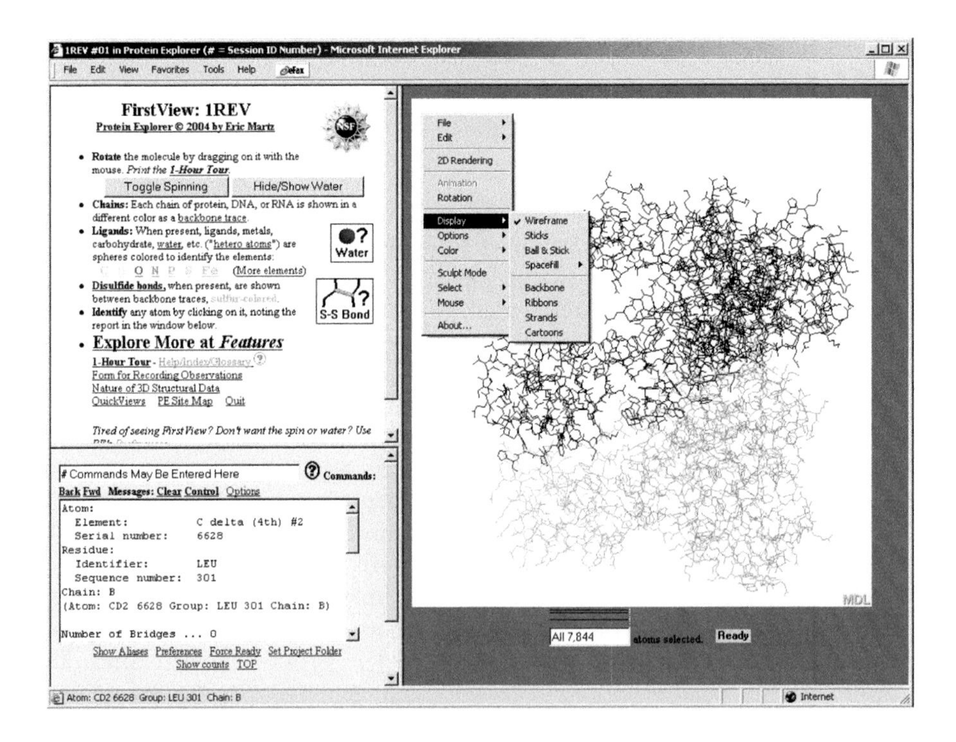

FIGURE 4.34
HIV reverse-transcriptase Protein Explorer View at PDB.

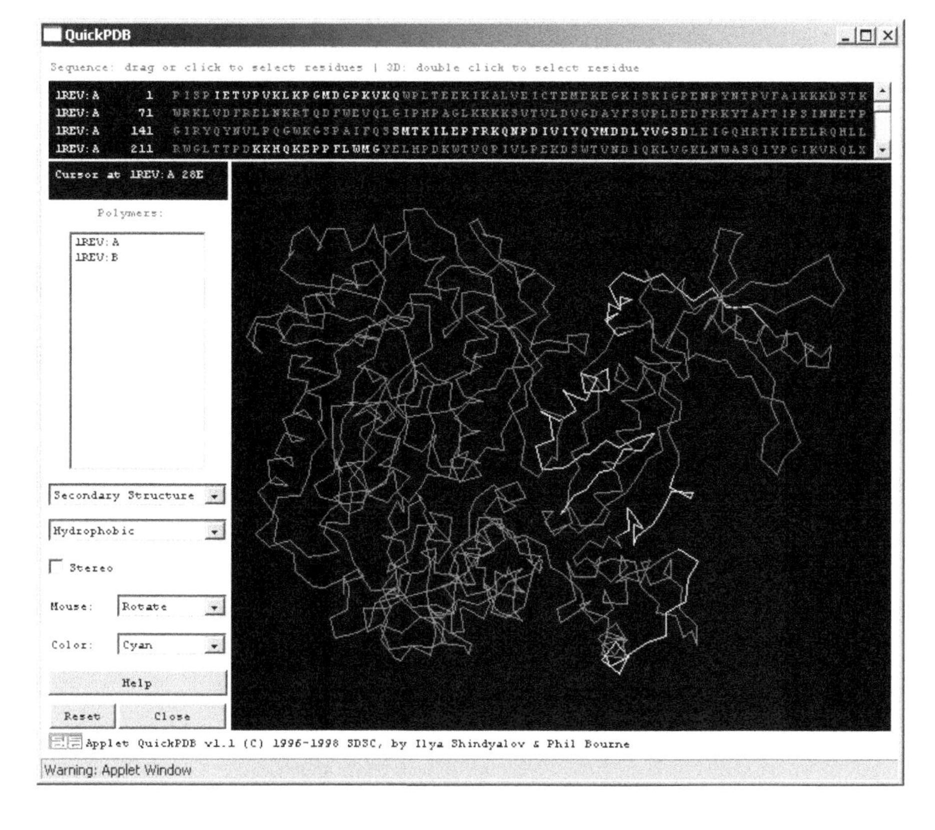

FIGURE 4.35
HIV reverse-transcriptase QuickPDB View at PDB. Highlighted sequencing in top window corresponds to the highlighted segments of the backbone structure in the main window.

Protein Explorer is the most inclusive structure viewer based on the Chime plug-in that allows many modifications of how the structure is displayed, from secondary structure composition to individual side chains and atoms. Quick PDB is a very fast way of getting an overview of the distribution of different types of amino acid based on chemical and physical properties including hydrophobicity, polarity, charge, size, aromatic and aliphatic characteristics. QuickPDB also allows highlighting the location of individual amino acids or longer peptide segments within the 3-D structure. This can be done by selecting a specific sequence, which colors the corresponding segment (see bright segments in Figure 4.35), or clicking on the structure, which selects a residue in both the structure window and the sequence window.

When structural information is not available, it has to be predicted, as outlined in the preceding section 4.3.1. Structural genomic approaches make use of homolog and paralog information at the sequence level to identify conserved motifs and domains, even if only covering a small portion of a novel protein, as a starting point to model the structure based on sequence similarity. Experimental structure determination and structure prediction go hand in hand and support each other by continuously adding new information that can help either experimental or predictive approaches to get a complete set of protein structure for any genome.

References

1. Badger, J. et al., New features and enhancements in the X-PLOR computer program. *Proteins*, 1999, 35(1), 25–33.
2. Sussman, J.L. et al., Protein Data Bank (PDB): database of three-dimensional structural information of biological macromolecules. *Acta. Crystallogr. D. Biol. Crystallog.*, 1998, 54(1 (Pt 6)), 1078–1084.
3. Benson, D.A. et al., GenBank. *Nucleic Acids Res.*, 1999, 27(1), 12–17.
4. Bairoch, A. and Apweiler, R., The SWISS-PROT protein sequence data bank and its supplement TrEMBL in 1999. *Nucleic Acids Res.*, 1999, 27(1), 49–54.
5. Richardson, D.C. and Richardson, J.S., The Kinemage: a tool for scientific communication. *Protein Sci.*, 1992, 1(1), 3–9.
6. Sayle, R.A. and Milner-White, E.J., RASMOL: biomolecular graphics for all. *Trends Biochem. Sci.*, 1995, 20(9), 374.

4.3.3 Rational Drug Design

Over the past several decades, the exponential growth in biological and biochemical data at the molecular level has been one of the greatest contributors to the transformation of pharmaceuticals from an empirical discovery science to a hypothesis driven approach based on rational or systematic design principles. A necessary ingredient in rational drug design is knowledge of one of the molecular structures involved—either the receptor protein or the drug molecule. In the absence of known structures—and this is true for many of the novel genes discovered through the genome projects as open or unidentified reading frames—structure prediction on the side of the receptor is the only viable alternative for such a rational approach.

Drug discovery makes use of both theoretical and empirical methods. To find a good (drug-) ligand for a protein or DNA surface (a receptor) experimental testing procedures using large chemical libraries with thousands of potential compounds can now be tested efficiently using high-throughput screening assays to search for candidates that could potentially serve as life-saving therapeutic agents. Facilitating these rapid but random screens are predictive approaches based on novel sequences. Genome projects facilitate the large-scale identification of promising targets.

Once a target is identified, two approaches can be followed. First, a gene can be cloned into an expression vector and used to make protein available for a ligand binding or functional assay in a cell line or animal model. This will generally lead to high-throughput screens and identification of lead compounds, small molecules with the potential to bind and affect the activity of the target. The lead compound structure can then form the bases of rational drug design to synthesize similar compounds with improved biochemical and pharmacological properties. Second, the gene sequence can be used to make a model of the protein structure or binding domain, which will give information about the structure a ligand will likely possess.

Rational drug design is based on concepts of chemical similarity and complementarity. Chemical similarity is measured by identifying distances between atoms on a receptor and a ligand. Based on the chemical properties of the interacting atoms (or group of atoms = functional groups) small differences in distance have a great influence on the "reactivity" of a ligand. Since proteins are fluid-like entities (alas highly viscous ones), their structures are very sensitive towards disturbances at their surface, e.g., a binding event. For small molecule ligands that exhibit similar binding but have different structures, disturbance of the receptor structure results in activation or inactivation of the receptor and the ligand thus behaves as agonist or antagonist, respectively.

Much of what we know about binding comes from natural and synthetic ligands that bind to the same ligand-binding site because the ligands have a similar structure called a pharmacophore. Most drugs are usually competitive inhibitors of natural substrates by mimicking part of all of their structure. Drugs activating receptors are called agonists, while competitive inhibitors are called antagonists.

Molecular similarity space and the structure of a ligand/receptor complex are two important features in drug design. The strength of an interaction (affinity) depends on the complementarity of the physico-chemical properties of atoms that bind, i.e., protein surface and ligand structure. Excluding catalytic mechanisms from the discussion, two classes of molecular properties important for binding can readily be distinguished:

Shape or volume
Surface potential

Talking about these properties, chemists refer to them as the molecular similarity space, which can now more precisely be described as:

Atom pair matching function → shape or volume → weak interactions
Charge matching function → surface potential → strong interactions

It is intuitive to think that binding has to do with complementarity of surface structures the way a key fits into its lock. The better the similarity or complementarity the stronger and more specific the recognition will be. In molecular modeling, complementarity is expressed as atom pair matching function. In biological molecules, the most common interactions are mediated by Van der Waals and hydrogen bonds. Both interactions are weak and are effective only over very short distances, i.e., they are short-range interactions

TABLE 4.7

Ligand–Receptor Complementarity

Physical Property	Receptor	Ligand	Interaction	Distance Relation
Shape or volume	convex–concave	concave–convex	induced dipole	short range ($1/r^6$)
Surface potential	positive–negative	negative–positive	charge	long range ($1/r$)
Hydrogen bonds	acceptor–donor	donor–acceptor	dipole	short range ($1/r^3$)
Solubility	apolar–polar	apolar–polar	induced dipole–dipole	short range ($1/r^6$) short range ($1/r^3$)

that can be easily broken by mechanical stress or increased temperature. Strong and highly specific molecular interaction in biological systems are the result of multiple such bonds constituting a binding site. This not only contributes to more binding energy, but the distribution of individual interaction types define the conformational specificity of binding.

Since binding sites can be made of multiple bonds, recognition by a ligand of its receptor-binding site can only happen, if at least one of the two partners is free to orient itself for optimal interaction. The rotational and translational movement of the ligand within the surface potential field of the receptor is an important attribute in understanding binding. A specific interaction is encountered when the orientation of the ligand fits complementary surface properties on the receptor.

The physical properties defining a contact surface are summarized in Table 4.7. A combination of any of these properties defines multivariate surfaces. This can obviously lead to complex surface structures or binding motifs, especially for large contact surfaces such as found between proteins, where one protein is the "ligand" and the other the "receptor." Protein–protein interactions are central to biological systems and relevant for any enzyme or receptor complex, cytoskeletal structures, or chromatin structures. Peptide ligands (e.g., neuropeptides, insulin, leptin) are among those agonists providing the largest variability in similarity space. Since protein surfaces are determined by their amino acid residues, binding surfaces can be mathematically described as sequence space. Sequence space forms the foundation of structural genomics approaches that attempt to calculate the protein folds of protein coding sequences in whole genomes. There are a few notable exceptions where backbone structures are involved in binding as is the case in histone–DNA interaction in a nucleosome core particle, ion binding sites in the selectivity filter of K-channels, or the main-chain substrate binding site on the active site of chymotrypsin/trypsin family of proteases.

In general, the surface topology of structurally and functionally related ligands that all exhibit effector quality (agonist or antagonist) can be merged and the contours of all ligands averaged into a union surface. This union surface of a ligand class is expected to be complementary to the

TABLE 4.8

Drug Information Databases

Database	Content	URL
ChemBank	Initiative for chemical genetics Collection of data about small molecules and resources for studying their properties, especially their effects on biology; >900,000 records	http://chembank.med.harvard.edu
ChemID*plus*	Chemical Information at the National Library of Medicine; 367,000 chemical records	http://sis.nlm.nih.gov/Chem/ChemMain.html
ChEBI	Natural or synthetic molecular entities used to intervene in the processes of living organisms	www.ebi.ac.uk/chebi/

surface mold of the corresponding binding site on the receptor or enzyme. In complex structures the distribution and combination of physical properties used to search for similarity (complementarity) is large. Sometimes one or more ligand structures differ significantly from each other, even if they bind to the same receptor. Superposition of their structures exposes a critical fragment with shared similarity among ligands. This conserved structural element of a class of antagonist/agonists is called the pharmacophore. In many cases ligand receptor interaction is not mediated by the entire ligand structure, but by ligand points or critical fragments of the pharmacophore. Thus when analyzing existing data of antagonist and agonist structures, it becomes clear why compounds belonging to seemingly different classes of chemicals can act on the same target proteins. Chemical ligand and drug databases (Table 4.8) are now common even in the public domain. They allow searching for ligand similarities based on structure or function and information related to known pharmacological and medical properties.

The pharmacophore of serotonin receptor agonists and antagonists has been well studied. Serotonin (5-hydroxytryptamine or 5-HT) is a neurotransmitter that acts on any of nine known serotonin receptors, which play important roles in neuronal signaling in the central and peripheral nervous system. 5-HT3 receptor antagonists have been shown to produce beneficial effects in animal models of cognitive and psychiatric disorders.[1] Whether 5-HT3 receptor antagonists may have similar profound effects in the treatment of anxiety, depression, or psychosis will be determined by the outcome of ongoing clinical trials. However, it is in the treatment of cancer chemotherapy induced emesis that 5-HT3 receptor antagonists have had their greatest impact. The cytotoxic agents used in cancer chemotherapy provoke the release of 5-HT from enterochromaffin cells in the peripheral vagal afferent fibers of the gastrointestinal tract initiating vomit reflexes (emesis). 5-HT3 receptor specific antagonists block this action and thereby greatly reduce the number of emetic episodes that occur during cancer chemotherapy. The marked clinical efficacy of 5-HT3 receptor antagonists such as ondansetron, granisetron, and tropisetron together with their lack of adverse side effects has greatly improved the treatment of cancer chemotherapy induced emesis.

5-HT3 receptors belong to the family of ligand-gated ion channels. When activated, ions flowing through these channels depolarize the cell membrane triggering action potentials and thus nerve conduction events. 5-HT3 receptor-mediated ion currents evoked by the full agonists 5-HT, quaternary 5-HT (5-HTQ), meta-chlorophenylbiguanide (mCPBG), and the partial agonists dopamine and tryptamine in whole-cell voltage clamp experiments can be used to characterize binding properties of these ligands such as affinity and specificity. Ligand-gated receptors typically switch into an inactive (desensitized) state within seconds of activation. Both serotonin and its synthetic analogues desensitize the 5-HT3 receptor completely with a steep concentration dependence and a potency order of: mCPBG > 5-HTQ >> 5-HT >> tryptamine > dopamine (see Table 4.9). The time course of recovery from desensitization depends on the agonist used.

A quantitative molecular pharmacophore model was derived to predict drug affinities for 5-hydroxytryptamine (5-HT3) receptors (Figure 4.36). The model was based on the molecular characteristics of a learning set of 40 pharmacological agents that had been analyzed previously in radioligand binding studies.

Molecules were analyzed for various structural features, i.e., the presence of a benzenoid ring and nitrogen atom, substitutions on the benzenoid ring, the location of the substitutions on the nitrogen, and the molecular characteristics of the most direct pathway from the benzenoid ring to the nitrogen. Weighting factors, based on published 5-HT3 receptor affinity data, were then assigned to each of 10 molecular characteristics. The following nine rules have been established for the 5HT-3A receptor pharmacophore structure[2]:

1. Contains aromatic ring structure (lower half of molecule); consistence with the hypothesis of Lloyd and Andrews which states that all central nervous system active drugs contain an aromatic ring
2. A tropane ring embedded nitrogen is present (see upper half of molecule shown above) and located at a nearest distance from the aromatic ring is not more than seven atoms from the aromatic ring
3. When aligning the tropane ring nitrogen in the same plane as the aromatic ring (torsion angle flexibility) the distance between the nitrogen and the aromatic ring center is 6.0 to 7.8 Å
4. Chemical substitutions of no more than 3 atoms are allowed at the nitrogen
5. The tropane ring structure itself does not tolerate substitutions larger than methyl groups (-CH_3); larger groups significantly reduce affinity

TABLE 4.9

Select Agonists for Serotonin Receptor 5-HT3 Subtype

Agonist	Affinity	Structure
5-HT	EC50 = 2.2 mM (N1E-115 cells)	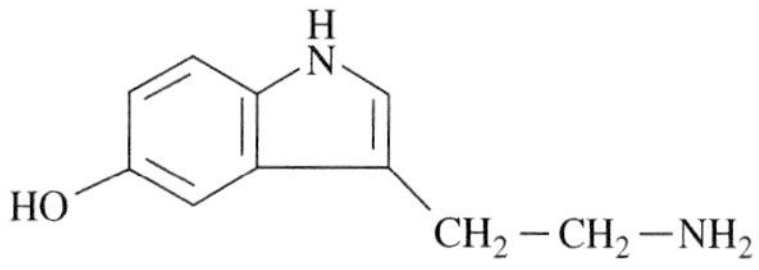
2-Me-5-HT	IC50 = 0.04 to 0.8 mM EC50 = 9.9 mM	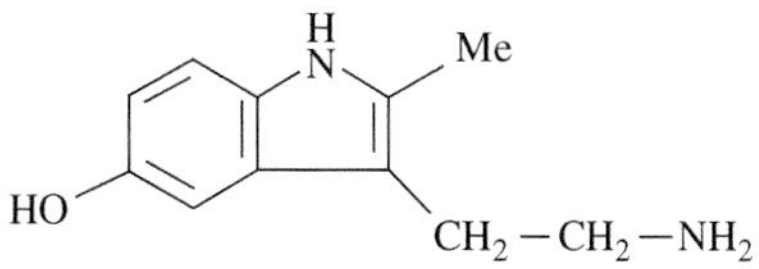
mCPBG	EC50 = 0.76 mM	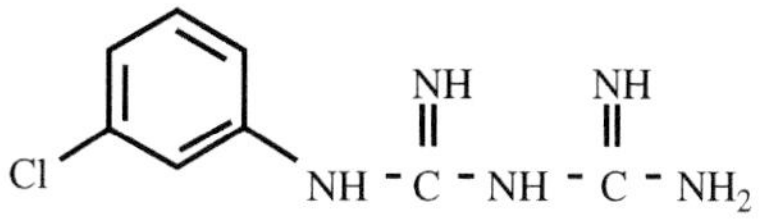
Phenyl-biguanide (PBG)	EC50 = 22 mM	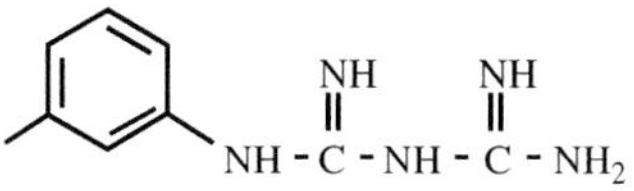
Quipazine	kDa = 0.62 nM	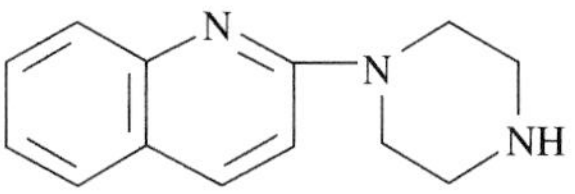
S 21007	EC50 = 27 mM IC50 = 2.8 nM	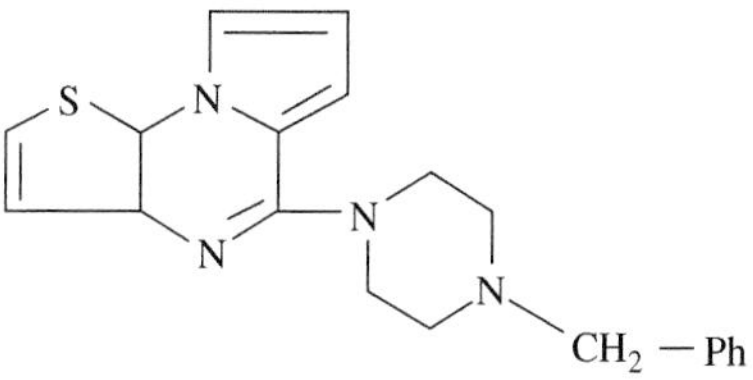

Abbreviations: 5-HT 5-hydroxytryptamine = serotonin; EC50 = ligand concentration at 50% efficacy (*in vivo*); IC50 = ligand concentration at 50% inhibition (functional assay); kDa = dissociation constant (ligand – receptor binding or affinity).

6. The linker structure between aromatic ring and ring nitrogen contains steric similarities that reduces flexibility (carbonyls or C=O bonds)
7. The first and second atom from the aromatic ring in the linker is never a tetrahedral carbon (no torsion angle flexibility; see point 6)
8. The third atom from the aromatic ring may be a tetrahedral carbon
9. Substitutions on the aromatic ring must be able to adopt a co-planar conformation

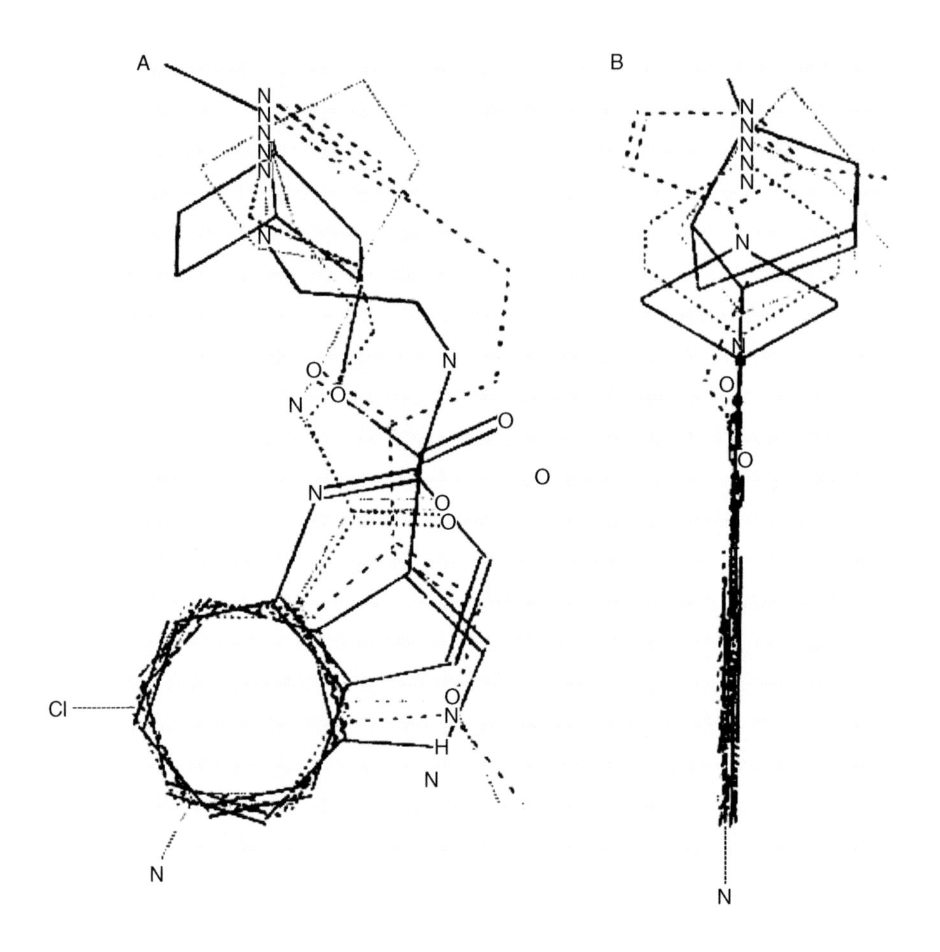

FIGURE 4.36
Pharmacophore structure of 5HT-3 receptor ligands. © Copyright 1998–2002, Molecular Pharmacology online by American Soc. for Pharmacology and Experimental Therapeutics.

The naturally occurring alkaloid atropine demonstrates how a "small" but significant violation of the "rules" in its structure explains its very low affinity for the serotonin receptor 3A subtype. Atropine is an antagonist of muscarinic cholinergic receptors and is used before eye examinations to dilate (open) the pupil. It also is used to relieve pain caused by swelling and inflammation of the eye, or to relieve symptoms of organophosphate poisoning (nerve gas, insecticides) by opening lung airways to allow breathing and preventing nausea, vomiting, abdominal cramping, low heart rate, and sweating.

In atropine, the first atom from the aromatic ring is a tetragonal carbon and thus allows torsion angle flexibility (see rules 6 and 7) explaining that the structure can deviate from being co-planar with the aromatic ring. These two atoms are the main difference as compared to the highly specific antagonist ICS 205-930. The molecular weight and chemical formula are nearly identical although the affinity of atropine for the 5HT-3A receptor is 1000 times lower than that for ICS 205 930.

The case of atropine and the 5HT-3A pharmacophore structure requirements demonstrate the influence of molecular motion on binding. The flexibility of the atropine molecule between its aromatic and tropane ring structures essentially reduces the chance in atropine of superimposing the tropane ring and aromatic ring, a gross-structure requirement for 5HT-3A antagonist binding. Further structure-activity relationship studies (SARS) show that all known receptor antagonists exhibit at least one degree of freedom. This hinders potential screening of ligand-receptor topology, which is best achieved with a rigid molecule as template for rational drug design. This importance of molecular motion in ligands (and receptors) is further demonstrated in thermodynamic analysis of drug binding.

For the search of new and effective agonists and antagonists, computer modeling has become an invaluable tool because powerful processors readily calculate the properties necessary to define a chemical similarity space. Not only can they be used to design new structures, or modify known structures of agonists/antagonists, they are also useful to screen existing compound libraries for structural and chemical similarity. As it turns out, molecular modeling tools are better at simulating specificity (conformation) than affinity (energy) of interactions. Affinity, but not specificity, depends on the entropy contribution of the solvent during binding and unbinding. The large number of highly mobile water molecule drastically increases computational needs, thus, molecular modeling (static or dynamic) is usually performed "in vacuum" reducing the number of calculations by excluding solvent-solvent and solvent-solute interaction. In addition, unfavorable enthalpic components of dehydration and ligand–receptor binding can be offset by favorable entropic components stabilizing the ligand–receptor complex, an effect known as enthalpy–entropy compensation.

In rational drug design the enthalpy–entropy compensation is a challenge that needs to be overcome to significantly improve the prediction of binding affinity of novel drugs to novel targets (for a review see [3]). Nevertheless, successful design of drugs on enzymes with deep binding pockets occluding bulk water (i.e., ordered water structure in binding pocket, a favorable condition for high affinity binding) has been achieved. Examples of drugs developed based on rational drug design strategies are Nelfinavir, an inhibitor of HIV-1 protease (Agouron Pharmaceuticals) and the thrombin inhibitor Ro 46-6240 (Hoffman-LaRoche). Nonpeptidic, small-molecule mimics as inhibitors of protein–protein interaction have proven more difficult to design. Much of the solvent occlusion of peptide inhibitors is provided by the main-chain and C atoms (amino acid side chain carbon) adjacent to binding hot spots, which explains why side-chain modifications heave little effect on affinity.

Essential for the rational design of the protease inhibitor was the successful crystallization of the HIV protease and related host proteases (e.g., pepsin) with and without bound inhibitors. The HIV protease is a member of the family of aspartate proteases and related to the pepsin family of proteases. Pepsin is inhibited by pepstatin, the natural inhibitor of pepsin. The structure of pepsin and its binding of pepstatin are known and this information forms the basis of a successful design of a HIV protease inhibitor by using computer models to identify the best possible inhibitor structure.

The active site of aspartate proteases contains a pair of aspartate residues in close proximity with a water molecule hydrogen-bonded and oriented optimally to attack the scissile bond of the substrate. The aspartate pair is located at the domain interface in pepsin, a monomeric protein, and at the subunit interface in HIV protease, a homodimer. The catalytic sites in viral and cellular aspartic proteases are very similar, but the importance lies in minute differences in symmetry relations at the interface of domains in pepsin and subunits in HIV protease. On the basis of the difference in symmetry at the active site of pepsin and HIV protease inhibitors have been designed that show a much higher affinity for the viral protein than for the host protease.

An effective protease inhibitor exhibits not only a high affinity for the active site, but is chemically inert and contains a nonhydrolysable peptide bond. Substrate analog inhibitors have therefore been designed that function as peptido mimetic. The scissile amide bond of a peptide substrate is replaced by nonhydrolysable hydroxyl isosteres with tetrahedral geometry. As shown in Figure 4.37, this hydroxyl ethylene structure mimics the substrate intermediate tetrahedral geometry of a peptide substrate.

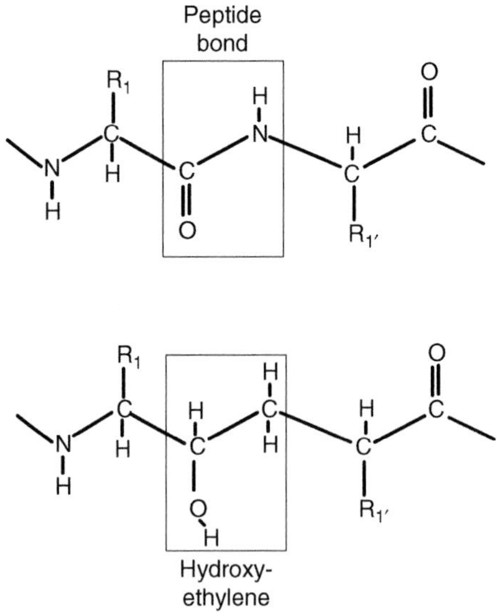

FIGURE 4.37
Peptide mimetic of a protease inhibitor. The hydroxy-ethylene structure cannot be broken by the protease.

The binding of a hydroxyethylene peptide mimetic is stabilized by the hydrogen bond formation of the hydroxyl of the backbone with the aspartate residues in the active site of the protease. The development of protease inhibitors has been accelerated by successfully using the concept that the best inhibitors are those that mimic the transition state structure of the substrate of proteases.

References

1. Eglen, R.M. and Bonhaus, D.W., 5-Hydroxytryptamine (5-HT)3 receptors: molecular biology, pharmacology and therapeutic importance. *Curr. Pharm. Des.*, 1996, 2(4), 367–374.
2. Schmidt, A. and Peroutka, S., Three-dimensional steric molecular modeling of the 5-hydroxytryptamine3 receptor pharmacophore. *Mol. Pharmacol.*, 1989, 36(4), 505–511.
3. Holdgate, G.A., Making cool drugs hot: isothermal titration calorimetry as a tool to study binding energetics. *Biotechniques*, 2001, 31(1), 164–166, 168, 170 passim.

4.4 Systems Biology

As the holistic saying goes, the whole is more than the sum of its parts. Systems biology deals with biological "wholes" treating an organism as a composite of molecular entities that interact in specific ways producing emerging properties that we describe as the organism's behavior, physiology, and metabolism. As the name suggests, a cell is defined as a physical system of components that all play an essential part. No single component can be removed without changing the state, behavior, or structural integrity of the system. Currently, systems biology deals mostly with interaction networks related to proteins, the proteome, and metabolic activity, the metabolome. Of course, the proteome and metabolome are only subsystems of a cell or organism, yet analyzing them independently of other systems components is a first good step in understanding the methodology of systems biology that also includes a great deal of bioinformatics.

4.4.1 Protein Interaction Networks

Protein interaction networks show protein components and their inter-relationships. These relationships are mainly evaluated through their effects on metabolites and signaling molecules. Interactions can also be defined through binding potentials between proteins. This binding can be categorized as follows. First, proteins that form subunits of stable protein complexes, and second, proteins that temporarily interact in reversible manners with the result of changing each other's activity. Examples of the former are cytoskeletal proteins, the subunits of hemoglobin tetramers or ligand receptor pentamers (acetylcholine receptor). Examples of reversible interactions are G-proteins binding to G-protein coupled receptors, or calmodulin binding to kinases and transporters and kinases binding to target proteins for the purpose of phosphorylation.

It should be noted, that many proteins are only temporarily expressed and that a complete set of proteins within a network thus undergoes dynamic changes. In addition, proteins represented by an interaction network will be present at different concentrations as the result of gene expression activity. Accordingly, some enzymes of metabolic pathways or kinases of regulatory pathways may be rate limiting for a pathway if expressed only at low levels. Protein interaction networks are at this stage rather static representation of the proteome of a cell or organism. Table 4.10 shows content and links to protein–network databases.

It is clear that much biochemical data is needed for useful biological interpretations of protein interaction networks, similar to the interpretation of gene network based on dynamic DNA microarray (gene expression) data. Further, the quality of the interactions needs to be confirmed and annotated to understand the strength of connectivity and the potential of regulatory or structural interactions. Experimental methods to corroborate protein complexes and interactions are difficult and time consuming (see Section 4.1 on hydrodynamic methods), but provide biologically relevant experimental data.

Techniques widely used to determine protein-protein interaction are immuno-coprecipitation (IP) and yeast two-hybrid system. Both techniques allow "pulling out" proteins that interact with a "bait" protein. With immuno-precipitation multiple proteins can be identified belonging to a single complex. Still, intracomplex contact site must then be established to unravel the details of interaction within a protein complex. IP is very sensitive, but time consuming and strictly depends on the ability to isolate and purify components. The yeast two-hybrid system is a systematic approach using molecular biology techniques. Here, both bait and target proteins are genetically engineered into a detection system composed of two halves of a transcription activator inducing an engineered beta-galactosidase (LacZ) reporter gene that is active against target-bait dimer formation. The results are easily read by observing a blue coloration of a cell containing the chromogenic substrate Xgal (5-bromo-4-chloro-3-indolylβD-galactoside). Thus a simple visual inspection of cell cultures yields information about existing protein–protein interaction.

Other sophisticated *in vivo* techniques rely on double labeling of proteins using fluorescence markers (often green fluorescence protein family) for coexpression colocalization in cells. As this technique has a rather low spatial resolution of several nanometers, direct interaction or close contact between labeled proteins can only be detected using the related fluorescence resonance energy transfer (FRET). Here, one label is activated and transfers its gained energy two a nearby second label, which now fluoresces. Secondary fluorescence intensity is a sensitive measure of distance between labels,

TABLE 4.10

Protein–Protein Interaction Databases

Database	Content	URL
UK HGMB Resource Center	A collection of protein interaction database sites	www.hgmp.mrc.ac.uk/GenomeWeb/prot-interaction.html
BIND (Biomolecular Interaction Network Database)	A database designed to store full descriptions of interactions, molecular complexes, and pathways	www.blueprint.org/bind/bind.php
DIP (Database of Interacting Proteins)	Catalogs experimentally determined interactions between proteins	http://dip.doe-mbi.ucla.edu/
GRID	A database of genetic and physical interactions for yeast, fly, and worm	http://biodata.mshri.on.ca/grid/servlet/Index
Yeast Protein Complex Database	Systematic analysis of multiprotein complexes in *Saccharomyces cerevisiae*, as a model system relevant to human	http://yeast.cellzome.com/
Kinase Pathway Database	An integrated database concerning completed sequenced major eukaryotes, which contains the classification of protein kinases and their functional conservation and orthologous tables among species, protein–protein interaction data, domain information, structural information, and automatic pathway graph image interface	http://kinasedb.ontology.ims.u-tokyo.ac.jp:8081/

usually in the Angstrom range (below 1 nm) indicating close interaction between the carrier proteins.

The nature of a protein interaction network has been described as scale-free meaning that its complexity does not depend on its size, i.e., the number of its components. In a scale-free network not every protein interacts with every other protein. Only a limited number of protein hubs or protein complexes play central roles in providing connectivity to a select few proteins, which provide connections to neighboring hubs. The observed average number of protein-protein interactions is two to three with hubs having many more interactions. Protein interaction networks also have a "small world" character where nodes are connected by relatively short paths. In cells, such paths represent metabolic and signaling pathways. In principle, the large-scale cellular networks are robust due to their hierarchical, scale-free organization.

WHAT ARE SCALE-FREE NETWORKS?

> A network's connectivity is distinguished by the probability $p(k)$ of a node having k links. If a network has random number of links between its elements, the probability $p(k)$ is highest for the mean number of links (i.e., $\langle k \rangle$). The probability decays exponentially for large k. A scale-free network, however, has many elements with only a few links (one, two, or three), with a small number of nodes having very large number of connections. The latter nodes are referred to as the hubs of a network. Such an uneven distribution of connectivity among elements has been found for gene, protein, and metabolic networks. Mathematically, these types of networks are scale free because the probability of a node having k connections has no well-defined peak typical for a Poisson distribution. Instead, for large k the probability decays as a power-law, $P(k) \approx k^{-r}$, appearing as a straight line with slope $-r$ on a log–log plot. (Adapted from Jeong, H., et al. *Nature,* 2000, 407(6804), 651–654.)

An interesting aspect of the small world character is the small network diameter; the average minimum distance between pairs of nodes (see box). Biological networks are dominated by hubs with short linear pathways between them accounting for the small network diameter. The network diameter for the human population is about six, as in "six degrees of separation" counting the average number of interpersonal connections needed to make a link between one person and any other person; I have a friend who knows a friend who knows a friend and so on. Interestingly, the metabolic network diameter measured in different bacteria with various complexities (genome size) is the same indicating that long linear metabolic and signaling pathways do not exist and are frequently dotted with hubs. The average path length in several bacteria is about three varying from two to five and is independent of the number of nodes. Bacteria with fewer nodes show slightly larger variation in pathway length. The number of nodes varies from about 200 to 800.[1]

Biological interpretation of protein interaction networks can be supported by correlation with gene expression data. Co-regulated genes and protein thereof found via proteomics to interact are likely to have biologically important interactions and functional integration in a cellular context. A big challenge of interpreting protein interaction networks is to screen out false positives from proteomics approaches such as the yeast two-hybrid system. The latter is prone to give physiologically nonrelevant interactions because of the forced coexpression of proteins. Apparent interaction within the heterologous expression system may not be relevant because those proteins may not be expressed at the same time *in vivo*. In addition, over-expression may result in nonphysiological levels giving positive signals through spurious but random associations at physiologically unrealistic high concentrations.

Bioinformatics approaches used to validating interaction networks will contribute to understanding and testing these networks for biological relevant information. Importantly, some aspects of networks that bioinformatics tools can help solve are:

Comparing protein complexes and proteome organization from different organisms

Reliability of experimental protein–protein interaction data

Modular organization (topology) of cellular networks

Developing computational methods for the prediction of protein interactions

Visualizing protein interaction networks

Correlation of protein–DNA and protein–protein interaction networks

For instance, in analogy to searching sequences and motifs in sequence and structure databases, a pathway Blast tool has been developed by the Whitehead Institute for Biomedical Research in collaboration with the University of California. As shown in Figure 4.38, short pathway structures can be used to search a complex network and identify similarity of interaction.[2] The application has been developed to compare networks across species to study evolutionary conservation of protein interaction networks. PathBlast is available at http://www.pathblast.org/. Queries are based on a series of known proteins by sequence, structure, or ID.

For biologists, one of the more intriguing features of protein networks is their robustness, i.e., insensitivity to random disturbances. In the context of

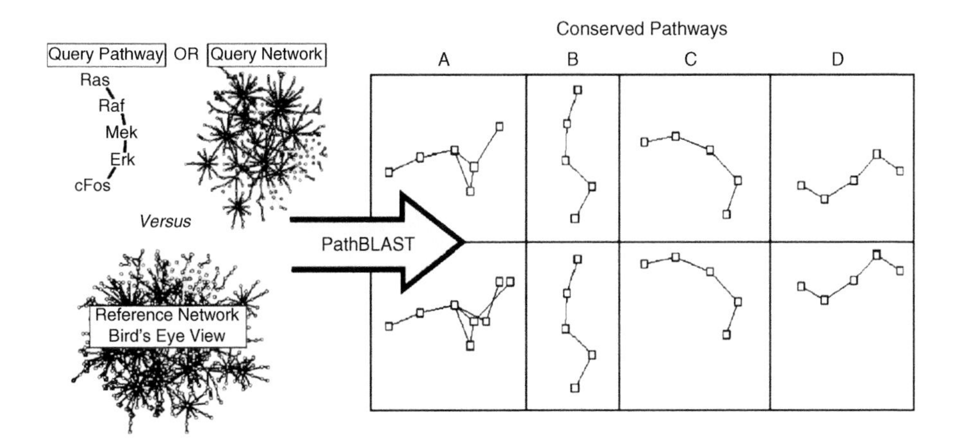

FIGURE 4.38
Identifying conserved protein interaction pathways with PathBLAST. PathBLAST operates in two modes, depending on whether the query is a single pathway or a whole network. In the first mode, single user-defined pathways are queried against a reference network of observed protein interactions from bacteria, yeast, fly, or worm. In the second mode, two large protein networks are aligned against each other to enumerate all of the pathways that are conserved between them. High-scoring pathway matches (A–D) are ranked by score and indicate pathways that are potentially conserved over evolution. The current focus of the PathBLAST website is on the first (more common) mode of query. (From Kelley, B.P., et al., *Nucl. Acids. Res.*, 2004, 32(suppl.2), W83–88. With permission.)

a protein interaction network, robustness can be measured by the change in network diameter upon removal of a node (protein element) through mutation resulting in an inactive protein or gene loss. While elimination of *N* number of proteins in hubs—the most connected proteins—the network diameter increases rapidly. However, randomly removing proteins does not affect, in average, network diameter. This robustness has been shown in silico using a metabolic network of *E. coli* strain MG1655[3] and is consistent with the observation that many single gene knock-outs are nonlethal and often do not affect growth rates appreciably.

References

1. Jeong, H. et al. The large-scale organization of metabolic networks. *Nature,* 2000, 407(6804), 651–654.
2. Kelley, B.P. et al., PathBLAST: a tool for alignment of protein interaction networks. *Nucl. Acids. Res.,* 2004, 32 (suppl.2), W83–88.
3. Edwards, J.S. and Palsson, B.O., The *Escherichia coli* MG1655 in silico metabolic genotype: Its definition, characteristics, and capabilities. *Proc. Nat'l Acad. Sci. USA,* 2000, 97(10), 5528–5533.

4.4.2 Metabolic Reconstruction

One of the essential functions of gene and protein interaction networks is cellular metabolism. Metabolism comprises the catabolic degradation of nutrients to extract energy and produce intermediary metabolites for the biosynthesis of cellular components and structures. Thus metabolic reconstruction's ultimate goal is to understand and model the flow of metabolites through a network of enzymes organized in metabolic compartments and connected via transport activity. The first metabolic pathways like glycolysis and Krebs cycle (citric acid cycle) have first been described in the 1930s. Today, a large number of metabolic pathways are known including the rich metabolic diversity found in plants and microorganisms able to synthesize a plethora of secondary metabolites (phytochemicals) and are the result of adaptations to diverse living conditions. Microorganisms express a variety of energy sources unavailable to animals and plants.

Bioengineers have for a long time modeled and modified small metabolic networks. Today's genome initiatives, however, push the task of modeling the metabolism of entire cells to new limits. It is an enormous task because it involves hundreds of components and many more interactions. Reconstruction and validation of genome-scale metabolic networks have been published for several model organisms including yeast and *E. coli*. The yeast in silico model iND750 includes 750 genes, 646 metabolites and 1149 reactions.[1] The model has been used to predict metabolic behavior for various carbon sources and gene knock-outs. A total of 4154 comparisons were performed of which

82.6% were confirmed largely from screening the existing literature for experimental results. Most false predictions were related to nuclear and mitochondrial compartment reactions. Pathways related to quinone and sterol synthesis, fatty acid and phospholipid biosynthesis, and the synthesis of branched chain amino acids had an error rate of more than 25% indicating that information about these pathways is incomplete or that the activity of a metabolic enzyme is also important for cellular processes related to biomass formation, e.g., DNA and protein synthesis for growth and cell structure maintenance.

The robustness of scale-free metabolic networks is of particular interest. An analysis of an *E. coli* metabolic flux model indicates the ability of cells of rerouting the flux of metabolites[2] and the effect of gene deletions on flux was correctly predicted for 86% of all cases tested.

Metabolic reconstruction requires a full understanding of the actual pathways. Several databases are available, of which Japan's KEGG database is one of the most elaborate and intuitive to find pathway information. Other databases and organizations have information similar to KEGG's. They are summarized in Table 4.11.

KEGG is a deductive database using a functional reconstruction model. The user can compute pathways and binary relations with the goal of computing wiring diagrams of genes and molecules (Table 4.12). This should lead to an understanding of cells as being a complex, self-assembling system whose components and the relationship between them are fully understood. In KEGG's words, the genome is simply a "warehouse of parts and all the regulatory signals in the genome are simply bar codes to retrieve them. In this view the blueprint of life is written in the entire cell as a network of molecular interactions." To gain an understanding of this network of molecular interactions, KEGG uses prediction tools for the computation of novel relationships based solely on the contents of its individual catalogs (the warehouse). These tools can be found through the "Search and compute with KEGG" link.

KEGG is part of the Japanese Human Genome Program at the Institute for Chemical Research, Kyoto University (www.genome.ad.jp:80/kegg/; also see Chapter 2.2.3 for details on using KEGG). Although the technical challenges and underpinnings of KEGG are the same as those of NCBI, KEGG addresses the complex interaction of proteins in cells from the perspective of metabolic interaction. Their philosophy includes finding answers to some of the common questions of modern molecular biology such as: what do we know about the relationship between the sequence of a gene and the function of the protein? What is the protein-folding problem in the cellular context? What are the challenges and problems facing the functional reconstruction problems, or how can we understand the relationship between the genome and the organism—its development and morphology? The goal of KEGG, therefore, is to build a functional map starting with available components within various molecular catalogs. The functional maps represent metabolic and regulatory pathways. The molecular catalogs include genome maps, base sequences, gene

TABLE 4.11

Pathways Databases

Database	Content	URL
KEGG (Kyoto Encyclopedia of Genes and Genomes)	Computerized current knowledge of molecular and cellular biology in terms of the information pathways that consist of interacting molecules or genes and to provide links from the gene catalogs produced by genome sequencing projects	http://www.genome.ad.jp/kegg/pathway.html
BioCyc Knowledge Library	A collection of Pathway/Genome Databases. Includes literature derived Pathway/ Genome Databases EcoCyc and MetaCyc and HumanCyc Also features *computationally* derived pathways based on Genome information	http://biocyc.org/
ExPASy Biochemical Pathways	Digitized version of wall charts courtesy Boehringer Mannheim et al., divided into Metabolic Pathways and Cellular and Molecular Processes, maintained by the Swiss Institute of Bioinformatics, Geneva, Switzerland	www.expasy.org/cgi-bin/search-biochem-index
MIPS	Comprehensive Yeast Genome Database (CYGD)	http://mips.gsf.de/genre/proj/yeast/index.jsp
UM-BBD	Microbial Biocatalysis and Biodegradatation–Microbial biocatalytic reactions and biodegradation pathways primarily for xenobiotic, chemical compounds	http://umbbd.ahc.umn.edu/
WIT (What is there?)	WIT2 provides access to thoroughly annotated genomes within a framework of metabolic reconstructions, connected to the sequence data; protein alignments and phylogenetic trees and data on gene clusters, potential operons and functional domains.	www-unix.mcs.anl.gov/compbio/
EMP(Enzymes and Metabolic Pathways)	EMP covers factual content of an original journal publication (30k+ records).	

allocations (physical map, inheritance map), and LIGAND databases (enzymes, compounds, and elements).

How can we reconstruct a biological organism in silico? KEGG's approach uses a hierarchical view of an organism, most easily viewed as atomic levels, molecular levels, and network levels (pathways). KEGG uses a system for data representation that structures a database according to the number of links between its components.

TABLE 4.12

KEGG Knowledge on Molecular Interaction Networks in Biological Processes

Pathway Element	Number	Name
Number of pathways	15,037	(PATHWAY database)
Number of reference pathways	229	(PATHWAY database)
Number of ortholog tables	85	(PATHWAY database)
Number of organisms	181	(GENOME database)
Number of genes	646,192	(GENES database)
Number of ortholog clusters	33,305	(SSDB database)
Number of KO assignments	4,429	(KO database)
Number of chemical compounds	10,823	(COMPOUND database)
Number of glycans	10,360	(GLYCAN database)
Number of chemical reactions	504	(REACTION database)

Elements of the catalog database include molecules (protein structures, metabolites), genes (sequences), and genomes (Figure 4.39). Pathway maps connect elements from the catalog database through binary relations, which are molecular interactions (structure) and genetic interactions (function). Interactions of more than two units are called networks and include metabolic pathways (molecular and genetic), genomes (linear and circular), hierarchies (classification, taxonomy), and neighbors (sequence similarities, structural similarities).

KEGG introduces the integration of genomic information with pathways to reflect the biological reality within a cell. This allows the individual scientist to search for proteins or genes of new or related pathways in model organisms other than the one being investigated. Missing structural information can quickly be obtained by finding homologs in "neighbors" whose structure has been solved. Novel pathways may be predicted by entering starting and end points of substrate and product and selecting an appropriate organism.

EcoCyc[5] is an *Escherichia coli* tailored metabolic database and part of the BioCyc Knowledge Library.[6] The bacterium *E. coli* is the ultimate laboratory test subject for geneticists, molecular biologists, microbiologists, and biochemists. It is also extremely important to human health and physiology, as it is part of our gastrointestinal system. Unfortunately, *E. coli* is also an opportunistic

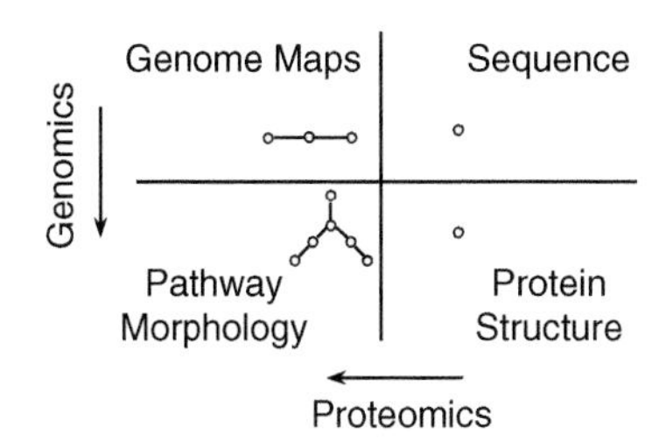

FIGURE 4.39
Database hierarchy.

pathogen; i.e., it can cause lethal infections if it enters our bloodstream. It is recognized by most people for its role in food poisoning caused by meat contamination, most often in undercooked hamburger meat. Its close genetic relationship to the bacteria *Salmonella typhimurium* (a problem found mostly in poultry) makes the integrative knowledge of its metabolism, genetics, and health problems a pressing yet fascinating issue.

EcoCyc (http://biocyc.org/) addresses the integration of the classic biochemical pathways (metabolism) of *E. coli* with its complete genome sequence information. For example, the metabolic pathways for amino acid synthesis in *E. coli* involve several enzymes that are often coregulated at gene expression levels. Thus, sensitive proteomics techniques should be able to see shifts in spot density, not only for one protein, but for several different ones. New pathways or homologous pathways in recently studied organisms with limited sequence information may be detected by proteomic means.

EcoCyc (similar to KEGG) uses chemical compound libraries that list the molecules involved in each biological reaction, the molecular weight of the compound and, in many cases, its chemical structure.

The EcoCyc KB (knowledge base) has a number of uses. It is an electronic reference source for *E. coli* biologists and for biologists who work with related microorganisms. Scientists can visualize the layout of genes within the *E. coli* chromosome, or of an individual biochemical reaction, or of a complete biochemical pathway (with compound structures displayed). The navigation capabilities allow the user to move from a display of an enzyme to a display of a reaction that the enzyme catalyzes, or of the gene that encodes the enzyme. The interface also supports a variety of queries, such as generating a display of the map positions of all genes that code for enzymes within a given biochemical pathway.

In addition to being a reference source for parts, the EcoCyc KB allows complex computations related to the metabolism, such as design of novel biochemical pathways for biotechnology, studies of the evolution of metabolic pathways, and simulation of metabolic pathways. The EcoCyc KB is also being used for computer-based education in biochemistry. The *E. coli* metabolic database nevertheless provides us with a considerable body of knowledge that is easily accessible from a PC terminal. As of March 2004, the following numbers of objects were contained as shown in Table 4.13.

What makes an organism alive? This question looms large in the minds of life scientists and the answer seems close in the minds of system biologists. The creators of EcoCyc talk about "an *in silico* model of *E. coli* metabolism that can be probed and analyzed through computational means." This suggests that experimenting on the model of metabolic pathways instead of the biochemical (wet bench) model will someday be as possible as a computer simulating nuclear tests thus replacing the actual tests, which are detectable seismographically by the enemy. Further, it suggests that the encyclopedia will someday be transformed into a workbench—an electronic laboratory rather than an electronic library. The motivation for many people is to get a predictive tool that can be used for

TABLE 4.13

BioCyc Statistics Version 8.0

EcoCyc KB Statistics		MetaCyc KB Statistics	
Pathways	178	Pathways	496
Reactions	3331	Enzymatic Reactions	4873
Enzymes	1043	Enzymes	1665
Transporters	183	Chemical Compounds	3051
Genes	4479	Organisms	231
Transcription Units	858	Citations	3771
Citations	8014		

KB = knowledge base

medical purposes, assessing risk and providing individualized health care and prevention.

References

1. Duarte, N.C. et al., Reconstruction and validation of *Saccharomyces cerevisiae* iND750, a fully compartmentalized genome-scale metabolic model. *Genome Res.*, 2004, 14(7), 1298–1309.
2. Edwards, J.S. and Palsson, B.O., The *Escherichia coli* MG1655 in silico metabolic genotype: Its definition, characteristics, and capabilities. *Proc. Natl. Acad. Sci. USA*, 2000, 97(10), 5528–5533.
3. Kanehisa, M., A database for post-genome analysis. *Trends Genet.*, 1997, 13, 375–376.
4. Kanehisa, M. and Goto, S., KEGG: Kyoto encyclopedia of genes and genomes. *Nucleic Acids Res.*, 2000, 28, 27–30.
5. Karp, P.D. et al., The EcoCyc database. *Nucl. Acids Res.*, 2002, 30(1), 56–58.
6. Karp, P.D. et al., The MetaCyc database. *Nucl. Acids Res.*, 2002, 30(1), 59–61.

5

The Bioinformatics Revolution in Medicine

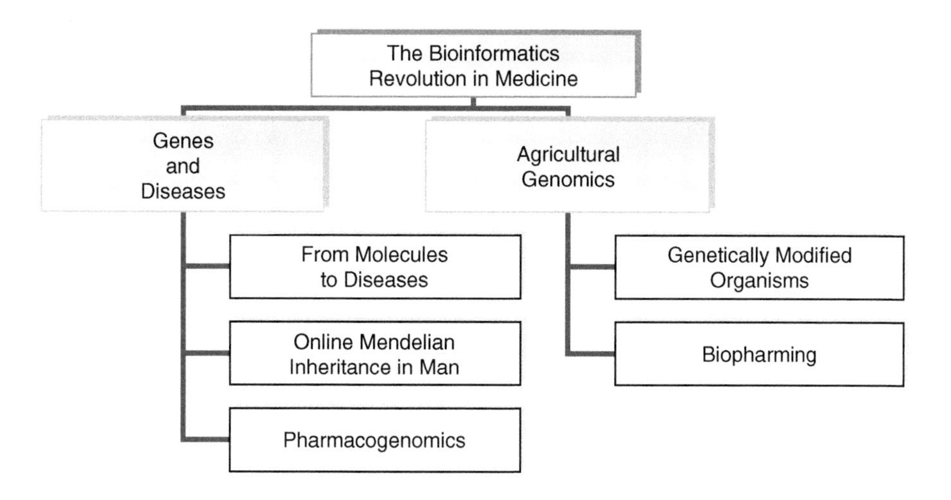

FIGURE 5.1
Chapter overview.

5.1 Genes and Diseases

5.1.1 From Molecules to Diseases

Physicians have all but abandoned the idea of a *vis vitalis*, a force of life that controls health and disease, life and death. Modern medicine, instead, is based on scientific principles and its practice is rooted in the idea that disease states are the result of genetic and metabolic defects and caused by toxins, drugs, or pathogens that invade the body and distort cellular homeostasis and growth. The successful use of *medicinal* drugs to intervene in diseased states, suppress infections, inflammations and pain, reverse cancer, and fight depression are all proof of principles of the molecular basis of disease. Pharmacology is the science that explains how drugs affect metabolism and physiology and that genetics has a role in how drugs work in people. The discovery that a lack of *essential* nutrients like vitamins, amino acids, fatty acids, and minerals stunts growth, causes gums to bleed or bones to easily fracture is consistent with the link between molecules and diseases.

But what makes one compound an essential nutrient and another a toxin? How does a missing, mutated, or extra copy of a gene cause a disease? How does an infection cause inflammation and pain? What causes cancer and autoimmune diseases? Understanding the molecular repertoire of our body and the molecular interactions that result in healthy and diseased states of cells and organs is key to modern medicine. Bioinformatics has become an important tool in making sense of genetic components of disease and will in the near future make an even bigger impact through functional and structural genomics. Networks of genes, proteins, and metabolic reactions are key pieces to solving the puzzle called the human body.

For a bioinformatician, the human body is a complex system characterized by hierarchical levels of structural organization: molecules, cells, tissues, organs, and the organism. Each level has its properties that cannot be found at a lower level, yet each level depends on the proper functioning of each lower level. The role of bioinformatics will be to systematically explore the information content of each level and use it to describe patterns of interaction that explain the transition from lower to higher level functions. In a systems view of the human body, a disease is a manifestation at the organism level of a defect at the molecular level, e.g., a mutation or misfolded protein. The human brain is probably the most compelling example of the difficulties of understanding how higher level properties emerge in complex systems. While we yet have to fully understand how diseases like Alzheimer's, Parkinson's, and Huntington's are caused by changes in the structure of a few—but important—proteins, we are far away from a cellular or molecular model of consciousness, precisely because our conscious experiences are of an immediate and personal nature that makes them appear so far removed from the chemical nature of molecules and cells.

The neurosciences have made great progress in characterizing the basic function of excitable tissues and their cells in muscles and brain (Table 5.1). At the tissue level, neurons form small groups of interconnected cells called

TABLE 5.1

Hierarchical Organization of Brain Structure and Function

Level	Structure	Emerging Systems Property
Organ	Brain	Cognition, motor control, consciousness
Tissue	Neural network	Chemical synapse, input/output connections, long-term potentiation
Cell	Neuron, membrane	Compartmentalization, action potential encoding, and propagation
Molecule	Gene, protein	Hereditary information, channel property

neural networks. At the cellular level, individual neurons carry action potentials that are the emerging property of ion channels embedded in their cell membranes. The finding that ion channels contribute to action potentials and that action potentials form the basic signaling mode of neurons linked together to form the brain, explains how mutations in genes coding for ion channels are known to cause neurological and muscle diseases. Thus, we have a good understanding of the chain of events from a molecular to a cellular, and from a cellular to a physiological level of the systems properties of the brain.

At the tissue level, neurons influence each other's action potentials and firing patterns through communication via chemical and electrical synapses. Modulating the strength of synaptic signaling between cells is recognized as an important mechanism of memory and learning. At the cellular level, action potentials are both the result and the cause of synaptic signaling. Action potentials are dynamic, fast-moving waves of small voltage changes (measured in millivolts) across the cell membrane. The "length" of the wave is measured in milliseconds. Their strength, duration, and firing frequencies (1 to 200 times per second; Hz) encode information carried by individual nerve cells. These firing patterns are the functional signature of individual neurons and differ for different types of cells. The electrical signals are generated by ions moving through various ion channels producing short-lived currents (and thus voltage changes) across cell membranes (see box). Defects at the molecular level cause changes in the kinetic behavior of ion channel proteins altering or preventing the firing patterns of neurons, and thus disrupt the communication patterns in the brain. Ion channels are also the basis of the control muscle contractions, and many mutations have been identified that affect both the proper working of the brain and peripheral organs: the heart and smooth and skeletal muscles.

The complexity of an organ like the brain arises, of course, because several different types of ion channels are needed to trigger and propagate action potentials in a neuron. Different neurons connected in a network have different combinations of channels. Subsequently, the particular cognitive function of a brain area is determined by many different proteins and thus depends on the activity of many different genes. The challenge is to understand how defects in one or more of these genes/proteins contribute to healthy and diseased states of the brain.

WHAT ARE ION CHANNELS?

Ion channels provide narrow, channel-like pathways used by ions to cross cell membranes. The membrane is an electrical insulator and blocks the movement of ions (positively or negatively charged molecules) in the absence of channels (or closed, inactive, defective). Ion flux can be measured as an electrical current. With extremely sensitive electronic amplifiers, the movement of a few thousand ions in every thousandth of a second can be measured and could reveal information about the electrical activity of a cell membrane or within a narrow patch thereof. Action potentials, the signaling mode of neurons, are the combination of at least three

> different classes of ion channels, meaning that for each of the different ions, a different protein has to be present for it to cross the membrane, each ion channel being responsible for the transport of one ion type. Many drugs act at the level of these ion channels and receptor proteins that control the activity of these channels. In addition, many potent and often lethal toxins from snake and spider venom interfere with the functional activities of these channels by directly interacting with these proteins by either blocking or activating channels. These toxins and drugs offer a unique pharmacological tool kit to localize, isolate, and functionally control specific channels types, even if they are only present at very low density in a preparation.

Identification of ion channels has greatly benefited from pharmacology. Ion channels are susceptible to naturally occurring toxins and drugs, such as calcium channels for snail toxins and caffeine, acetylcholine receptors for nicotine, and potassium channels for another snake toxin charybdotoxin. The effects of toxins are often paralysis, yet at low doses, they can have beneficial effects on neurological and cardiological disorders. As such, they have become primary tools to dissect the link between the molecular action of ion channels and the physiological output of an organ. Pharmacology of receptors and transporters has greatly enhanced our understanding of the mechanism of disease related to these proteins. As for hereditary diseases, they are the result of mutations in the genetic sequence of these same proteins altering their function, structure, or both.

A few select examples of known diseases linked to mutations (defects) in ion channels and related transporters and their susceptibility to toxins and drugs are listed in Table 5.2. Some of the inherited diseases linked to ion channels, pumps, and transporters, include cystic fibrosis (chloride channel), diastrophic dysplasia (sulfate transporter), Long QT syndrome (potassium channel), Menker's syndrome (copper transport), Pendred syndrome (thyroid-specific sulfate

TABLE 5.2

Transporters Linked to Diseases and Susceptibility to Toxins

Transport Protein	Hereditary Diseases	Susceptibility to Toxins and Drugs
Nicotinic acetylcholine receptor	Myasthenia gravis (slow channel syndrome)	Nicotine, alpha-bungarotoxin
Calcium channels	Malignant hyperthermia (MHS1); episodic ataxia; hypokalemic periodic paralysis	Omega-conotoxin (snail toxins), caffeine, Amlodipine (Norvasc)
Chloride channel ClC-2	Myotonia; epilepsy	NPPB*
CFTR	Cystic fibrosis	Bumetanide (bumex), NPPB[a]
Gap junction protein Cx43, Cx26	X-linked Charcot-Marie-Tooth disease; deafness	
Na(+), K(+)-ATPase	Polycystic kidney disease	Glucosides (digitoxin)
Potassium channels	Long-QT–syndrome	Charybdotoxin, propafenone

[a] 2-phenpropylamino-5-nitrobenzoic acid; 5-Nitro-2-(3-phenylpropylamino)benzoic acid; for more information go to the National Center for Biotechnology Information page on genes and diseases at www.ncbi.nlm.nih.gov/; follow link Genes&Diseases under Hotspots)

transporter), polycystic kidney disease (cell–cell interaction, membrane protein organization), Wilson's disease (copper transport, ATPase), and Zellweger syndrome (PXR1, peroxisome protein import receptor, peroxisome biogenesis disorder). A more inclusive list of genetic diseases can be found at the Genes and Disease website at The National Center for Biotechnology Information. Genes and Disease is part of the National Library of Medicine (NLM) Bookshelf, a collection of searchable biomedical topics ranging from cancer medicine, to neurochemistry to immunology (www.ncbi.nlm.nih.gov/entrez/query.fcgi?db = Books).

Like the brain, the heart is a complex organ susceptible to disease, and structural abnormalities in the excitable tissues of the heart predisposes affected persons to an accelerated heart rhythm. This can lead to sudden loss of consciousness and may cause *sudden cardiac death* in teenagers and young adults who are faced with stresses ranging from exercise to loud sounds. Sudden cardiac death is a *polygenic* disease, and multiple mutations have been identified in potassium, calcium, and sodium channels, as well as cytoskeletal proteins anchoring channels to the membrane. Long QT syndrome is one of several *myopathies* contributing to sudden cardiac death (Figure 5.2).[1]

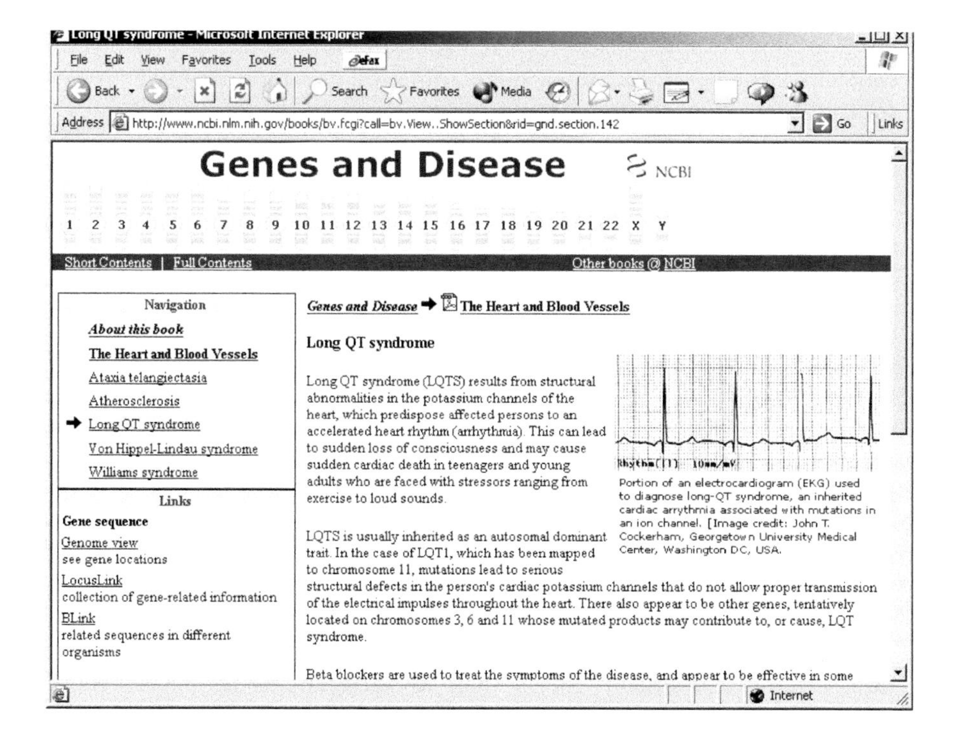

FIGURE 5.2
Long QT syndrome: Genes and Diseases. A database at NCBI for genes involved in hereditary diseases; shown is the page for long QT syndrome, which affects heart beat regulation.

TABLE 5.3

Genetic Polymorphism of Sudden Cardiac Death (SCD) Due to Cardiac Arrhythmias

Condition	Gene	Protein	Function	OMIM[a]
Long QT syndrome 1	KCNQ1	K+ channel	Channel activity	+192500
Long QT syndrome 2	KCNH2	K+ channel	Channel activity	+152427
Long QT syndrome 3	SCN5A	Na+ channel	Channel activity	+603830
Long QT syndrome 4	ANK2	Ankyrin-B	Channel location	+600919
Long QT syndrome 5	KCNE1	K+ channel	Channel activity	+176261
Long QT syndrome 6	KCNE2	K+ channel	Channel activity	+603796
Cardiomyopathy[b] 1	MYH7	Myosin heavy chain	Contractility	+160760
Cardiomyopathy 2	TNNT2	Troponin T2	Contractility	+191045
Cardiomyopathy 3	TPM1	Tropomyosin 1	Contractility	+191010
Cardiomyopathy 4	MYBPC3	Myosin-binding protein C	Contractility	+600958

[a] OMIM, Online Mendelian Inheritance in Man
[b] Cardiomyopathy, familial hypertrophic cardiomyopathy

As summarized in Table 5.3, *familial hypertrophic cardiomyopathy* is a genetic variant caused by defects in muscle cell proteins affecting contractility. Sudden cardiac death illustrates the problem identifying the genetic underpinnings of a complex disease. Dozens of different proteins are involved in controlling the heartbeat, and a single mutation present in just one type can contribute to irregularities leading to death.

Polygenic genotypes are studied using *quantitative trait loci* (QTL) analysis. QTL establishes a percentage of how much a specific allele contributes to a trait in a population, its *penetrance*, be that a normal physiological process or disease state. Of the known mutation associated to the Long QT syndrome, many are found in potassium-selective ion channels.

What is the importance of these channels? Potassium is found mostly inside cells, forming a 20-fold concentration gradient toward the outside. The effect is that whenever a potassium channel opens, potassium ions move from the inside of the cell to the outside, moving positive charges to the extracellular space. This movement will happen as long as the channel is open. The net effect of potassium channel activity is the repolarization (restoration) of the resting potential of muscle cells. In the Long QT syndrome, this process is markedly slowed when a mutation causes potassium channels to be less active than normal. As a result, the time lapsing between the Q and T wave in an electrocardiogram is extended, thus the name *long* QT syndrome.

How many potassium channel genes are there? The current number of cloned and sequenced genes is more than 50, a number that includes genes from different organisms—from man to bacteria. Figure 5.3 shows the EntrezGene web site for KCNQ1, a voltage-gated potassium channel that carries mutations identified in Long QT syndrome 1. The growing numbers of sequences allow interesting interpretations that are based simply on their sequence differences with well-elaborated information about their functions.

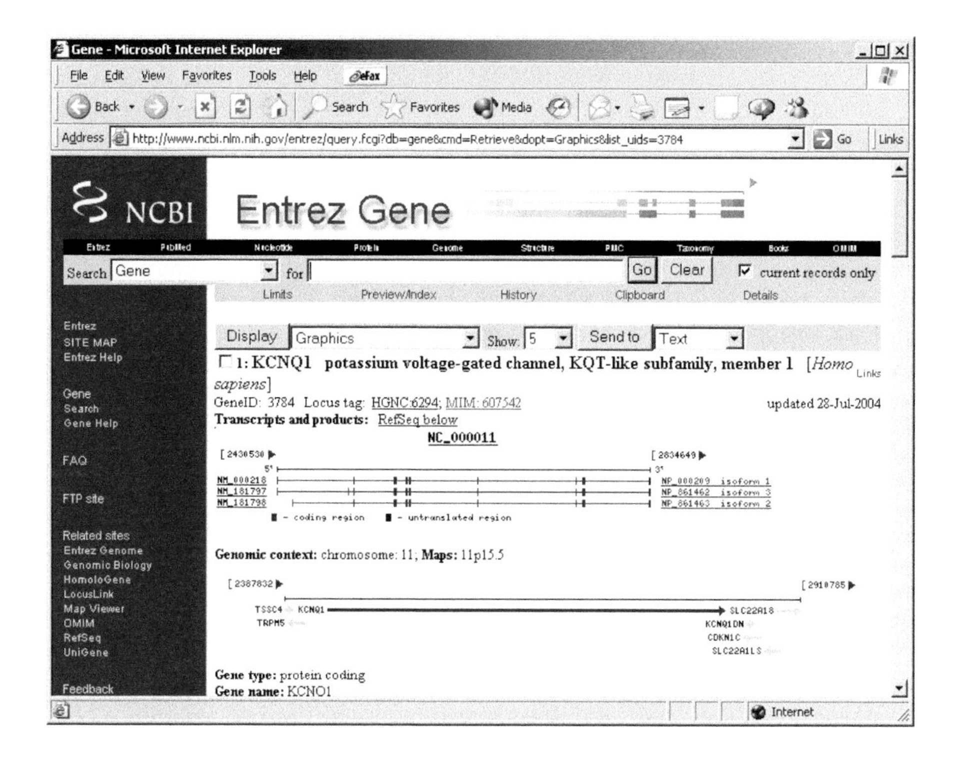

FIGURE 5.3
Entrez Gene for voltage-gated potassium channel KvLQT1 (KCNQ1). This ion channel contains a mutation responsible for the disease known as Long QT syndrome, which produces human cardiac arrhythmias (see Table 5.2).

Once again, the relationship between sequence, structure, and function is enormously helpful in our understanding of the physiology, including diseases, of these channels.

References

1. Yang, W.P. et al., KvLQT1, a voltage-gated potassium channel responsible for human cardiac arrhythmias. *Proc. Natl. Acad. Sci. USA.*, 1997, 94(8), 4017–4021.

5.1.2 Online Mendelian Inheritance in Man (OMIM)

The relationship between genes and physiology, and also genes and pathophysiology, is well known from family histories. It is not surprising, then, that an often-mentioned reason for sequencing the human genome and the genomes of many other organisms is the potential implication on understanding, preventing, and curing human diseases. A comprehensive listing and description of the currently known genetic diseases can be found at the Online Mendelian Inheritance in Man (OMIM) database at NCBI.[1] This database is a "catalog of human genes and genetic disorders maintained by Johns Hopkins University. The database contains textual information, pictures, and reference information." Table 5.4 summarized the database content as of July 1, 2004. It is structured into genes of autosomal, sex-linked, and mitochondrial diseases, and distinguishes the level of information associated to a gene or disease regarding sequence, phenotype, and Mendelian inheritance patterns.

Most genes listed in OMIM tend to have high penetrance or *population-attributable fraction*, PAF, which is a measure of the importance of a locus (allele) from a population genetics, and thus public health, point of view. Although population genetics is a century-old science, only the recent advent of high-resolution maps for most human chromosomes increased the marker density for QTL analysis for the identification of minor genes affecting a phenotype. Prior to whole genome information, the influence on the course of a disease came from family studies where markers of diseases show a clear Mendelian pattern of heredity, as the term "Mendelian inheritance" indicates. In other words, these genotypes are identified based on classical population genetics, which is typically feasible for monogenic traits. Most phenotypes and disease are polygenic. They are complex traits and affected by multiple genes, or better yet their gene products.

Nevertheless, an analysis of the genes listed in OMIM will promote a better understanding of the molecular mechanisms underlying diseases. Whole genome sequencing, the mapping of genetic markers at high density, and functional gene expression studies will further the identification of novel genes,

TABLE 5.4

OMIM Statistics of a Catalog of Human Genes and Disorders

OMIM Statistics	Autosomal	X-Linked	Y-Linked	Mitochondrial	Total
* Gene with known sequence	9226	414	41	37	9719
+ Gene with known sequence and phenotype	352	37	1	0	390
# Phenotype description, molecular basis known	1422	131	0	23	1577
% Mendelian phenotype or locus, molecular basis unknown	1277	128	4	0	1409
Other, mainly phenotypes with suspected Mendelian basis	2199	154	2	0	2355
Total	14476	864	48	61	15445

Source: From www.ncbi.nlm.nih.gov/entrez/query.fcgi?db=OMIM; July 1, 2004. With permission.

particularly those alleles of polygenic traits with a penetrance or PAF below 5%. Comparative genomics will play an important role, too, because it helps identifying genetic elements on genomes such as gene clusters, rearrangements, and synteny, all of which will facilitate the use of animal models in diseases.

One particular disease that captures the imagination of everyone is cancer, and the fight against it has been institutionalized many decades ago. Measured against the tremendous resources put toward the goal of eradicating cancer, biomedical research has been fairly ineffective, with successes against some, but not most, forms of cancer. The realization of years of research and treatment is that cancer is a heterogeneous disease that has no single cause despite its common phenotype of unchecked tissue growth. The genome projects have revitalized the hope of understanding cancer at the molecular level and the misregulation of genetic programs controling cell growth and differentiation gone awry. Hence the attempt of a systematic analysis of the genetic "anatomy" of cancer.

The National Cancer Institute coordinates the Cancer Gene Anatomy Project and is the major contributor of EST library relevant to cancer tissue. Such data will help understand changes in gene expression between healthy and diseased cells "allowing the user to find *in silico* answers to biological questions in a fraction of the time it once took in the laboratory." Many genetic abnormalities have been linked to cancer including chromosome break points and SNPs.

Table 5.5 summarizes the tools available at the Cancer Gene Anatomy Project database. The database allows searching for markers in tissue-specific manner to identify potential genes involved in tumor growth. From a bioinformatics point of view, cancer has a robust metabolism that rejects intervention making it an autonomous local tissue. Understanding both the local autonomy and metabolic robustness thus might bring new insights into finding the Achilles heel of this disease.

TABLE 5.5

The Cancer Gene Anatomy Project (CGAP)

Search Category	Resource Available
Genes	Gene information, clone resources, SNP500Cancer, GAI, and transcriptome analysis
Tissues	cDNA library information, methods, and EST-based gene expression analysis
Pathways	Diagrams of biological pathways and protein complexes, with links to genetic resources for each known protein
RNAi	RNA-interference constructs, targeted specifically against cancer relevant genes
Chromosomes	FISH-mapped BAC clones, SNP500Cancer, and the Mitelman database of chromosome aberrations
SAGE Genie	Serial analysis of gene expression using SAGE data
Tools	Direct access to all analytic and data mining tools developed for the project

Source: From Web Site http://cgap.nci.nih.gov/. With permission.

The Cancer Gene Anatomy Project database includes tools to find genes, cDNA libraries, and SNPs, and to examine gene expression (e.g., SAGE Genie) and chromosomal cytogenetic maps and recurrent aberrations (rearrangements, breaks, translocations etc.) that are thought to cause cancer formation. As an important addition, the database provides precalculated pathway maps taken from the KEGG and BioCarta databases.

One of the promising novel strategies of studying disease using bioinformatics tools is identifying differences in gene expression by comparing healthy and diseased tissue, or cell lines, as disease models. In SAGE Genie, a tool to search for known genes expressed in normal and cancer tissues and cell lines, the important difference is the number of sequence tags associated to a SAGE library, indicating different gene expression. SAGE Genie is a database tool that allows gene-oriented searching with a graphical identification of differences in the number of associated tags found in any given tissue or cell line derived from such a tissue. As shown in Color Figure 5.4

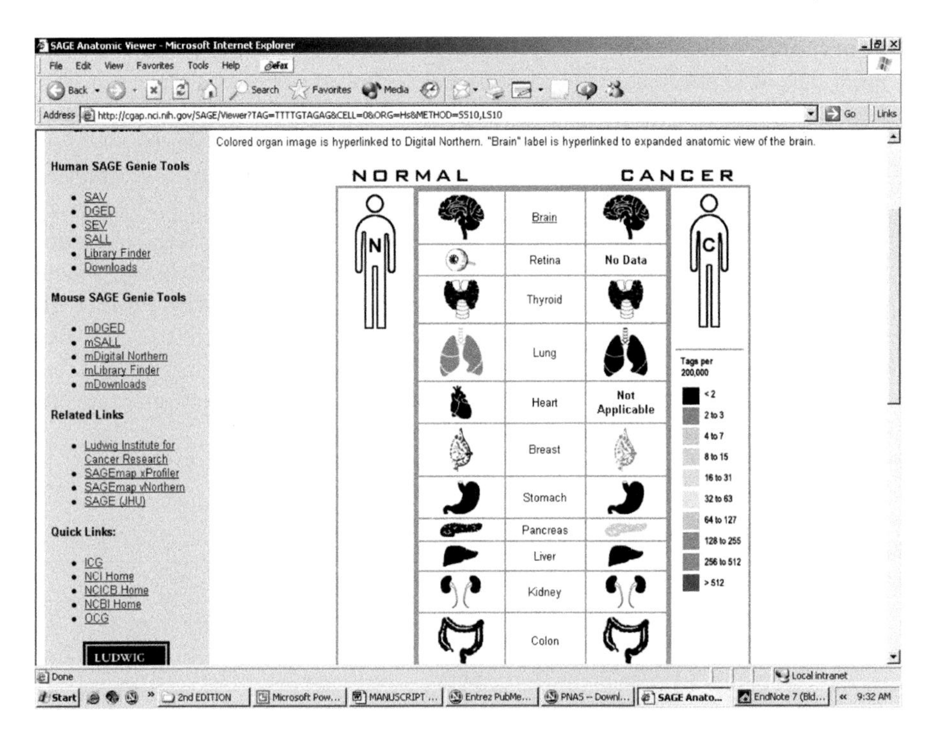

FIGURE 5.4 (See color insert following page 44)
Human SAGE Genie tools for p53 (TP53) Search query: TTTTGTAGAG. Search query: TTTTGTAGAG, tissues only. Colored organ image is hyperlinked to Digital Northern. "Brain" label is hyperlinked to expanded anatomic view of the brain. SAGE is a gene expression tool Part of the National Institute of Health, U.S. Department of Health and Human Services; http://cgap.nci.nih.gov/.[2] Differences in shading indicated differences in number of tags identified in normal and cancer tissues in respective organ tissue samples. (Shading scale on right: top [blue] less than 2 tags/200,000 bp, bottom [red] more than 512 tags/200,000 bp.) For an introduction to SAGE (serial analysis of gene expression) see Section 3.3.1.

a search for the SAGE tag TTTTGTAGAG associated the oncogene p53, a tumor suppressor protein. It is thought to activate genes coding for inhibitors of cell growth and invasion (tissue growth). In cancer cells, levels of p53 are often elevated, possibly because the cell tries to overcome the reduced ability of p53 mutants to bind to DNA regulatory sites.

Thus, finding differences in gene expression of p53 is a way of identifying putatively transformed (malignant) cells based on sequence tags with mutations linked to cancer. Color Figure 5.4 shows differences in tag numbers for lung, breast, and pancreatic tissue. Light blue and red shades indicate more tag copy numbers, i.e., higher gene expression and thus potentially cancerous tissue. Healthy tissue is marked by dark blue shades and quantified as 1 or 0 tags per 200,000 bp sequenced.

Tissue samples are also analyzed histologically for signs of tumor growth. Comparing the number of tags found in normal and cancer breast tissue (Figure 5.5) shows that among normal tissue samples, 3 out of 10 show increased levels of tags (max 6), while 6 out of 18 cancer tissues show increased p53-associated tag numbers. The percentage is virtually identical. It is important to realize that different cancer types are included in the comparison, and a more detailed look may reveal that p53 mutants are

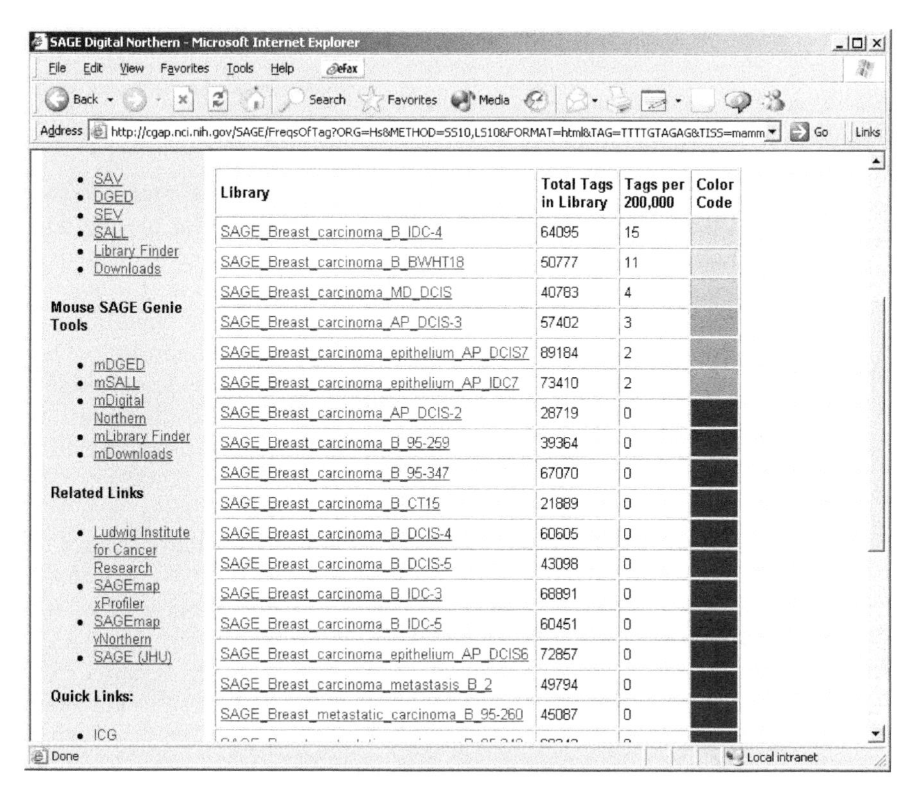

Library	Total Tags in Library	Tags per 200,000	Color Code
SAGE_Breast_carcinoma_B_IDC-4	64095	15	
SAGE_Breast_carcinoma_B_BWHT18	50777	11	
SAGE_Breast_carcinoma_MD_DCIS	40783	4	
SAGE_Breast_carcinoma_AP_DCIS-3	57402	3	
SAGE_Breast_carcinoma_epithelium_AP_DCIS7	89184	2	
SAGE_Breast_carcinoma_epithelium_AP_IDC7	73410	2	
SAGE_Breast_carcinoma_AP_DCIS-2	28719	0	
SAGE_Breast_carcinoma_B_95-259	39364	0	
SAGE_Breast_carcinoma_B_95-347	67070	0	
SAGE_Breast_carcinoma_B_CT15	21889	0	
SAGE_Breast_carcinoma_B_DCIS-4	60605	0	
SAGE_Breast_carcinoma_B_DCIS-5	43098	0	
SAGE_Breast_carcinoma_B_IDC-3	68891	0	
SAGE_Breast_carcinoma_B_IDC-5	60451	0	
SAGE_Breast_carcinoma_epithelium_AP_DCIS6	72857	0	
SAGE_Breast_carcinoma_metastasis_B_2	49794	0	
SAGE_Breast_metastatic_carcinoma_B_95-260	45087	0	

FIGURE 5.5
p53 tag distribution in cancer breast tissue samples.

involved in some, but not all, cancers, and that the presence of a mutation does not always translate into a disease phenotype.

Differences in the database may also stem from false positives and false negatives associated to very short tag sequences (here 10 base pairs). A search using the longer 17 bp tag sequence TTTTGTAGAGATGGGGT (tissues only) shows fewer results because of the more stringent query sequence, and a striking difference for breast cancer tissues (the only organ for which enough data is available) with 5 out of 7 breast carcinomas studied having increased number of p53-associated tags. Thus, mutant p53 are likely involved in transformation of healthy breast tissue into malignant carcinomas.

References

1. Online Mendelian Inheritance in Man, OMIM (TM). McKusick–Nathans Institute for Genetic Medicine, Johns Hopkins University (Baltimore, MD) and National Center for Biotechnology Information, National Library of Medicine (Bethesda, MD), 2000. World Wide Web URL: http://www.ncbi.nlm.nih.gov/omim/.
2. Boon, K., et al., An anatomy of normal and malignant gene expression. *Proc. Natl. Acad. Sci.*, 2002, 99(17), 11287–11292.

5.1.3 Pharmacogenomics

Over 50 years of clinical studies have shown that the way we react to some drugs depends on ethnicity and can be inherited. Molecular biology analysis and genomic sequencing identified genetic polymorphism in drug-metabolizing enzymes, transporters, and target proteins as the basis of individual differences. *Pharmacogenomics* combines these insights from pharmacology with the genomic biology to facilitate understanding of how multiple genetic determinants affect drug efficacy and how knowledge on genetic polymorphism can be used to individualize drug therapies. The known examples of drugs affected by genetic polymorphism relate to proteins that are usually monogenic and thus have a high penetrance, i.e., a single gene largely determines the efficacy of a drug. Thus, drugs are either not effective because they cannot recognize a mutated target, are excreted due to better recognition by a multi-rug resistance transporter (p-glycoprotein), or have toxic effects due to lack of metabolic degradation by the cytochrome P450 monooxygenases (CYPs).

The challenge is to understand and identify the majority of cases which are polygenic, and where each gene contributes only a fraction to the overall efficacy of a drug. Genome-wide approaches are now possible thanks to the availability of whole-genome scans (SNP haplotype maps) and functional genomics techniques (bioarrays). Whole-genome approaches attempt to find new associations using genetic markers on completed genome maps that alter the nucleotide sequence of the coding region, or alter gene expression or splice-variant formation. For genetic polymorphism of drug-metabolizing enzymes, focusing on pathways is a logical approach. Catabolic reactions usually involve a set of enzymes, and reduced detoxification activity may be the result of changes in more than one protein of a pathway. Genetically, such pathways can be characterized as a network of genes that are coregulated, and together create a single phenotype (e.g., toxicity).

Genes and their associated pharmacological data can be searched using the Pharmacogenetics research network database at www.pharmGKb.org. Genes that can be found include those for which information regarding pharmacological data and drug interaction is available. This includes data relating to phenotype, genotype, and mechanisms of action, e.g. that of cholesterol-lowering statins, blood pressure-controling angiotensin-converting enzyme (ACE) inhibitors, or beta-blocker binding to adrenergic receptors.

Knowing the many genes (i.e., proteins) involved in pathways related to disease and identifying sequence differences (mutations) in individuals will eventually lead to an understanding of how certain drugs are more effective in some patients but not in others, and why some people come down with serious side effects. The goal of pharmacogenomics is to map these pathways and individual differences from a genome, transcriptome, and proteome point of view. Obviously, understanding the gene and protein interaction networks and the corresponding signaling and metabolic pathway flows is a necessary first step in establishing a personalized medicine.

5.2 Agricultural Genomics

Biotechnology defines a wide range of applications in genetic engineering, enzymology, bioengineering, and drug discovery. While it impacts medicine, biotechnology also has a large impact on modern agriculture in such diverse areas as water quality and supply, pesticide reduction and pest control, food quality, crop resistance and productivity, biomass increase, and renewable energy sources such as hydrogen gas from algae through the development of genetically modified organisms (GMOs).

The contribution of bioinformatics to biotechnology has been drastically changed with the advent of modern genome projects. Both plant and livestock genomes have been sequenced and are being sequenced to completion as summarized in Table 5.6. For a comprehensive link to all plant genomes, see NCBI's "Plant Genome Central" at www.ncbi.nlm.nih.gov/genomes/PLANTS/PlantList.html.

An important feature of genome biology is comparative analysis. Thus, genome projects now include more and more animal and plant species, allowing an unprecedented look into the biology of culturally important organisms. As for humans, the study of disease is important for the health

TABLE 5.6

Agriculturally Important Organisms with Known Genomes

Group	Genomes	URL
Plants	NCBI Plant Genome Central	http://www.ncbi.nlm.nih.gov/genomes/PLANTS/PlantList.html
	AatDB—*Arabidopsis thaliana*	http://www.arabidopsis.org/
	BeanGenes—*Phaseolus* and *Vigna*	http://beangenes.cws.ndsu.nodak.edu/
	ChlamyDB—*Chlamydomonas reinhardtii* (green algae)	http://www.biology.duke.edu/chlamy/
	CoolGenes—cool season food legumes	http://www.ukcrop.net/perl/ace/search/CoolGenes
	CottonDB—*Gossypium hirsutum*	http://cottondb.tamu.edu/
	GrainGenes—wheat, barley, rye, and relatives	http://wheat.pw.usda.gov/index.shtml
	MaizeDB—maize	http://www.maizegdb.org/
	MilletGenes—pearl millet	http://jic-bioinfo.bbsrc.ac.uk/cereals/millet.html
	RiceGenes—rice	http://bioserver.myongji.ac.kr/ricemac.html (Korea)
		http://rgp.dna.affrc.go.jp/index.html (Japan)
	SolGenes—Solanaceae	http://grain.jouy.inra.fr/cgi-bin/webace/webace?db=solgenes
	SorghumDB—Sorghum bicolor	http://grain.jouy.inra.fr/cgi-bin/webace/webace?db=sorghumdb
	SoyBase—soybeans	http://soybase.agron.iastate.edu/
	TreeGenes—forest trees	http://dendrome.ucdavis.edu/treegenes.html
Livestock	Multiple at NCBI	http://www.ncbi.nlm.nih.gov/Genomes/index.html
	Multiple at Roslin Inst., UK	http://www.thearkdb.org/
	BovGBASE—bovine	http://www.genome.iastate.edu/
	ChickGBASE—poultry	http://www.genome.iastate.edu/chickmap/dbase.html
	PiGBASE—swine	http://www.genome.iastate.edu/maps/pigbase.html
	SheepBASE—sheep	http://www.projects.roslin.ac.uk/sheepmap/front.html

Note: To access any known genome, visit NCBI Genomic Biology www.ncbi.nlm.nih.gov/Genomes/index.html.

of both animals and plants, but ultimately for the benefit of human health through studying animal models of human diseases. The Online Mendelian Inheritance in Animals (OMIA) database provides information on known diseases in animals which are related to human diseases. Being able to conduct animal experiments on diseases relevant to humans offers a great advantage. Animal models allow controlled studies to be performed, including the intentional induction of a disease.

The database lists 1996 disorders, of which 762 are potential models for human diseases obtained from some 13 species including cattle, chicken, goat, sheep, horse, and pig, but also cat, dog, emu, fox, quail, rabbit, and turkey. In total, the database contains information on 206 species (Figure 5.6).

Studying diseases in animals has the advantage of being able to conduct experimental approaches to disease models in a carefully controlled environment. Although results from animal studies don't always translate into human studies, mechanistic insights always help improve experimental design. This is particularly interesting in animals with evolutionarily related pathways and conserved gene sequence and function. The OMIA database is a good first step in identifying genes from ongoing or previously published results on animal diseases with links to potential human homologs.

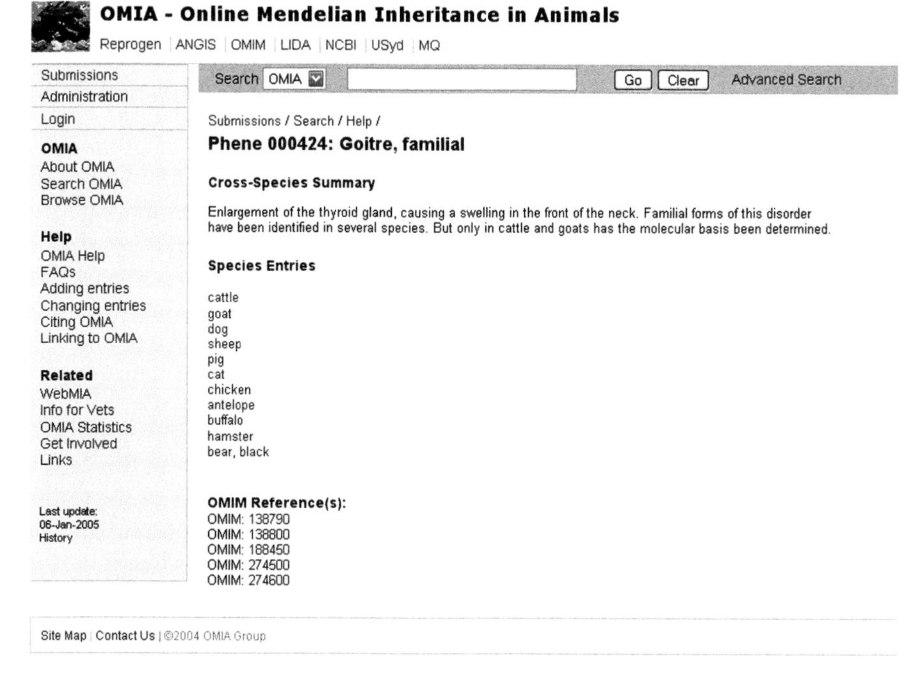

FIGURE 5.6
Online Mendelian Inheritance in Animals (OMIA). The Online Mendelian Inheritance in Animals database provides information about diseases in animals that are also relevant for human medicine. Shown here is the entry page for familial goiter, a disease causing the enlargement of the thyroid gland. As can be seen, goiter is a disease commonly found in mammals. Reference links to the Online Mendelian Inheritance in Man (OMIM) database are provided. (URL www.angis.org.au/omia/).

5.2.1 Genetically Modified Organisms

Any organism that receives an extra functional gene through recombinant DNA technology (gene splicing), usually from an unrelated organism, qualifies as a genetically modified organism. While genetic modifications technically include mutations and deletions (including knockouts), the term "genetically modified organisms" has come to mean the addition of foreign genes (e.g., from a bacteria) into animals and plants used in agriculture and ultimately destined for human consumption. GMO applications are still rare, and usually are used to solve issues with food growth (resistance) and storage (delayed ripening), rather than to improve the quality of food. The latter remains a technically feasible yet politically difficult goal.

For instance, in 2000 Ingo Potrykus and colleagues at the Swiss Federal Institute of Technology invented "golden rice" by creating a rice strain with the engineered capability to produce beta-carotene, or provitamin A.[1] The World Health Organization estimates about 250 million people globally are deficient in vitamin A, increasing their risk of blindness, immune problems, and other serious conditions. Ready availability of beta-carotene-rich food is an important step in eliminating deficiencies and preventing disease.

Golden rice is not just a scientific tool to produce a provitamin-fortified food. Its creation exemplifies the hopes, limitations, and politics of modern agriculture. While engineering a vitamin source into the main staple of a large segment of the world's population is conceptually a sound idea, it alone cannot solve vitamin A deficiencies. A daily portion of golden rice provides less than 10% of a person's vitamin A requirements.[2] However, the proof-of-concept of engineering functional food has been demonstrated by this and related projects such as beta-carotene-enriched mustard seeds.

Aside from the issue of usefulness, the golden rice project raised a particular scientific issue. Rice plants already have the full genetic complement to synthesize beta-carotene expressed in their leaves, so why the need to *add* a recombinant gene into the rice genome? The answer: unlike its natural counterpart, bioengineered beta-carotene is expressed in all parts of the rice plant, including its seeds, because the engineering process bypasses developmental regulation of beta-carotene synthesis in the plant. To understand this lack of regulation of recombinant genes, scientists will have to get a better understanding of gene regulatory networks through functional genomics and the systems biology of development of multicellular organisms. Insights from these plant systems will also benefit human genetics and medicine. After all, plants are complex multicellular organisms and use developmental programs similar to those of humans. They will only differ in their network components.

As for rice, beta-carotene is used as a pigment in photosynthetic membranes, extending the absorption spectra of visible light for the plant beyond that provided by chlorophyll, but it also protects the plant lipids against oxidative damage from excess photon radiation. It does not use the beta-carotene in its endosperm, the rice grain, as is the case in other plant species like yellow maize, sorghum, or carrots. Beta-carotene is a typical *secondary metabolite* that plants synthesize for protective and defensive purposes. From an evolutionary point of view, the protective mechanism against photon damage of cell membranes in plant leaves explains its *nutraceutical* value for humans as antioxidant. Of course, the main importance of beta-carotene is its function as a vitamin, an essential component of metabolic processes. Processed beta-carotene (vitamin A) comes in several metabolic forms (the retinal family), functioning as master regulator (retinoic acid) of many genes controlled by hormonal signals and as a photo pigment in animal retinas as a major component of the eyesight in humans.

References

1. Ye, X. et al., Engineering the provitamin A (carotene) biosynthetic pathway into (carotenoid-free) rice endosperm. *Science*, 2000, 287(5451), 303–305.
2. Brown, P., GM rice promoters "have gone too far." *The Guardian* (U.K.), February 10, 2001.

5.2.2 Biopharming

Biotechnology companies have adopted molecular biology techniques, bioinformatics, and modern genomics in the agricultural sector with great commitment. Improvement of crops and livestock through specific and speedy genetic engineering has created great hopes, although successes have been rare and have been achieved slowly.

Initially, the expectation of having no longer to depend on conventional breeding techniques with multigenerational backcrossing or hybridizations to eliminate undesirable traits was well founded. However, the limitations of using transgenic techniques have quickly turned out to be harder to overcome than expected, largely because most desirable traits have complex genotypes. With many traits and diseases being polygenic, insertion and deletion of individual genes has rarely resulted in the expected results because many genes have a low penetrance for a phenotype.

Bioinformatics with the rapidly increasing whole genome sequences available are starting to change the old approach to basic science. Earlier, transgenic methods have relied on finding promising intervention points, i.e., genes in metabolic pathways or signaling pathways that lead to altered phenotypes (traits) through quantitative trait loci (QTL) analysis which identifies alleles and their effect or contribution to a phenotype. QTL depends largely on easy-to-detect genetic markers such as STSs, ESTs, and SNPs. In the absence of whole genome sequences, marker density is usually too low to use QTL successfully. With genome projects, marker density has improved and novel gene targets for transgenic intervention in marker-assisted breeding. Thus, genomics is transforming agriculture, however slowly, where entire chromosomal regions can now be associated with desirable traits such as growth rate, leanness, feed intake, litter size, and disease resistance in livestock (see Table 5.7).

Transgenic methods using bacterial and viral vectors such as *Agrobacterium tumefaciens* or the *tobacco mosaic virus* for plants and *lenti virus* in animals are at the center of modern breeding techniques. They allow further precision of targeted mutagenesis, much more efficient than radiation hybridization, where x-ray-induced random mutations are screened and selected for through conventional breeding and backbreeding (hybridization techniques). Transgenic methods also allow production of human biomedical products in plants and animals, mostly through the expression of proteins and peptides for therapeutic and diagnostic purposes. These

TABLE 5.7

Applications of Agricultural Genomics

Group	Organisms[a]	Agricultural Traits	Biomedical Traits	Model Organism
Crop	Corn, soybean, cotton, potato, cereals, fruits	Growth performance, grain quality, seed oil content and composition, fruit ripening, pest, herbicide resistance, drought resistance, increased sweetness, protein quality, feed quality	Therapeutics, vaccines, industrial enzymes, heat-stable enzyme	Arabidopsis
Forestry	Trees	Wood density, fiber length, lignin content, delayed flowering, disease and pest resistance		Poplar, Arabidopsis
Livestock	Pig, cattle, chicken, sheep, goat, rabbit, horse, fish	Growth, yield, feed intake, body composition and leanness, fiber and hair quality, reproduction and litter size, disease resistance and immune function	Veterinary medicine, genetic research, modification of milk, xeno-transplantation	Mouse, rat, fruit fly, *C. elegans*, zebra fish

[a] See the National Center for Genome Resources http://www.ncgr.org/; the Arabidopsis information resource http://www.arabidopsis.org/; and NCBI's Plant Genome Central http://www.ncbi.nlm.nih.gov/genomes/PLANTS/PlantList.html for genome information.

TABLE 5.8

Biopharming with Plants

Product Type	Entity	Applications
Therapeutics / pharmaceuticals / diagnostics	Collagen	Bone crafts, corneal implants, tissue engineering, drug delivery
	Antithrombin III	Anticoagulation
	Aprotinin	Anti-inflammatory, surgery
	Gastric lipase	Pancreatic insufficiency
	Lactoferrin	Antibiotic, anti-inflammatory
	Brazzein	Protein sweetener
Vaccines	Respiratory syncytial virus	Antiviral
	Hepatitis B	Antiviral
	Gastroenteritis virus	Antiviral
	HIV	Antiviral
Antibodies	IgGs	Herpes simplex 2, colon cancer
	Single domain antibody	Substance P
Industrial enzymes	Trypsin	Processing of pharmaceuticals, proof-of-concept
	Avidin	Diagnostic reagent
	Beta-glucuronidase	Visual marker in genetic screening

Source: From Horn, M E., et al., *Plant Cell Rep.*, 22(10), 711–720. With permission.

biopharming processes, as shown in Table 5.8, depend on the precise insertion of recombinant human genes that can be expressed in the host organism. In animals, human proteins are expressed in the milk, where the products are readily orally available. When expressed in plant hosts, the engineered proteins can be expressed in leaves, tubers (roots), and fruits for ready harvesting.

The advantage of using plants is their lower production costs and the potential absence of human pathogens that are a health risk from engineering human proteins in animal systems. Health risk has stopped harvesting animal organs for xenotransplantation (from pig and baboon) despite engineered surface antigens from human donors, which reduces the risk of immune rejections. Bacterial systems have been used as the first cells to express human proteins for biomedical and industrial applications. While industrial applications are fairly successful, the biomedical use of biopharming human therapeutics is largely hampered by the need for clinical trials to demonstrate safety and effectiveness. Obviously, such barriers are not important when producing enzymes for diagnostic or bioengineering purposes strictly for laboratory use.

The use of genetic engineering in agriculture has a political component, because genetically modified organisms that are used for human consumption experience strong regulatory barriers and consumer rejections, largely outside the U.S. Safety concerns and hostile market conditions increase the costs of biotechnology in agriculture, costs that are absent from conventional breeding. Traditional husbandry, of course, is costly for the much longer time periods required to breed desirable traits over many generations, time periods that can span decades for larger animals and plants, whose generation times are measured in years rather than days.

This particular time constraint in conventional breeding is required because the trait is a phenotype such as litter size, milk production, or disease resistance that often only can be observed in adult animals. Consider in particular forestry, where breeding trees with better fruit yields or denser wood with more consistent fiber growth can take many years until it can be measured. Bioinformatics has changed the equation. Using genetic markers and biochemical screening tests allows a breeder to test the phenotype at the level of the genome, i.e., breeding becomes a genetic marker-driven process. This can be fast, as the presence of desirable markers (QTLs) can be identified in the embryo, where selection can now be performed much earlier. Genomics thus not only helps in identifying more markers, but also drives selection based on marker identification, rather than waiting for a phenotype to be established.

With the help of whole genome sequence information, the use of model organisms accelerated gene discovery and the identification of promising gene candidates for genetic improvement of crops or livestock. In addition, model organisms such as Arabidopsis can be used as actual models for studying the genetic modification first in the model plant in forestry. For instance, delaying flowering in the mustard plant has been studied, a process that is thought to be important in poplar wood growth, which stops when flowering is induced. Genetic animal models can also be used to study human disease mechanisms.

References

1. Horn, M.E. et al. Plant molecular farming: systems and products. *Plant Cell Rep.*, 22(10), 711–720, 2004.

Appendix A

Glossary of Biological Terms

allele the genetic variant of a gene. A gene can be found in different variants in a population, even in the same individual. Alleles are responsible for the different traits of certain characteristics, such as eye and hair color in animals, and flower and seed color in plants. Alleles are also responsible for genetic diseases.

amino acid small molecules with various chemical properties forming the building blocks of proteins.

archaea a prokaryotic form of life that forms a domain in the tree of life. There are three domains: bacteria, archaea, and eukarya. One defining physiological characteristic of archaea is their ability to live in extreme environments. They are often called extremophiles and thrive in high salt, at high or low temperature, high pressure, or high or low pH.

bacteria a prokaryotic form of life. Unlike archaea, bacteria are more versatile and are responsible for many diseases in animals and plants.

base distinct chemical structures found in nucleic acids and part of nucleotides. The bases of nucleotides form the signature letters allowing sequence information to be stored in DNA and RNA strands.

bioinformatics computational analysis of biological information such as nucleic acid and protein sequences and protein structures.

cDNA complementary DNA obtained from a messenger RNA template through a process called reverse transcription.

cell basic, self sustaining unit of living organisms. Composed of cell membrane as outer boundary surrounding the cytoplasm and internal organelles that carry out specialized functions (see mitochondrion).

cell culture artificially (*in vitro*)-maintained cell population in growth medium containing specifically isolated cell types which grow indefinitely; used to express recombinant DNA or proteins for physiological studies simulating experiments that would have been done in living organisms.

chromosome structurally independent unit of a genome. Prokaryotes have usually one large circular chromosome and one or more small circular extrachromosomal DNA (plasmids). Eukaryotic cells have often several to several dozen chromosomes. Humans have 46 chromosomes comprising two sets of 23 chromosomes (see also karyotype). Each set constitutes the complete human Genome coding for 20,000 to 25,000 genes.

cloning, clone 1. organism: reproducing a genetically identical offspring; 2. gene: duplicating a nucleic acid sequence without introducing mutations.

contig continuous DNA sequence obtained from individually sequenced DNA fragments that contain overlapping sequences at their ends.

cytoplasm main compartment of a cell (excluding organelles) in which most metabolic processes occur.

DNA deoxyribonucleic acid; is part of chromosomes and contains the genetic information in all organisms.

domain 1. an independently folding or functioning part of a protein; 2. in taxonomy and evolution the highest and most inclusive division of all forms of life; the three domains are archaea, bacteria, and eukarya.

enzyme protein that catalyzes a chemical reaction.

eukaryote group of organisms that contain organelles within their cytoplasm, specifically a nucleus containing all chromosomal material.

evolution (biological evolution) the perpetual change of the genetic composition of living organisms.

exon gene sequence on chromosome that belongs to the coding sequence; exon sequences are interrupted by introns.

gene hereditary unit of life on DNA in chromosomes; individual genes code for proteins or RNA molecules; in eukaryotic cells, most genes are organized into exon and intron segments, with the former containing the coding sequence.

gene pool collection of all genes or coding sequences within a population of an organism.

genetic code the "language" by which genetic information is stored on chromosomes. Consists of four "letters" or bases: A (adenine), G (guanine), C (cytosine), and T (thymine), where triplets form codons; each codon represents an amino acid.

genome total genetic content of an organism, both structurally and functionally.

genotype actual set of alleles (genetic information) on an individual's 46 chromosomes; differs from individual to individual, except in clones (e.g. twins).

homology evolutionary derived structural similarity between functional units such as genes or body parts.

hydrodynamics the physical study of biological macromolecules in solution; allows determination of molecular weight and size of protein complexes under physiological conditions.

in silico a computer model of a biological system used to perform an "experiment;" a predictive model of a biological system.

intron chromosomal DNA sequences interrupting the coding sequence of genes (called exons).

karyotype chromosome set of an organism; species specific; chromosomes are arranged by size and characterized by their banding pattern.

library 1. DNA library: collection of DNA fragments of various origin; 2. chemical library: collection of compounds generated by random, combinatorial synthesis strategies.

ligand a biologically active molecule that recognizes and binds to a unique receptor protein, which is activated or inactivated as a result.

membrane semipermeable cell envelope made of phospholipids and membrane proteins; while phospholipids provide stability of membranes, proteins provide transport and signaling processes across this otherwise impermeable structure.

metabolism chemical reactions in cells for the degradation and biosynthesis of molecules. Chemical energy is extracted from nutrients and used to synthesize macromolecules, promote transport, signaling and growth.

mitochondrion organelle in eukaryotic organisms responsible for oxygen-dependent energy metabolism.

molecular biology the science of studying the genetic composition and mechanism of living organisms at the molecular level. It historically refers to the understanding and manipulation of genes (DNA). The molecular studies of all other organic molecules like proteins, fats, and carbohydrates is called biochemistry.

mutation change in the base or nucleotide sequence in a gene or chromosomal structure.

nanometer, nanotechnology one nanometer is one billionth of a meter; nanotechnology pertains to molecular devices with dimensions in the nanometer range.

neuron nerve cell in the brain responsible for electrical and chemical signal transmission.

nucleic acid biological macromolecule important for storage of genetic information; is composed of nucleotides which determine the sequence of genes; comes in two common forms, DNA (deoxyribonucleic acid) and RNA (ribonucleic acid).

nucleotides building blocks of DNA and RNA; composed of base, ribose (sugar), and phosphate groups.

organelle small, membrane-encapsulated particle in the cytoplasm of eukaryotes that carry out highly specialized functions.

ortholog a homologous sequence that is derived from a common ancestor found in individuals of different species.

paralog a homologous sequence that is derived from gene duplication and found within the same organism.

pathway an organizational map of a complex biological process or network of interactions; most commonly known as metabolic pathway.

PCR, polymerase chain reaction an enzyme-mediated DNA amplification mechanism which allows sequence specific selection of the DNA to be amplified.

pharmacophore the pattern of binding sites of a class of ligands that bind to the same receptor.

polymorphism sequence differences among individuals found on specific chromosome locations within a population.

protein biological macromolecules that carry out most functional activities in cells (cell structure, regulation, digestion, and biosynthesis); classes of proteins include enzymes, hormones, receptors, and antibodies.

proteome, proteomics the full set of proteins expressed in a cell or organisms; proteomics is the study of all proteins used by an organism or cell.

receptor protein that serves as binding site for signaling molecules such as growth factors, hormones, or neurotransmitters (see also ligand).

recombinant DNA genetically modified or structurally altered nucleic acid sequence (usually a gene) within a vector or host DNA.

RNA, ribonucleic acid a form of nucleic acid that comes in three distinct polymeric types: messenger RNA which mediates the translation of DNA into amino acid sequences; transfer RNA which couples amino acids with the corresponding codon on the messenger RNA; rRNA which is part of ribosomes, the cytoplasmic particles catalyzing protein biosynthesis.

taxonomy a system for naming, ranking, and classifying organisms.

vector DNA small piece of DNA containing regulatory and coding sequences of interest; vector DNA functions to insert and amplify a gene into a target genome.

Appendix B

Bioinformatics Web Sites

Databases

EMBL	www.ebi.ac.uk
EBI	www.ebi.ac.uk
NCBI	www.ncbi.nlm.nih.gov
Entrez	www.ncbi.nlm.nih.gov/Entrez
DDBJ JapanGenomeNet	www.genome.ad.jp
KEGG	www.genome.jp/kegg
BioCyc	http://biocyc.org
ChemBank	http://chembank.med.harvard.edu
ChemIDplus	http://sis.nlm.nih.gov/Chem/ChemMain.html
ChEBI	www.ebi.ac.uk/chebi
PDB	www.rcsb.org/pdb
Worldwide Protein Data Bank	http://www.wwpdb.org/

Genome Browsers

NCBI Map Viewer	www.ncbi.nlm.nih.gov/mapview
Ensembl Genome Browser (EBI)	www.ensembl.org
UCSC Genome Browser	http://genome.ucsc.edu/cgi-bin/hgGateway?org=human
Apollo Genome Browser	www.bdgp.org/annot/apollo
Genome Tracking Database	http://maine.ebi.ac.uk:8000/services/cogent

Bioinformatics Projects

The Whole Brain Atlas	www.med.harvard.edu/AANLIB
The Human Brain Project	www.gg.caltech.edu/hbp
International HapMap Project	www.hapmap.org
Tree of Life Project	http://tolweb.org/tree
Gene Ontology Consortium	www.geneontology.org
Pharmacogenetics Research Network	www.pharmGKb.org

Index

C

D

E

L

M

N

O

P

Q

R

S

T

U

V

W

X

Y

Z